The Impact of Nitrogen Deposition on Natural and Semi-Natural Ecosystems

ENVIRONMENTAL POLLUTION

VOLUME 3

Editors

Editorial Board

The Impact of Nitrogen Deposition on Natural and Semi-Natural Ecosystems

edited by

Simon J. Langan

Macaulay Land Use Research Institute,
Aberdeen, Scotland,
United Kingdom

KLUWER ACADEMIC PUBLISHERS
DORDRECHT / BOSTON / LONDON

Library of Congress Cataloging-in-Publication Data

The impact of nitrogen deposition on natural and semi-natural
ecosystems / edited by Simon J. Langan.
p. cm. -- (Environmental pollution ; v. 3)
Includes index.

1. Nitrogen compounds--Environmental aspects. 2. Atmospheric
deposition. I. Langan, Simon J. II. Series: Environmental
pollution (Dordrecht, Netherlands) ; v. 3.
TD196.N55I44 1999
577.27'5711--dc21 99-29975

ISBN 978-90-481-4025-1

Published by Kluwer Academic Publishers,
P.O. Box 17, 3300 AA Dordrecht, The Netherlands.

Sold and distributed in North, Central and South America
by Kluwer Academic Publishers,
101 Philip Drive, Norwell, MA 02061, U.S.A.

In all other countries, sold and distributed
by Kluwer Academic Publishers,
P.O. Box 322, 3300 AH Dordrecht, The Netherlands.

Printed on acid-free paper

Printed in the Netherlands.

CONTENTS

PREFACE

It has been recognised that nitrogen deposition is playing an increasingly important role in the acidification and eutrophication of ecosystems, both terrestrial and aquatic. The member states of Europe have recently negotiated and agreed an acid emission abatement policy for sulphur, based on the critical load approach. Such an approach was adopted and implemented relatively quickly. It was largely made possible through the synthesis of substantial amounts of process-based scientific study over the preceding decade. With the general agreement concerning abatement of sulphur emissions, attention is very rapidly turning to the role of atmospheric nitrogen, not only in Europe and N. America but also across Asia. The question arises as to whether the same critical load and modelling approach should be used to influence policies designed for nitrogen emission abatement. To look at some of the data, processes and issues involved in making this assessment, a two day workshop was held in Aberdeen, Scotland during September 1996. The origins of this book stem from that meeting.

The workshop entitled 'Nitrogen Deposition and the Acidification of Natural and Semi-Natural Ecosystems' was attended by over one hundred scientists and policy advisors from thirteen countries. The chapters of this book represent an overview of the science presented. The book in its entirety provides a synthesis of our knowledge of the impact of nitrogen deposition on extensively managed and natural ecosystems of the northern temperate latitudes.

We hope the book will provide some focus and inspiration for researchers involved in problems associated with increased N deposition. For students the book should provide an overview as to what and who has done what research in each of the ecosystem elements considered. Finally for the policy maker the book provides a review of the key scientific issues, how far these have been understood and what is left to be researched and of course funded.

Simon Langan
February 1999

ACKNOWLEDGEMENT

In preparing this book I have been assisted by numerous people and organisations. Firstly we would wish to acknowledge colleagues and scientists who attended the Aberdeen workshop in 1996. Secondly our principal sponsors the Scottish Office, Department of Agriculture, Environment and Fisheries Department but also support from the Department of the Environment, Air Quality Division (now part of the Department of Environment, Transport and Regions). Finally my thanks go to colleagues at MLURI, in particular Ed Paterson, Jeff Wilson and the Director, Professor Jeff Maxwell. I would also like to thank Ian Baird for assistance with copy preparation and page make-up.

Simon Langan
Macaulay Land Use Research Institute
Craigiebuckler
Aberdeen Scotland

NITROGEN DEPOSITION: SOURCES, IMPACTS AND RESPONSES IN NATURAL AND SEMI-NATURAL ECOSYSTEMS

M. HORNUNG [1] AND S. J. LANGAN [2]

[1] *Merlewood Research Station, Institute of Terrestrial Ecology, Grange Over Sands, Cumbria, LA11 6JU, UK*

[2] *Macaulay Land Use Research Institute, Craigiebuckler, Aberdeen, AB15 8QH, UK*

1.1 Introduction And Background

Anthropogenic emissions of sulphur and nitrogen compounds, principally derived from the burning of fossil fuels, have lead to regional changes in atmospheric and precipitation chemistry. The fate and environmental consequence of these changes on ecosystem functioning has attracted considerable research and aroused public concern at the international, national and local levels. Throughout the 1970s and 1980s this research and concern was primarily directed at the effects associated with the increased deposition of sulphur. This largely reflected sulphur as the dominant component of atmospheric pollution. Research in most countries in this period was therefore largely focused on the effects of sulphur and acidity, although nitrogen (N) inputs were also quantified in many field-based studies. The results of the research underpinned the development of agreements to limit sulphur emissions and was fed into the setting of critical loads for sulphur (Bull, 1995). The impacts of the reduction in industrial activity in western Europe in the 1980s, combined with the effects of the implementation of sulphur protocols lead to a sharp reduction in sulphur emissions through the 1980s and early 1990s.

In contrast N emissions have steadily increased since the late 1940s. This increase continued through the 1980s and early 1990s, at a time when S emissions were declining sharply. The contribution atmospheric inputs of N make to total acid inputs have therefore increased. The extensive research programme in the Netherlands on acidification during the 1980s also highlighted the large effects of NH_y emissions from intensive agriculture. In addition concerns about eutrophication of ecosystems due to N inputs had increased, thus broadening the effects debate and the rationale for controls. The balance of the research programmes in a number of countries changed, therefore, in the late 1980s and 1990s to give greater emphasis to N emissions and effects.

Research on the impact and fate of atmospherically deposited N has largely concentrated on a range of extensively managed ecosystems, such as forests, and natural ecosystems of high conservation interest. For these ecosystems, characterisation and quantification of the nitrogen transfers and transformations between plant foliage, soils and surface waters under different pollution inputs have been undertaken. Finally in an attempt to identify the extent of any undesirable effects these increased N inputs may have, the spatial representation of ecosystem 'tolerances' have been calculated, set and mapped as critical loads (see Section 1:6 below).

As concern about the impacts of transboundary pollution developed, the first protocol to limit N emissions was formulated in 1988, under the auspices of the United Nations Economic Commission for Europe, and implemented in 1991 (the Sofia protocol). This first protocol focused entirely on NO_x and aimed for a reduction in emissions from 1994 to 1987 levels. A second protocol is currently being discussed and aims to be effects based with consideration of proposed abatements judged against critical loads and exceedance. This protocol may well consider total N emissions, oxidised and reduced and take into account the role of NO_x in ozone formation.

Against this background, the current chapter provides an introductory overview, inevitably selective, of some developments over the last few years and identifies gaps in our knowledge, particularly those influencing the calculation and setting of critical loads. Many of the topics introduced here are dealt with in much greater detail in the subsequent chapters.

1.2 Emission Sources And Trends Over Time

Emissions of N to the atmosphere have increased steadily since the Second World War with the most dramatic increases occurring since the 1950s. Galloway (1996) estimates that in 1990 20Tg N year was being produced globally through the combustion of fossil fuels, representing more than double natural emissions (lightening ~8, biomass burning ~3 and microbial processes ~2 Tg N yr-1). Subdivided on a continent basis, Galloway *et al.* (1995) show the emissions from energy production to be dominated by the developed regions of the world (Table 1.1), most noticeably N. America and Europe.

Table 1.1: Present day emissions of nitrogen (Terragrams N yr^{-1}) on a continent basis, after Galloway *et al.* 1995.

Continent	*Emissions*
United States and Canada	7.6
Europe	4.9
Australia	0.3
Japan	0.8
Asia	3.5
South America	1.5
Africa	0.7
Former Soviet Union	2.2

Of the range of oxidised and reduced species of N introduced into the atmosphere within the northern hemisphere the most reliable data are available for NO_x emissions and show continuing increases until the 1990s with a plateau in the mid-1990s but model predictions suggest a further increase towards the millennium and beyond (IPCC,1992, Graedel *et al.*, 1995). The main sources of NO_x emission are industry, power generation and motor vehicles. The earlier post-war increases in emissions were largely derived from industry and power generation but the role of vehicle emissions has gradually increased and now represents the most significant contributor which will largely determine whether or not emissions stabilise or increase further (INDITE, 1994). The diffuse nature of vehicular emissions contrasts sharply to those of SO_2 which predominantly originate from a relatively few stationary point sources. Reliable data on NO_x emissions are available for most European countries but the degree of spatial disaggregation of these data varies considerably between countries.

Data on emissions of NH_y are much less reliable and this has precluded the calculation of trends in emissions over time. Present day emissions over Europe have been reconstructed on the basis of animal numbers and their management for production (ApSimon *et al.*, 1987, Pain *et al.*, 1998), the dominant source of emissions. The last five years has seen a considerable effort to improve data on emissions at national levels and to improve spatial disaggregation of the data. The available data and models are reviewed and examined in greater detail by Metcalfe *et al.*, in Chapter 2.

There are also significant emissions of N gases, N_2O, N_2 and NO from soils (Firestone, 1982) as a result of denitrification processes. These are of more relevance to concerns about global warming and ozone production and consumption than to discussions of acidification or N enrichment. However, there is a link between the magnitude of N inputs to soils, as atmospheric deposition and/or fertilizers, and emissions of N gases via denitrification; this linkage is discussed briefly below. The understanding of both the denitrification processes and the quantification of emissions from soils have increased considerably over the past 10 years.

1.3 Deposition

The residence time of NO_2 in the atmosphere is 1 to 2 days compared to that of NH_3 which is 1 to 2 hours. NO_2 emissions are therefore transported further, giving rise to transboundary pollution while the majority of NH_3 emissions are deposited locally and hence largely confined within the country of origin. Thus, in the UK while a larger amount of N emitted is in oxidised forms, the greatest quantity of deposited N is in reduced forms (INDITE, 1994). The importance of interactions between SO_x and NH_y deposition, sometimes referred to as co-deposition, has become clear over the last few years. These interactions may have major effects on deposition of NH_y and SO_x both locally and nationally. The reduction of SO_2 concentrations over recent years may be resulting in a smaller proportion of NH_y emissions being deposited close to source, effectively increasing the residence time of the NH_y. Will the continuing decline in SO_2 concentrations lead to increased transboundary pollution from ammonia?

The processes controlling NO_x deposition seem reasonably well understood and the models for the calculation of deposition relatively robust. There are, understandably requests for finer resolution deposition data from the effects research community and work is in progress to produce data at finer scale grids at both the European and national level. Thus, EMEP data will soon be available on a 50 km grid and data for the UK is planned at 5 km.

Information for NH_y is much less reliable but at the European scale and for most individual countries research is in progress to improve the situation. These finer scale datasets are important for work and assessment of local and regional impacts. It is clear that the majority of the NH_y emitted from intensive animal units is deposited close to the source, probably within 2 km.

In other regions of the world such as N. America and there is a paucity of spatially distributed data of sufficient integrity to construct and model concentration and flux calculations at such a fine resolution. Although as the appreciation of the role and magnitude of N emission and deposition increases, there is a growing awareness of the need to establish appropriate inventories and monitoring networks for emission and deposition.

1.4 The Fate And Impact Of Changing Inputs Of Atmospheric Nitrogen

The fate of deposited N is determined by a complex series of processes occurring within soil and vegetation components. Whilst the main processes have been recognised for many years, recent research has begun to clarify the controls on rates and directions of these processes.

Plant uptake. The last 10 years have seen clear demonstration of direct foliar uptake of both ammonium and nitrate. Thus, for example, Thoen *et al.*, (1991 and 1996) showed that an amount equivalent to 40% of the N incorporated into the annual growth by spruce (Picea abies) was taken up through the foliage. A linked response between foliar and root uptake has also been demonstrated with, for example root uptake 'suppressed' when there is significant foliar uptake and foliar uptake 'suppressed' when there is a large supply of available N in the rooting zone (Rennenberg *et al.*, 1996a and b).

The foliar uptake pathway is not generally incorporated into dynamic N models used to assess likely responses to pollutant inputs or to determine at what N loading saturation of the system will take place. This gives rise to the question: Should this uptake pathway be treated in a different way to root uptake in models? The different uptake pathways should have no impact on calculation of critical loads of nutrient N using the mass balance approach. The key input to the equation is net N uptake; usually determined in forest systems as the N removed in biomass at harvesting. However, the pathway may need incorporating explicitly into models which consider sources and sinks of acidity so that it can be treated differently to root uptake of N.

There is now clear evidence of organic N uptake by mycorrhizae (eg Michelsen *et al.* 1996). Much of the current work on this uptake pathway to date has focused on plant species typically found in highly stressed environments, such as the Arctic fringe and there is a need to extend this work to other environments.

Direct phytotoxicity. Direct effects of NO_x or NH_y have been shown in experimental conditions (eg Van der Eerden, 1982, Van der Eerden and deWit, 1987, Saxe and Muralin, 1989, Dueck *et al.*, 1988) and critical levels (see Section 1:6) have been set for both NH_y and NO_x (Ashmore and Wilson, 1992, Federal Environment Agency 1996). Damage is rarely seen with ambient concentrations but NH_y damage has been reported from sites close to large agricultural point sources. Although direct effects of these pollutants are clearly possible, the major impacts on plant-soil systems are soil mediated.

Impacts on plant species composition. In natural and semi-natural ecosystems N is most commonly the limiting nutrient and natural plant communities are adapted to conditions of low N supply. Enhanced N deposition to these systems results in changes in the competitive interactions of species, favouring the more N responsive species. Ellenberg (1988) has ranked many of the plant species of central Europe in terms of nutrient, primarily N, requirements; enhanced N inputs would favour the high ranking species. Grime *et al.*, (1988) have classified plant species in terms of their growth strategies into either competitors, ruderals and stress tolerators. Increased N inputs favour the competitors to the disadvantage of stress tolerators; many of the plant species which are of high conservation interest are stress tolerators. There is now a considerable body of European literature, at the site scale and from national surveys, which shows increases in competitors/N demanding species over the last 20 years. Perhaps the most dramatic and best documented changes are the replacement of shrub heaths with grasses in the Netherlands. The relevant work is reviewed in greater depth by Aerts and Bobbink in Chapter 4. Parallel changes towards increases in N responsive, competitor species have also been reported from woodlands, montane vegetation and calcareous grassland (Bobbink *et al.* In press). However, the research has also highlighted the importance of interactions between N inputs and land management, and between N inputs, other environmental stresses and pest outbreaks in determining impacts. Plant species changes have been used as key indicators in setting critical loads.

Links with plant N status. Relationships have been shown between N content of foliage of shrub heaths and bryophytes and current atmospheric inputs of N (Pitcairn *et al.*, 1995, Woolgrove and Woodin 1996, Baddeley *et al.*, 1994). Thus, Pitcairn *et al.* (1995) showed higher concentrations in *Calluna* foliage from southern Scotland and England than from the more remote areas of Scotland receiving smaller inputs. The foliar N content also increased with altitude, paralleling the atmospheric inputs. Morecroft and Woodward (1996) have also shown that the foliar N content of transplanted *Alchemilla alpina* increased with altitude reflecting an increase in atmospheric inputs. Increased foliar N concentrations have been linked to the increased possibility of damage from pathogens and insect grazers and increased sensitivity to environmental stresses. Thus, relationships have been shown between larval growth rates and adult weights of heather beetle and *Calluna* N status (Power *et al.* 1998, Heil and Diemont 1983). Caporn *et al.* (1994) have also shown changes in frost hardiness of *Calluna* related to N inputs while impacts on drought sensitivity have been demonstrated for a number of species (eg Power *et al.* 1998, Neighbour *et al.* 1988, Lucas 1990).

Causal links between increased N deposition and mycorrizal changes and vulnerability to secondary stress such as frost are sparse. That which is available is reviewed by Aerts and Bobbink in Chapter 4.

Impacts on forest growth. Many forest systems are N limited and enhanced atmospheric N inputs to these systems will initially have a fertilizer effect. Analysis of regional data on forest production in Europe has shown significant increases in forest growth and production over the last 40 years (Spiecker *et al.*, 1996). While part of this increase is likely to be due to improved silviculture, it is thought that a significant part is due to the fertilizer effect of increased atmospheric inputs of N. There is a conflict in many semi-natural forest systems in Europe between the 'benefits' of increased N deposition as seen in the increased production and the adverse impacts in terms of changes in the species composition of ground flora as a result of the alteration in competitive interactions.

Fate of N inputs to soils and impacts on NO_3 fluxes to waters. Nitrification and immobilisation are key process controlling the response of ecosystems to enhanced N inputs. Dise and Wright (1995) have brought data together from catchment and plot based budget studies in forests and shown that up to inputs of c.9 kg N ha^{-1} yr^{-1} the input N is retained, immobilised within the system. Above inputs of c.25 kg N ha^{-1} yr^{-1} leaching was enhanced in roughly direct proportion to inputs and between inputs of 15 and 25 kg N ha^{-1} yr^{-1} there was a considerable variety of response, with some sites showing enhanced leaching and others continuing to immobilise inputs. There is clearly also a difference in response depending on whether the inputs are as NH_y or NO_x. Thus, Emmett *et al.* (1988) have shown at a site in North Wales, part of the NITREX network (Wright and van Breeman, 1995), that ammonium inputs of up to 75 kg N ha^{-1} yr^{-1} are largely retained within the system while nitrate inputs are essentially leached. A clearer understanding is needed of the factors controlling the response overall to the inputs but more especially to the variation in response to inputs of between 15 and 25 kg N ha^{-1} yr^{-1}. Specifically, what triggers the change in a given system from immobilisation to nitrification?

Much of the data and studies undertaken on this aspect are covered by Wilson and Emmett in Chapter 5. The carbon:nitrogen (C:N) ratio is one of the most common parameters used to characterise soil organic materials and to predict/interpret nitrification potential. This is a relatively crude measure but it is simple to determine and has proved robust. Kriebnitzsch (1978) examined the rate of nitrate production in incubated forest humus layer material in relation to C:N ratio. In materials with a C:N ratio >25, there was almost no nitrate production; any ammonium produced is clearly immobilised in the microbial biomass. More recently, Matzner and Grosholz (1997), Gundersen *et al.* (1998) and Wilson and Emmett (Chapter 5) have brought together data from forest studies and shown a relationship between C:N ratio and nitrate leaching. This suggests that C:N ration in the forest floor or A horizon could be used to indicate the likely response of systems to pollutant N inputs. Similarly, Yesmin *et al.* (1996) have shown an inverse relationship between N in peat soils and C:N ratio of the surface horizon. The surface layers becoming relatively enriched in N in response to N deposition although the work suggests that the history of N inputs at a given site is an important factor controlling response to further additions.

Impacts on N emissions from soils. The increase in emissions of N to the atmosphere from industry, vehicles and agriculture has been paralleled by increased use of N fertilizers. Thus, in the UK application of N fertilizer nationally increased from c 600 thousand tonnes in 1960 to c1.6 million tonnes in 1984, when it plateaued for a number of years and more recently has shown signs of decline. Thus, fertilizer applications to land in the UK are of the same order as emissions of oxidised N to the atmosphere. The increased application of fertilizer N has a feedback to N emissions from soils as recent research has shown clear links between the rates of N inputs to land, as fertilizer and/or atmospheric deposition, and N emissions from soils. The emissions from soils are a significant source term. They are dominated by emissions from the soils of intensive agriculture but low levels of emissions also occur from forest soils and from semi-natural ecosystems. Until recently the focus of research was on emissions of N_2 and, more particularly the radiatively active N_20. However, recent research has also shown significant emissions of NO from fertilized grassland on gley soils in southern England (Skiba *et al.* In press). However, it is clear that much of this NO is 'trapped' in the vegetation canopy.

The major emissions of N through denitrification at national and European scales originate from lowland agricultural soils receiving fertilizer N inputs and there has been considerable improvement in the quantification of these losses over recent years. Although the emissions from natural and semi-natural systems are relatively small these small annual outputs can be significant when calculating critical loads. There are relatively few data for semi-natural ecosystems. Values for denitrification losses incorporated into the critical load mass balance equations are either calculated, using simple models, or a default value based on a review of published data by Ineson and Sverdrup (1992). There is a clear relationship for semi-natural systems between the magnitude of atmospheric inputs and denitrification outputs. Denitrification losses from forest soils vary through the forest cycle and with the type of forest management, generally peaking after clearcutting eg (Ineson *et al.*, 1991, Dutch and Ineson 1990). This variation needs to be incorporated into the calculation of critical loads over a full management rotation.

1.5 The N Saturation Concept

Most ecosystems in the northern hemisphere are considered to be N limited with the result that the addition of N has stimulated an increase in growth. However, with continued additions of N, ecosystems become unable to utilise excessive inputs and disruption to ecosystem functioning may result. This has given rise to the concept of nitrogen saturation. Some of the issues surrounding this concept have been dealt with by Skeffington and Wilson (1988). Aber *et al.* (1989) building on earlier works suggested that within northern temperate forest ecosystems different stages of the development of N saturation could be linked with characteristic attributes and changes to the nitrogen cycle. The developments associated with this work are reported by Wilson and Emmett in Chapter 5.

At a catchment scale, Stoddard (1994) has illustrated using both site and regional data from across the United States that it is possible to identify four stages in the development of N saturation related to increasing nitrogen deposition. This development is reflected by characteristic changes in surface water nitrate concentrations at seasonal and long-term time scales. An increasing number of studies are also starting to suggest that with increased N inputs there is an increased leaching and flux of organic nitrogen. The spatial and temporal elements of this are reviewed in depth by Chapman and Edwards in Chapter 6.

1.6 Critical Loads

The encapsulation of the key processes and links between emission source, deposition and impacts on receptors is increasingly being considered by the work on critical loads. At its simplest level critical loads can be considered as a precautionary threshold above which any stress, such as increased deposition of pollutants will give rise to some negative effect. The most commonly cited definition of critical loads is that of Nilsson and Grennfelt (1988):– 'A quantitative estimate of the exposure to one or more pollutants below which significant harmful effects on specified elements of the environment do not occur. According to present knowledge.' This is graphically presented in Figure 1.1. This concept has been developed and widely applied in international discussions on the second sulphur protocol and it is likely the second international UNECE nitrogen protocol will adopt the same approach. The concept has permitted policy makers to consider the potential impact of atmospheric inputs of sulphur and N as an acidifier and in eutrophication and more recently the potential for combined effect.

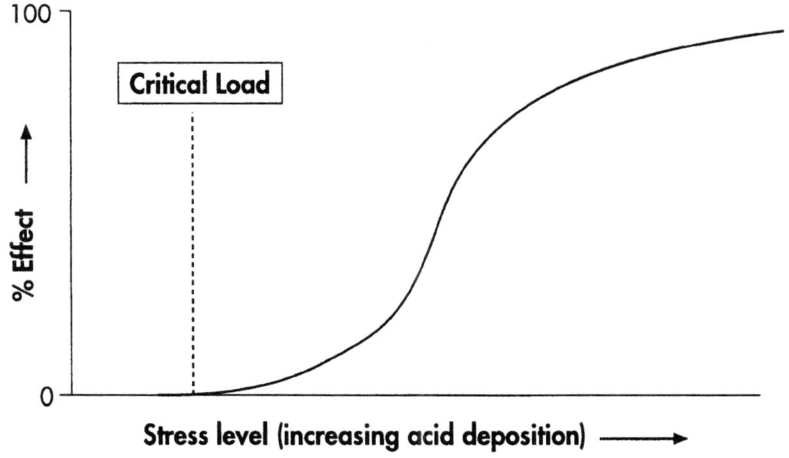

Figure 1.1: Representation of the dose response relationship within a critical load framework.

Similarly the concentration (as opposed to the load) of atmospheric pollutants above which direct adverse effects occur on plants, such as plant tissue, are defined in terms of critical levels. The work on critical levels is not dealt with further in this book and the reader is referred to the reviews and work reported in UNECE 1993 and Ashmore and Wilson (1992) and Federal Environment Agency (1996).

Critical loads can be set using a variety of approaches and data sets. Generally speaking, there are three levels of approach to calculating critical loads. At the simplest, with relatively low data requirements is the level 0, empirical approach; level 1 in which sources and sinks are balanced in a mass balance approach and finally, level 2 in which data intensive dynamic models are used.

Bobbink and Roelofs (1995) and Bobbink *et al.* (1996) have published a range of empirically derived critical loads for natural and semi-natural ecosystems and freshwaters based on long-term monitoring, observations across pollution gradients or manipulative experiments. The published values correspond to the deposition at which changes in plant competition gives rise to changes in species composition. These effects are usually considered to be the result of 'eutrophication' of the system as a result of enhanced N inputs.

Recent developments in the formulation of the critical load mass balance equations and dynamic modelling approaches are presented and discussed by Posch and De Vries in Chapter 7. These methodologies still have a number of major uncertainties. Thus, for example the mass balance equation includes a term for an 'acceptable level of N immobilisation', expressed in kg ha^{-1} yr^{-1}, under long term equilibrium conditions. There is considerable debate about how this value should be defined and quantified. Long term mean rates of accumulation can be calculated for undisturbed natural systems, eg undisturbed soils developed on glacial moraines of known age, from total soil N content divided by the period of soil formation. These calculations generally give low rates of accumulation, less than 1 kg ha^{-1} yr^{-1}. Current or recent rates of N accumulation can be calculated from ecosystem N budgets, or catchment based calculations, or from repeated analysis and quantification of soil N stores over a period of years. Such calculations tend to give much larger values than the long term means noted above. Are these current/recent rates 'acceptable rates' under equilibrium conditions? There is a view that the system will eventually become destabilised or be disturbed and that there will then be a large release of N. Further consideration of the contributions of soil processes to N fluxes are reviewed by Williams and Anderson in Chapter 3.

Outputs of N from soils through denitrification form an N loss from the system and losses via this pathway are incorporated into the calculation of critical loads. They potentially increase the critical load by a value equal to the annual denitrification outputs; in terms of critical loads for nutrient N the denitrification losses might been seen as a positive advantage. However, the emissions as N_2O also contribute to global warming and NO is important in atmospheric transformations. These effects should be taken into account in multi-impact models or assessments.

As noted above, negotiations of the second N protocol have been initiated with a target date for agreement of 1998. Consideration of critical loads, current and modelled future exceedances will probably be fed into these negotiations. Integrated modelling assessment is also looking at the relative pay off of N reductions as opposed to further reductions in S emissions and critical load methodologies have been modified to allow

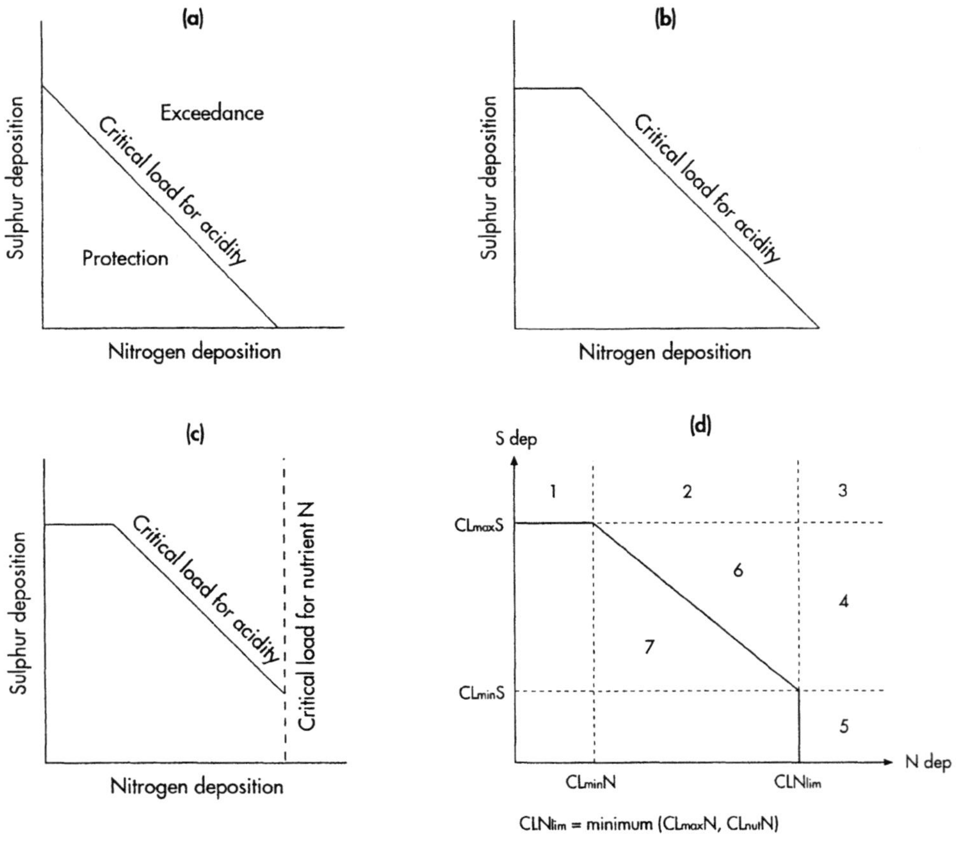

S dep = sulphur deposition
N dep = nitrogen deposition
$CL_{max}S$ = maximum critical load of sulphur
$CL_{min}S$ = minimum critical load of sulphur
$CL_{min}N$ = minimum critical load of nitrogen
$CL_{max}N$ = maximum critical load of nitrogen
$CL_{nut}N$ = critical load of nutrient nitrogen
CLN_{lim} = limiting critical load of nitrogen

1 only sulphur reduction will offer protection
2 sulphur reductions may subsequently offer the option of reducing sulphur or nitrogen
3 sulphur and nitrogen must be reduced before there is any option of reducing either sulphur or nitrogen
4 nitrogen reductions may subsequently offer the option of reducing sulphur or nitrogen
5 only nitrogen reduction will offer protection
6 either sulphur or nitrogen reductions may offer protection
7 ecosystem protection where critical loads are not exceeded

Figure 1.2: The assessment of the potential impact from atmospheric deposition of sulphur and nitrogen using a critical load function: a) a simple acidification function without N processes, b) an acidification function with N removal included, c) the acidification function with nutrient N critical load, d) the solid line joins all possible pairs of values of critical loads of sulphur and nitrogen.

simultaneous consideration of acidification and eutrophication effects and of S and N through the use of the critical load function (Bull, 1995). Without the dual role of N deposition as a nutrient and an acidifier the function could be drawn as a linear relationship between S and N (Figure 1.2a). However, consideration of immobilisation processes constrain the potential for damage at low deposition inputs (Figure 1.2b) and at higher deposition loads ecosystem response will largely be a function of excess nutrient supply (Figure 1.2c). Thus it is possible to use this approach to consider options in reducing pollutant deposition relative to ecosystem damage (Figure 1.2d). By plotting the deposition relative to the selected critical load receptor it is possible to consider the incidence of damage (represented by critical load exceedance) for different receptors and for different geographical regions. In situations where deposition is < the critical load, (7), no ecosystem damage is predicted to occur. Areas in which deposition is in the areas 1 to 6 in Figure 1.2d, critical load exceedance and ecosystem damage will prevail.

To change this situation requires emission abatement and the various options for this are self explanatory in the Figure. Perhaps not readily apparent from the figure, and a limitation to setting critical loads on a mass balance, steady state approach, is the inability to consider the temporal element of both inputs and ecosystem response. To meet this requirement the use of dynamic models to represent the key transformations and interactions between the various ecosystem sources sinks and pools for N are needed. Given our current understanding, the representation of N dynamics within existing models is relatively simple.

For the scientist, the challenges lie in formulating the method, data, processes and indicators on which such assessments and improvement in their predictive capability can be made. These challenges and how they are being met provide the basis for the subsequent chapters. These are laid out according to emissions, sources and deposition (Chapter 2); soil–vegetation mediated processes (Chapter 3); impacts: non forest (Chapter 4), forests (Chapter 5); surface waters (Chapter 6); critical loads and modelling (Chapter 7) and future directions (Chapter 8).

References

Aber, J.D, Nadelhoffer, K.J., Steudler, P. and Melillo, J.M. (1989) Nitrogen saturation in northern forest ecosystems. *Bioscience*, **39**, 378–386.

ApSimon, H., Kruse, M. and Bell, J.N.B. (1987) Ammonia emissions and their role in acid deposition. *Atmospheric Environment*, **21**, 1939–1946.

Ashmore, M.R. and Wilson, R.B. (eds.). (1992) *Critical levels of air pollution for Europe*. Department of the Environment, London.

Baddeley, J.A., Thompson, D.B.A. and Lee, J.A. (1994) Regional and historical variation in the nitrogen content of Racomitrium–Lanuginosum in Britain in relation to atmospheric nitrogen deposition. *Environmental Pollution* **84**, 189–196.

Bobbink, R. and Roelofs, J. G. M. (1995) Nitrogen critical loads for natural and semi-natural ecosystems: The empirical approach. *Water, Air and Soil Pollution*, **85**, 2413–2418.

Bobbink, R., Hornung, M. and Roelofs, J.G.M. (In press.) The effects of air-borne nitrogen pollutants on species diversity in national and semi-natural vegetation – a review. *Journal of Ecology*.

Bull, K.R. (1995) Critical loads – Possibilities and constraints. *Water, Air and Soil Pollution*, **85**, 201–212.

Dise, N.B. and Wright, R.F. (1995) Nitrogen leaching from European forests in relation to Nitrogen deposition. *Forest Ecology and Management* **71**, 153–161.

Dueck, T.A., Dil, E.W. and Pasman, F.J.M. (1988) Adaptation of grasses in the Netherlands to air pollution. *New Phytologist* **108**, 167–174.

Dutch, J. and Ineson, P. (1990) Denitrification of an upland forest site. *Forestry*, **63**, 363–377.

Caporn, S.J.M., Carroll, J.A. and Lee, J.A. (1994) Effect of N supply on frost hardiness in Calluna vulgaris (L.) Hull. *New Phytologist*, **128**, 461–468.

Emmett, B.A., Reynolds, B., Silgram, M., Sparks, T.H. and Woods, C. (1998) The consequences of chronic nitrogen additions on N cycling and soil water chemistry in a Sitka spruce stand, North Wales. *Forest Ecology and Management* **101**, 165–175.

Ellenberg, H. (1988) *Vegetation ecology of Central Europe (4th edition).* Cambridge University Press, Cambridge.

Firestone, M.K. (1982) Biological denitrification, in *Nitrogen in Agricultural Soils, Agronomy No 22*, (ed F.J. Stevenson), American Society of Agronomy, Madison, Wisconsin.

Federal Environment Agency. (1996) *Mapping critical levels/loads*. Federal Environmental Agency, Berlin.

Galloway, J.N., Schlesinger, W.H., Levy, H., Michaels, A. and Schnoor, J.L. (1995) Nitrogen fixation: Anthropogenic enhancement–environmental response. Global *Biogeochemical Cycles*, **9**, 235–252.

Galloway, J.N. (1996) Anthropogenic mobilisation of sulphur and nitrogen: immediate and delayed consequence. *Annual reviews in Energy Environment*, **21**, 261–292.

Graedel, T.E., Benkovitz, C.M., Keene, W.C., Lee, D.S. and Marland, G. (1995) Global emissions inventories of acid related compounds. *Water, Air and Soil Pollution*, **85**, 25–36.

Grime, J.P., Hodgson., J.G. and Huny, R. (1988) *Comparative plant ecology: a functional approach to common British species*. London, Unwin Hyman.

Heil, G.W. and Diemont, W.H. (1983) Raised nutrient levels change heathland into grassland. *Vegetatio* **53**, 113–120.

Hornung, M., Dyke, H., Hall, J.R. and Metcalfe, S.E. (1997) The critical load approach to air pollution control. *Issues in Environmental Science and Technology*, **8**, 119–140.

INDITE 1994. *Impacts of nitrogen deposition in terrestrial ecosystems*. Report prepared for the Department of the Environment. HMSO, London, England.

Ineson, P., Dutch, J. and Killham, K.S. (1991) Denitrification in a Sitka spruce plantation and the effect of clearfelling. *Forest Ecology and Management*, **44**, 77–92.

Ineson, P. and Sverdrup, H. (1992) Compuscript on nitrogen denitrification. In: *Critical loads for nitrogen – a report from a workshop held at Lokeburg Sweden 6–10 April 1992*. (eds.) P.Grennfelt and E. Thornelof Nord 1992: 41, Nordic Council of Ministers, Copenhagen.

IPCC (1992) Climate Change 1992. *The Supplementary Report to the IPCC Scientific Assessment*. (ed. J.T. Houghton, B.A. Callander and S.K. Varney). Cambridge University Press.

Kriebnitzch, W.U. (1978) Stickstoffnachlieferung in saure Waldboden Nordwestdeutschlands. *Scr. Geobot.*, **35**, 293–302.

Lucas, P. (1990) The effects of prior exposure to sulphur dioxide and nitrogen dioxide on the water relations of Timothy grass (Pheum pratense) under drought conditions. *Environmental Pollution*, **66**, 117–138.

Matzner, E. and Grosholz C. (1997) Beziehung zwischen NO_3-Austragen, C/N-Verhältnissen der Auflage und N-Eintrogen in Fichtenwald (Picea abies Karst.) Ökosystemen Mitteleuropas. *Forstw. Cbl.* **116**, 39–44.

Gundersen, P., Emmett, B.A., Kjonass, O.J., Koopmans, G.J. and Tietema, A. (1998) Impact of nitrogen deposition on nitrogen cycling in forests: a synthesis of NITREX data. *Forest Ecology and Management* **101**, 37–55.

Morecroft, M. D. and Woodward, F.I. (1996) Experiments on the causes of altitudinal differences in the leaf nutrient contents, size and d13C of Alchemilla alpina. *New Phytologist*, **134**, 471–479.

Michelsen, A., Schmidt, I.K., Jonasson, S., Quarmby C. and Sleep, D. (1996) Leaf ^{15}N abundance of subarctic plants provides field evidence that ericoid, ectomycorrhizal and non-and arbuscular mycorrhizal species access different sources of soil nitrogen. *Oecologica*, **105**, 53–63.

Muller, B., Touraine, B. and Rennenberg, H. (1996) Interactions between atmospheric and pedospheric nitrogen nutrition in spruce (Picea abies L. Karst.) seedlings. *Plant Cell and Environment*, **19**, 345–355.

Neighbour, E.A., Cottam, D.A. and Mansfield, T.A. (1988) Effects of sulphur dioxide and nitrogen dioxide on the control of water loss by birch (Betula spp.) *Environmental Pollution*, **54**, 149–157.

Nilsson, J. and Grennfelt, P., (eds.) 1988 *Critical loads for sulphur and nitrogen. Report from a workshop held at Skokloster, Sweden.* March 1988, Published by Nordic Council of Ministers, Copenhagen.

Pain, B.F., Weerden Van Der T. J., Chambers, B.J., Phillips, V.R. and Jarvis, S.C. (1998) A new inventory for ammonia emissions from UK agriculture. *Atmospheric Environment*, **32**, 309–313.

Pitcairn, C. E. R., Fowler, D. and Grace, J. (1995) Deposition of fixed nitrogen and foliar nitrogen content of bryophytes and Calluna vulgaris (L) Hull. *Environmental Pollution*, **88**, 193–205.

Power, S.A., Ashmore, M.R., Cousins, D.A. and Sheppard, L.J. (1998) Effects of nitrogen addition on the stress sensitivity of Calluna vulgaris. *New Phytologist*, **138**, 663–673.

UNECE (1993) *Manual on methodologies and criteria for mapping critical levels/loads and geographical areas where they are exceeded.* Texte 25/93 Umweltbundesamt, Bitzmarkplatz 1, 1000 Berlin 33, Germany.

Rennenberg, H., Schneider, S. and Weber, P. (1996b) Analysis of uptake and allocation of nitrogen and sulphur compounds by trees in the field. *Journal of Experimental Botany,* **47**, 1491–1498.

Rennenberg, H., Herschbach, C. and Polle, A. (1996a) Consequences of air pollution on shoot–root interactions. *Journal of Plant Physiology,* **148**, 296–30.

Saxe, H. and Murali, S. (1989) Diagnostic parameters for selecting against novel spruce Picea abies decline. II. Responses of photosynthesis and transpiration to acute nitrogen oxides exposures. *Physiology Plant,* **76**, 349–55.

Skeffington, R.A. and Wilson, E.J (1988) Excess Nitrogen deposition: Issues for consideration. *Environmental Pollution,* **54**, 159–184.

Skiba, U., Sheppard, L.J., Pitcairn C.E.R., Leith, I.D., Crossley, A., van Dijk, S., Kennedy, V.H. and Fowler, D. (In press) Soil nitrous oxide and nitric oxide emissions as indicators of the exceedance of critical loads of atmospheric nitrogen deposition in semi-natural ecosystems. *Environmental Pollution.*

Spiecker, H., Mielikainen, K., Kohl, M. and Skovsgaard, J.P. (eds.). (1996) *Growth trends of European forests, studies from 12 countries.* Springer Verlag, Berlin. 372 pp.

Stoddard, J.L., 1994 Long term changes in watershed retention of Nitrogen: its causes and aquatic consequences, in *Environmental chemistry of lakes and reservoirs, Advances in Chemistry Series No. 237.*, (ed. L.A. Baker) American Chemical Society, pp 223–284.

Thoene, B.,Schroder, P., Paper, H., Egger, A. and Rennenberg, H. (1991) Absorption of atmospheric NO_2 by spruce (Picea abies L. Karst) trees. 1. NO_2 influx and its correlation with nitrate reduction. *New Phytologist,* **110**, 327–338.

Thoene, B., Rennenberg, H. and Weber, P. (1996) Absorption of atmospheric NO_2 by spruce (Picea abies) trees. 2. Parameteriation of NO_2 fluxes by controlled dynamic chamber experiments. *New Phytologist,* **124**, 257–266.

Woolgrave, C.E. and Woodin, S.J. (1996) Current and historical relationships between the tissue nitrogen content of a snowbed bryophyte and nitrogenous air-pollution. *Environmental Pollution,* **91**, 282–288.

Van Breeman, N. and Verstraten, J.M. (1991) Soil acidification/ N cycling, in *Acidification research in the Netherlands.* (eds. G.J. Heij and T. Schneider). Elsevier: Amsterdam.

Van der Eerden, L.J. (1982) Toxicity of ammonia to plants. *Agricultural Environment,* 223–225.

Van der Eerden, L.J. and de Wit, A.K.H. (1987) Effecten van NH_3 en NH4 op plantern en vegetaties; relevantie van effectgrenswaarden. *Acute en chronische effecten van NH3 (en NH_4) op levende organismen.* A.W. Boxman and F.J.M. Greelen. Lab. Voor Aquatiche oecologie, K.U. Nimegei.

Yesmin, L., Gammack, S. M. and Cresser, M. S. (1996) Medium-term response of peat drainage water to changes in nitrogen deposition from the atmosphere. *Water Research,* **30**, 2171–2177.

SPATIAL AND TEMPORAL ASPECTS OF NITROGEN DEPOSITION

S. E. METCALFE [1], D. FOWLER [2], R. G. DERWENT [3],
M. A. SUTTON [2], R. I. SMITH [2] AND J. D. WHYATT [4]
[1] *Department of Geography, University of Edinburgh, Edinburgh EH8 9XP, UK*
[2] *ITE Edinburgh Research Station, Bush Estate, Penicuik, Midlothian EH26 0QB, UK*
[3] *Meteorological Office, Bracknell, Berkshire RG12 2SZ, UK*
[4] *Department of Geography, Lancaster University of Hull, Lancaster LA1 4YB*

2.1 Introduction

It has been recognised that nitrogen deposition is playing an increasingly important role in the acidification and eutrophication of both terrestrial and aquatic ecosystems, (Ambio, 1997). Over North America and north-west Europe, while emissions of sulphur dioxide (SO_2) have fallen considerably over the last 20 years, emissions of oxides of nitrogen (NO_x) have not. Atmospheric inputs of acidity from nitrogen compounds now exceed those from sulphur in many areas. Measurements of nitrate (NO_3) concentrations in systems as diverse as precipitation in New Hampshire (Fay *et al.*, 1986), Norwegian lakes (The Norwegian State Pollution Control Authority, 1987) and the Greenland ice cap (Mayewski *et al.*, 1990) all indicate a steady increase, while sulphate has generally declined since the 1970s.

In this chapter we review our understanding of the chemistry of atmospheric nitrogen compounds and describe the nature of the sources and the processes which remove the gases and particulate forms of nitrogen from the atmosphere. The emissions, chemistry and deposition processes are applicable globally, although the different characteristics of the physical and chemical climate determine the relative importance of the different deposition pathways in the different countries. To illustrate the processes, data for the UK are used and the scientific and political context for the paper is that of Europe. However, the same primary pollutants are emitted to the atmosphere in similar quantities in North America and SE Asia on spatial scales which are similar to those of Europe. Thus, while the specific data used to illustrate the points are for the UK and Europe the principles apply globally, as illustrated for North America and SE Asia by Galloway and Rodhe (1991). In this respect, it is surprising that the level of scientific and political interest in deposited nitrogen should be so much greater in Europe than in North America. The rapidly developing economies of SE Asia are creating similar problems of long range transport and deposition of pollutants. It appears likely that during the next decade, global interest in the effects of deposited pollutants will move from the regions in which abatement strategies are gradually reducing the problem to areas in which the magnitude of the problem is growing rapidly, notably in Asia.

2.2 Atmospheric Chemistry of Nitrogen Compounds

2.2.1 INTRODUCTION

An understanding of the atmospheric chemistry of nitrogen compounds is essential to an understanding of the potential impacts of nitrogen deposition. This is because the nitrogen compounds emitted in to the atmosphere are not necessarily those nitrogen compounds which are readily deposited, nor those with important impacts on natural and semi-natural ecosystems. Atmospheric chemistry plays an important role in processing the nitrogen-containing constituents emitted into the atmosphere into those pollutants which can cause harm to target ecosystems in natural and semi-natural environments.

In this section, some of the key atmospheric chemical processes involved in the transformation of nitrogen-containing trace constituents into ecosystem damaging pollutants are first reviewed. Having laid out the transformation pathways involved, attention is then given to timescales, since in a constantly changing atmosphere, timescale and spatial scale are inextricably linked. Atmospheric chemistry controls not only the nature of the potentially damaging pollutants produced, but where and how these pollutants are formed in and removed from the atmospheric circulation.

As will be shown in later sections, deposition is a vitally important atmospheric removal process because it controls the ultimate spread of the pollutants away from their sources. It may however lead to ecosystem contamination and ultimately to ecosystem damage. The interplay between atmospheric chemistry and deposition controls the regions where damaging impacts may be anticipated. Estimates of emissions of some atmospheric nitrogen compounds are discussed in more detail in Section 2.3.

2.2.2 FATE AND BEHAVIOUR OF THE ATMOSPHERIC NITROGEN COMPOUNDS

Nitrogen (N_2): Nitrogen is the main constituent of the atmosphere and is the species which is used as the reference against which all other concentrations are expressed. Its concentration is therefore, by definition constant at 78.084 % by volume. The total dry atmospheric mass of N is 5.132 x 10^{21} g (Trenberth and Guillemot, 1994).

Nitric oxide (NO): Nitric oxide (NO) is by far the most important nitrogen-containing species emitted into the atmosphere, on a mass basis, from human activities involving combustion of fossil fuels (e.g. in motor vehicles, power stations, in the home and in industrial processes). Few adverse environmental impacts are associated directly with NO and most concern is associated with its atmospheric transformation products.

The main fate of NO is to react with the ozone (O_3), ubiquitously present in the lower atmosphere in reaction (1). This is the case for the NO generated by motor traffic, power stations, microbial processes in soils, lightning, biomass burning and high flying aircraft. This reaction (1) has been well studied under laboratory conditions and is one of the best understood of all the atmospheric chemical processes:

$$NO + O_3 = NO_2 + O_2 \qquad (1)$$

Under typical atmospheric conditions, the reaction takes place within a few seconds and either leads to the complete conversion of all of the O_3 to nitrogen dioxide (NO_2) with an excess of unreacted NO or to the conversion of all of the NO to NO_2 with an excess of unreacted O_3. In highly polluted atmospheres, or close to individual pollution sources, the former behaviour is typically observed because although ozone is widely distributed in the lower atmosphere, its concentration is not usually high compared with that of NO in urban areas or close to major roads and hence ozone concentrations become rapidly depleted.

In stagnant weather conditions, particularly during cold, wintertime conditions, a further oxidation route converting NO into NO_2 may come into play through reaction (2):

$$NO + NO + O_2 = NO_2 + NO_2 \qquad (2)$$

Because of its second order dependence on NO concentrations, this process only becomes of any significance when NO concentrations are exceedingly high. This process explains the occurrence of the unprecedented concentrations of NO_2 observed in some urban areas during wintertime (Derwent *et al.*, 1995). In one such episode in central London during December 1991, hourly peak concentrations of NO_2 as high as 423 ppb were reported in association with NO concentrations in excess of 1 ppm (Bower *et al.*, 1994).

In addition to its reactions with O_3 and oxygen (O_2), the other important atmospheric reactions of nitric oxide are with the hydroperoxy (HO_2) free radical in reaction (3) and with organic peroxy (RO_2) radicals in reactions (4) and (5):

$$HO_2 + NO = NO_2 + OH \qquad (3) \text{ and}$$

$$RO_2 + NO = NO_2 + RO \qquad (4) \text{ and}$$

$$RO_2 + NO = RONO_2 \qquad (5)$$

All of the peroxy radical reactions exemplified by reactions (3) and (4) above result in the conversion of NO to NO_2. This conversion process is an essential element of the photochemical generation of ozone which is of crucial importance in photochemical smog formation in the polluted urban boundary layer, in the formation of greenhouse gases in the lower atmosphere and in the chemistry of the stratospheric ozone layer.

Nitric oxide can react at high concentrations in laboratory systems to form higher oxides of nitrogen such as N_2O_3. These reactions have no importance in the ambient atmosphere. Almost all of the adverse consequences following from the emission of NO to the atmosphere from human activities result from its relatively rapid and efficient conversion to NO_2 in reactions (1) to (4) above.

Nitrogen dioxide (NO_2): Nitrogen dioxide is the first and most immediate reaction product of the atmospheric oxidation of the NO emitted by human activities. During daylight, the main fate of the NO_2 so formed is to absorb solar ultraviolet radiation (Leighton, 1961) and to undergo photolysis, reforming the NO and O_3 from which it was made, through reactions (6) and (7):

$$NO_2 + \text{radiation (wavelengths 200–420 nm)} = NO + O \tag{6}$$

$$O + O_2 + M = O_3 + M \tag{7}$$

where M represents an oxygen O_2 or nitrogen N_2 molecule which acts as a so-called 'third body'. The rate of photolysis of NO_2 depends on the solar actinic irradiance and this in turn depends on the height of the sun in the sky and hence time-of-day and season as well as on the amount and height of any cloud and haze which may obscure the sun. For much of the daylight portion of the year, the lifetime of NO_2 is only a matter of minutes before it is photolysed back to NO.

The reaction of NO with O_3 and the photolysis of NO_2 form a cycle of chemical reactions which occur rapidly over the timescales of minutes in the sunlit atmosphere. Together, this cycle of reactions generates a mixture of NO and NO_2 in the sunlit atmosphere from initially either pollutant present singly. This ensures that under most atmospheric conditions, NO and NO_2 coexist as a mixture. The mixture is conveniently referred to as NO_x=NO+NO_2. This reaction system is the so-called 'photostationary state' system (Brimblecombe, 1986) which when established, leads to a rather constant relationship, equation (8), between the concentrations of [NO], [NO_2] and [O_3] in the sunlit atmosphere:

$$\frac{[NO] \times [O_3]}{[NO_2]} = J / k_1 \tag{8}$$

where J is the photolysis rate coefficient for reaction (6) and k_1 is the first order rate coefficient for reaction (1).

Under summertime conditions, in the mid-afternoon at mid-latitudes and clear sky conditions, J will typically have a value of about $7 \times 10^{-3}\ s^{-1}$ and k_1 about $4 \times 10^{-4}\ ppb^{-1}\ s^{-1}$. This leads to the relationship:

$$\frac{[NO] \times [O_3]}{[NO_2]} = 18 \tag{9}$$

which implies: $[NO_2] = 2 \times [NO]$ at about 30–40 ppb [O_3] and

$= 5 \times [NO]$ at about 90 ppb [O_3] in the sunlit atmosphere.

In wintertime conditions and at night time, the photolysis on NO_2 does not occur to the same extent as during the summer and NO_2 becomes a rather more stable product of the atmospheric oxidation of NO. Under these wintertime conditions, almost any concentrations of NO and NO_2 can coexist and there is no simple relationship such as the 'photostationary state' relationship between them. In polluted atmospheres, then, there is a tendency for all of the O_3 to be depleted by reaction (1) since NO concentrations are generally greater than those of O_3. In rural atmospheres, because ozone concentrations are generally greater than those of NO, the opposite tendency usually occurs and reaction (1) goes to completion, leaving the NO greatly depleted.

Higher Oxides of Nitrogen: There are two higher oxides of nitrogen which are important in the atmosphere and these are NO_3 and N_2O_5. Both are formed by the reaction of ozone with NO_2 and both are only of importance in the night-time atmosphere. The formation of the higher oxides begins with the complete oxidation of NO to NO_2 by reaction with ozone which can only occur at night time. The NO_2 then reacts with any remaining ozone to form the nitrate radical NO_3 in reaction (10):

$$NO_2 + O_3 = NO_3 + O_2 \tag{10}$$

NO_3 radicals show significant reactivity with a range of common atmospheric constitutents (Wayne *et al.*, 1991. Reaction with NO_2 generates nitrogen pentoxide, N_2O_5:

$$NO_2 + NO_3 + M = N_2O_5 + M \tag{11}$$

N_2O_5 is thermally unstable and falls apart reversing its formation reaction.

None of this night-time chemistry involving the higher oxides of nitrogen is of any importance during daylight for several reasons. NO_3 and N_2O_5 are themselves photolytically unstable:

$$NO_3 + \text{radiation (wavelengths 400–660 nm)} = NO + O_2 \tag{12}$$

$$NO_3 + \text{radiation (wavelengths 400–660 nm)} = NO_2 + O \tag{13}$$

$$N_2O_5 + \text{radiation (wavelengths 200–385 nm)} = NO_2 + NO_2 + O \tag{14}$$

Also, during daylight, NO_3 reacts rapidly with the NO ubiquitously present from NO_2 photolysis, inhibiting the formation of the higher oxides of nitrogen:

$$NO + NO_3 = NO_2 + NO_2 \tag{15}$$

The night-time formation of the higher oxides of nitrogen is of environmental importance since it is one of the more important conversion routes of NO_x to nitric acid (HNO_3) and aerosol particle nitrate. This former route involves the reactions of the higher oxides of nitrogen with cloud droplets or the water taken up by aerosol particles:

$$N_2O_5 + H_2O = 2HNO_3 \tag{16}$$

The reaction of N_2O_5 with sea salt aerosol particles (a) and leads to the formation of aerosol nitrate and the displacement of chloride from these coarse particles (Finlayson-Pitts and Pitts, 1986).

$$N_2O_5 + NaCl_{(a)} = NaNO_{3(a)} + Cl\,NO_2 \tag{17}$$

Once nitrogen-containing trace gases have been incorporated into the atmospheric aerosol particles, it is then possible to explain the presence of oxidised nitrogen compounds in rain, through rain-out of the aerosol nitrate (see Section 2.4.1).

The nitrate NO_3 radicals are also of some importance in night-time chemistry through their addition reactions to olefinic hydrocarbons, many of which are derived from natural biogenic emissions (Atkinson, 1994).

Nitrous oxide (N_2O): Nitrous oxide (N_2O) is the oxide of nitrogen present in the atmosphere at the highest concentration, pole-to-pole and from the surface to the middle of the stratospheric ozone layer, except in the most polluted of urban atmospheres. The main sources of nitrous oxide are undoubtedly associated with soil processes, and with the stratospheric photolysis sink. These were in equilibrium prior to the industrial revolution, maintaining a concentration of 270 ppb. The main sources are in the soils of the wet tropical forests and in the soils of the temperate forest and grasslands. By the early 1990s, the global background surface concentration of N_2O had increased to about 311 ppb, with northern hemisphere concentrations typically 1 ppb higher than southern hemisphere values (IPCC, 1990).

N_2O is extremely unreactive chemically in the lower atmosphere and undergoes only two chemical processes of any significance and then only in the upper atmosphere. These two processes are photolysis in reaction:

$$N_2O + \text{radiation (wavelengths 173–240 nm)} = N_2 + O^1D \tag{18}$$

and destruction by excited oxygen atoms in reactions (19) and (20)

$$N_2O + O^1D = NO + NO \tag{19}$$

$$N_2O + O^1D = N_2 + O_2 \tag{20}$$

The NO liberated in the above process is the main source of oxidised nitrogen species in the stratospheric ozone layer (Brimblecombe, 1986). Ultimately, these oxidised nitrogen species are converted to nitric acid (HNO_3) and are transported through the stratosphere and into the troposphere by the atmospheric circulation. This stratosphere–troposphere flux of nitric acid ensures a small baseline loading of nitrate in rain and snow throughout both hemispheres.

Nitrous oxide has an exceedingly long atmospheric lifetime of about 120 years, which is controlled by its stratospheric photolysis. The mean global N_2O concentration has been rising at about 0.8 ppb/year throughout the 1980s (IPCC, 1995). Pre-industrial concentrations have been established accurately through the analysis of air bubbles trapped in Antarctic ice cores. Over the past 2000 years, levels remained constant at about 275 ppb. The increase away from the pre-industrial baseline and the present upwards trend in N_2O levels are unambiguously linked to human activities. Atmospheric N_2O levels are now currently higher than they have ever been since the beginning of recorded time.

Oxyacids of Nitrogen: Three of the many oxy-acids of nitrogen are commonly present in the atmosphere: nitrous acid (HONO); nitric acid (HNO_3) and peroxynitric acid (HO_2NO_2). Nitrous acid is formed in ppb quantities from NO_2 on surfaces, particularly soil surfaces, in the presence of water vapour or moisture. It is photochemically-labile and disappears rapidly in the matter of a few minutes immediately after sunrise, through the reaction:

$$\text{HONO} + \text{radiation (wavelengths 305–390 nm)} = \text{OH} + \text{NO} \quad (21)$$

Nitric acid on the other hand is relatively more stable to photolysis because its ultraviolet absorption spectrum is shifted towards much shorter wavelengths. At the surface, photolysis times are days to weeks decreasing to days in the stratospheric ozone layer. The photolysis of nitric acid may be represented as:

$$HNO_3 + \text{radiation (wavelengths 200–335 nm)} = OH + NO_2 \quad (22)$$

Nitric acid is the major atmospheric conversion product of NO_x and is formed through the reaction of NO_2 with hydroxyl (OH) radicals:

$$OH + NO_2 + M = HNO_3 + M \quad (23)$$

This reaction is an important process throughout the sunlit atmosphere. Daytime oxidation rates may approach 10–30% per hour under summertime conditions and 1% per hour or less during wintertime at high latitudes.

Nitric acid is a highly reactive gaseous species which is readily taken up by soil, vegetation and water surfaces. Dry deposition lifetimes are a few hours to tens of hours and so dry deposition is the main removal process for nitric acid in the atmospheric boundary layer. Nitric acid is also taken up into pre-existing aerosol particles and cloud droplets and reacts with sea-salt particles forming nitrate aerosol particles (Finlayson-Pitts and Pitts, 1986):

$$NaCl_{(a)} + HNO_3 = NaNO_{3(a)} + HCl \quad (24)$$

where $_{(a)}$ refers to the aerosol phase.

Nitric acid also reacts with gaseous ammonia to form ammonium nitrate which may be reversibly taken up into pre-existing aerosol particles or taken up into cloud droplets:

$$NH_3 + HNO_3 = NH_4NO_{3(a)} \quad (25)$$

$$NH_4NO_{3(a)} = NH_3 + HNO_3 \quad (26)$$

Peroxynitric acid (HO_2NO_2) is a highly unstable oxyacid which falls apart in a matter of seconds, liberating the species from which it was formed:

$$HO_2 + NO_2 + M = HO_2NO_2 + M \quad (27)$$

$$HO_2NO_2 + M = HO_2 + NO_2 + M \quad (28)$$

In the colder regions of the upper atmosphere, the thermal decomposition becomes particularly slow and appreciable concentrations (at the ppt–ppb level) of peroxynitric acid can build up. Peroxynitric acid can undergo photolysis and reacts with hydroxyl radicals so that in the stratospheric ozone layer it can become an important temporary reservoir species:

$$HO_2NO_2 + \text{radiation (wavelengths 200–335 nm)} = H_2O + NO_2 + O \quad (29)$$

$$OH + HO_2NO_2 = H_2O + NO_2 + O_2 \quad (30)$$

Organic Nitrogen Compounds: The atmosphere contains trace concentrations at the ppt–ppb levels of organic nitrogen compounds which are formed by photochemical reactions involving organic compounds and the oxides of nitrogen. These organic nitrogen compounds include alkyl nitrates, peroxyalkyl nitrates and peroxyacyl nitrates.

Ammonia (NH_3): Ammonia is the most important alkaline gas present in the atmosphere. Because of its short lifetime and because its main sources are at the earth's surface, it is present at appreciable concentrations only in the atmospheric boundary layer and in the ppt–ppb concentration range. The main source of ammonia is farming, although quantifying ammonia emissions is problematic (see Section 2.3.2 for a more detailed discussion relating to the UK).

Because of its high solubility, ammonia is readily taken up by damp soils and vegetation and by water bodies. Dry deposition is therefore an efficient process, giving ammonia an atmospheric lifetime of a few hours or less in the atmospheric boundary layer.

Because of its alkaline nature, ammonia is readily taken up by any pre-existing aerosol particles of sulphuric acid (Erisman *et al.*, 1988). This uptake is likely to be virtually instantaneous, whilst the particles remain acidic. However, once complete conversion of sulphuric acid to ammonium sulphate occurs, the particles will cease to be acidic and no further uptake of ammonia will occur. The atmospheric boundary layer, at any one time, therefore tends to contain either neutral ammonium sulphate with an excess of acidic sulphuric acid particles or neutral ammonium sulphate with an excess of gaseous ammonia. Since ammonia sources are, by and large, land-based, the former condition tends to occur over the oceans and the latter over the land, particularly farmland close to major point emission sources.

Nitrogen-containing aerosol particles: The atmosphere contains aerosol particles which cause haziness and turbidity and reduce the visibility of distant objects and the horizon. These aerosol particles are extremely variable in terms of their physical and chemical properties and their concentrations. They vary in nature from small solid particles with diameters of nanometres through to large mist droplets, with diameters of a few tenths of a millimetre.

A substantial fraction of the mass of the atmospheric aerosol is present as ammonium sulphate aerosol. This ammonium sulphate is largely removed from the atmospheric circulation by rain and this removal process gives ammonium sulphate an atmospheric lifetime of several days, at least up to 5–10 days. As a result ammonium sulphate aerosol particles may be transported for hundreds and thousands of kilometres away from pollution sources.

Ammonium sulphate aerosol particles are formed in a multi-stage process which is initiated by the oxidation of sulphur dioxide to form acidic sulphuric acid particles or cloud droplets. These acidic aerosol particles or cloud droplets can then take up alkaline ammonia from the atmosphere to form ammonium sulphate aerosols or droplets. Some of these cloud droplets may pass out of the cloud and become aerosol particles. Ammonium sulphate is usually present in the atmosphere as a sub-micron aerosol with particles in the fine particle size range covering 0.2–0.5 microns.

Ammonium sulphate is not the sole nitrogen-containing component of the atmospheric aerosol because of additional contributions from ammonium nitrate and sodium nitrate. Ammonium nitrate is formed by the condensation of ammonia and nitric acid onto pre-existing aerosols and so is largely present in the fine particle size range. However, sodium nitrate aerosol is formed by chemical reactions involving sea-salt aerosol particles and so is present largely in the coarse particle size range covering 1–5 microns.

Nitrate and Ammonium in Rain and Snow: It has been understood since the beginnings of agricultural science over a century ago, that rain and snow are contaminated with nitrogen-containing compounds in the form of nitrate and ammonium. The wet removal of these compounds is discussed further in Section 2.4. However, the ultimate source of this material is the ammonia and NO_x initially present in the atmosphere.

Understanding of the complex chemistry of the various forms of atmospheric nitrogen has improved considerably in recent years as a result of detailed laboratory and field experiments. Given the very short lifetimes of some N species, the incorporation of many reactions is outside the capabilities of many models, especially those concerned with making estimates of deposition at regional scales. Inevitably, these models have to employ some simplified representation of what is actually known in far greater detail. Two long range transport models used to estimate N deposition across Europe and for the UK specifically are discussed in more detail in Sections 2.5 and 2.7 below.

2.3 Emissions

Estimates of emissions of nitrogen compounds are made for a variety of reasons e.g. for modelling the effects of radiatively active gases and aerosols (IPCC, 1996) or modelling the long-range transport of potentially acidifying pollutants (Berge *et al.*, 1995). There are a variety of methods by which emissions inventories may be compiled, basically falling in to two categories: top-down, where totals are disaggregated to smaller spatial scales using statistics such as population and electricity generation; and bottom-up, where local data sets and emissions factors are used to aggregate up (Lindley *et al.*, 1996). Which of these methods is used will depend on both the availability of local data and resources. It is extremely unlikely that emissions inventories compiled by different groups, using different methodologies will reach the same totals. Global inventories in particular show wide variations (see Lee *et al.*, 1997 for a discussion of global NO_x emissions and their uncertainties) and patterns of emissions change through space and time (Galloway, 1995).

The detailed quantitative description of the emission and deposition budgets of fixed nitrogen for the UK is helpful to identify the major pathways, their controls and the spatial patterns. For a wider perspective it is helpful to consider the global picture, in which anthropogenic emissions of oxidized nitrogen at 21 Tg y^{-1} are approximately a factor of two larger than the less well known natural inputs from soils (as NO) and lightning (Galloway, 1995). So that while in industrial countries the emissions are dominated by anthropogenic sources, this is also the case globally and with rapid industrial development throughout the world, anthropogenic emissions will steadily account for a growing fraction of the atmospheric input.

In the case of NH_3 emissions, the complexity and spatial variability of the sources make global budgets very uncertain (Dentener *et al.*, 1994) but the emissions are again believed to be dominated by anthropogenic sources.

On continental scales emissions of NH_3-N and NO_x-N are of a similar magnitude in Europe and North America, while for SE Asia emissions of ammonia are almost an order of magnitude larger than those of oxidized nitrogen (as nitrogen). It is also in SE Asia that, as a consequence of rapid growth in population and industrial development, the emissions of ammonia are projected to increase most in the next two decades.

2.3.1 EMISSIONS IN THE EMEP AREA

Under the terms of the 1979 Convention and Resolution on Long-Range Transboundary Air Pollution (LRTAP), all the countries within the UNECE area are obliged to submit emissions data to the ECE secretariat. These data are then passed on to EMEP (the Co-operative Programme for Monitoring and Evaluation of the Long-Range Transmission of Air Pollutants in Europe) to allow the modelling of long-range transboundary fluxes. EMEP MSC–W (Meteorological Synthesising Centre–West), in Oslo, plays a major role in developing this emissions database and in quality control. Data are submitted in a variety of formats: totals; by source category; by source height; gridded and with projections. The supply of data from the different signatories is quite variable and where no data are supplied, estimates are taken from emissions calculated under the CORINAIR programme run by the EU (now under the auspices of the European Environmental Agency). It should be noted that the emissions totals for particular pollutants for individual years vary between EMEP reports as methods of estimation change. Originally, gridded emissions data across the EMEP area were held at a 150 km x 150 km resolution, although the current aim is to collect data on 50 km x 50 km grids for the period since 1990. Emissions of NO_x and NH_3 on the 150 km x 150 km grid in 1994 are shown in Figure 2.1, based on data in Barrett and Berge (1996). The total emission of NO_x as NO_2 across the EMEP domain is 20,111 k tonnes (about 6114 k tonnes N) of which 19,477 k tonnes comes from land-based sources. The belt of high NO_x emissions (Figure 2.1a) runs from central England eastwards into Belgium, the Netherlands and Germany. The EMEP area total for NH_3 in 1994 is 7486 k tonnes (6161 k tonnes N), all the emission is from land-based sources (see Figure 2.1b).

When the official country emission totals submitted to EMEP are compared with the CORINAIR90 estimates, the differences are quite small for NO_x (generally less than 10%), but large for NH_3, the worst cases being Hungary (65% difference) and the UK (35% difference). The emissions data held at MSC–W are used in the EMEP model (see Section 2.5) and in the UK scale HARM model (see Section 2.7). NO_x emission estimates (in k tonnes of NO_2) over the period since 1980 are given in Table 2.1 for a number of EMEP member countries and for the United States and Canada (Berge, 1997; United Nations 1997). Comparable emissions data for NH_3 are not available. NO_x emissions across Europe between 1980 and 1990 showed few clear trends. Overall, emissions began to decline after 1990, but there are exceptions (eg in Spain) and in some cases, the decline in emissions is probably due to major economic restructuring following the break up of the former Soviet Union. The North American figures show very similar patterns, but with much smaller reductions in emissions being predicted by 2010. The policy on emissions within Europe and a modelling assessment of possible future N deposition are discussed in Sections 2.6 and 2.7 below.

Table 2.1: NO_x emissions (in k tonnes) for a range of countries

	1980	1985	1990	1994	2010*
Czech Republic	937	831	742	435	398
Germany	3334	3276	2640	2210	2130
The Netherlands	583	576	575	542	120
Spain	950	839	1178	1223	892
USA	21120	20568	21040	21423	20071
Canada	1959	2038	2104	2026	2085

2010* emissions estimated

2.3.2 EMISSIONS IN THE UNITED KINGDOM

Estimates of UK emissions of air pollutants are made for the UK Department of Environment, Transport and the Regions (DETR) National Atmospheric Emissions Inventory by the National Environmental Technology Centre (NETCEN) at Culham. NETCEN use a bottom-up approach and their methodology is set out in Salway *et al.* (1996). The inventory (NAEI) includes emissions estimates for NO_x (measured as NO_2), N_2O and NH_3. Large point sources are dealt with on an individual basis, while other data are compiled into 10 km x 10 km grids. Data are collected for a wide range of source types, but are also put in to the UNECE/CORINAIR categories. Long term trends in the emissions of acidifying pollutants are discussed in RGAR (1998).

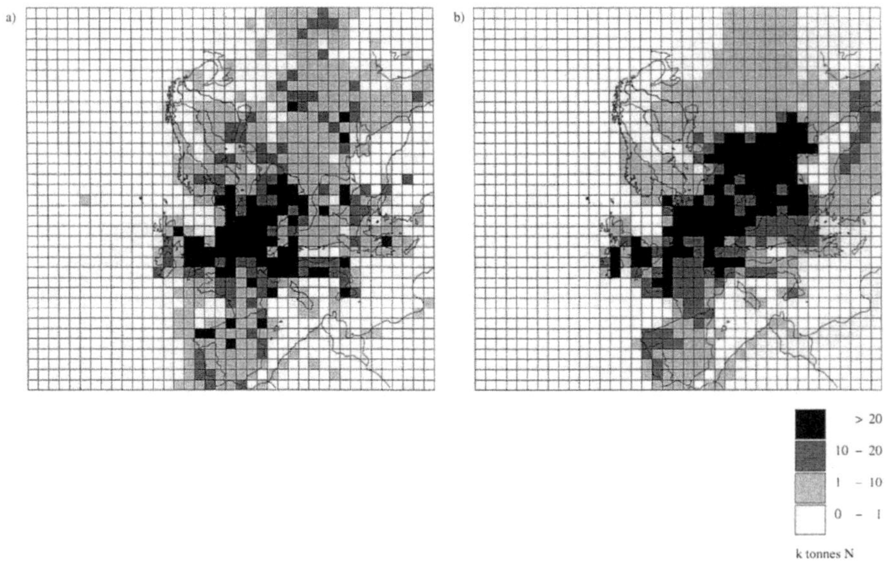

Figure 2.1: (a) EMEP area NO_x and (b) NH_3 emissions (in k tonnes N) for 1994 (from Barrett and Berge, 1996)

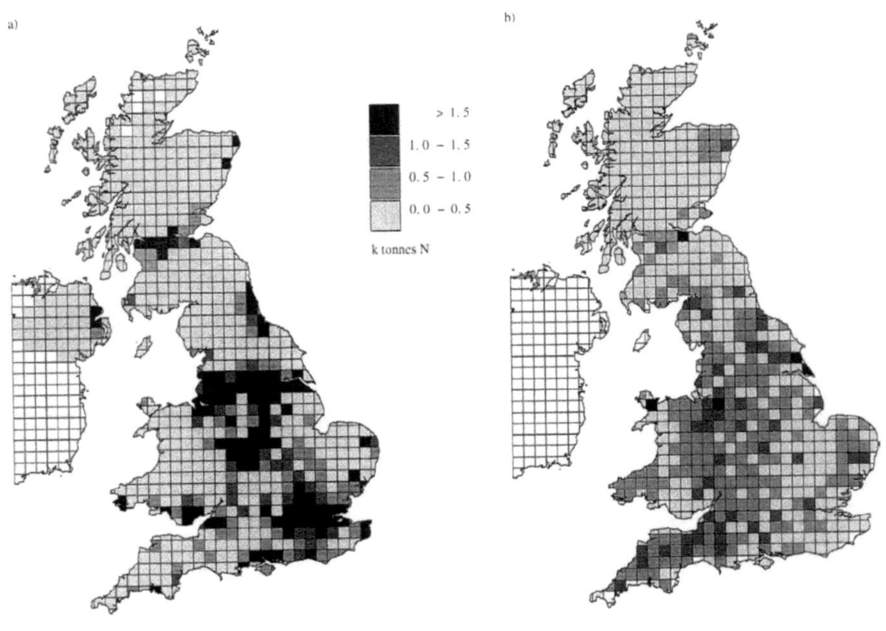

Figure 2.2: (a) UK NO_x emissions for 1994 on a 20 x 20km grid and (b) UK NH_3 emissions based on the Department of Environment, Transport and Regions official estimate of 1995.

UK NO_x as NO_2 emissions in 1994 were 2200 k tonnes (669 k tonnes N), of which 56% came from transport and 24% from the electricity supply industry (ESI). The estimate is thought to be accurate to ± 30%. The spatial distribution of these emissions on a 20 km x 20 km grid is shown in Figure 2.2a. The emissions show a strong association with major urban and industrial conurbations such as Greater London, Birmingham, Merseyside, Tyne and Teesside and through the central valley of Scotland. Since 1970, the trend in NO_x emissions has been relatively flat, with the peak in 1989 of 2754 k tonnes. The nature of the emission sources has, however, undergone a significant shift from point sources to mobile, low level sources (transport). In 1979, for example, ESI contributed 37% of the emissions and all transport 40%. Emissions from industry have fallen by 53% since 1970. The current estimate for NO_x emissions in 2010 is 118k tonnes (RGAR, 1998), of which all transport will provide 47% and the ESI 18.5%. The changing nature of NO_x sources has considerable implications for modelling N (see Section 2.5) and for emission regulation.

The UK's total N_2O emission in 1993 was estimated to be 78 k tonnes, with 81% coming from industrial sources and 13% from agriculture (Salway *et al.*, 1996). Although the percentage contribution was small, an increasing contribution from cars fitted with catalytic convertors was noted by the authors. Emissions of N_2O from soils are highly uncertain (Skiba *et al.*, 1994) and may be considerably higher than the published estimate.

The latest official estimate of UK NH_3 emissions is 320 k tonnes (260 k tonnes N) for 1993 (DoE, 1995) and the gridded data for Great Britain are illustrated in Figure 2.2b. The map shows generally higher emissions over the lowlands than the uplands. The highest emissions occur in the southwest, along the Welsh borders and in parts of north-west England. Emissions in Scotland are generally low except in the central valley. Although it is clear that cattle are the dominant source of NH_3 (> 50%), the actual emission total is highly uncertain, Sutton *et al.* (1995) suggest ± 54%. Making estimates of NH_3 emissions is extremely complex. The uncertainty arises through difficulty in obtaining representative emission factors for each of the main sources (dairy cows, other cattle and livestock). There is a huge variability in farming practice as well as in the stages in processing annual waste that present an opportunity for release of NH_3 to the atmosphere. Over grasslands and croplands bidirectional fluxes may occur, with deposition and then emission occurring at the same place over time (see Section 2.4.3). Particularly when NH_3 emissions estimates are used by atmospheric modellers interested in fluxes over 10s to 100s km, it may be necessary to reconsider what is meant by an 'emission' (RGAR, 1998). A number of different estimates have been made for the UK over recent years and are discussed in INDITE (1994) and RGAR (1998). The current value for the UK of 260 kt NH_3-N per year is substantially smaller than the value 450 kt NH_3-N estimated by Sutton *et al.* (1995) and that used by EMEP (1996). It is inevitably very difficult to identify trends in NH_3 emissions though time. Dragostis *et al.* (1996) suggest an increase of about 5% in ammonia emissions in GB between 1969 and 1988.

The provision of reliable emission inventories in terms of totals, allocations to source categories and spatial distribution is of great importance to those involved in atmospheric pollution modelling. The precision with which such inventories can be compiled varies markedly between different pollutants depending upon the nature of the sources (point vs area, static vs mobile), the pollutants atmospheric lifetimes and our understanding of the processes involved (see Section 2.2).

2.4 Nitrogen Deposition In The UK

The removal of nitrogen from the atmosphere may be divided into two broad categories: wet and dry deposition. Each category may be sub-divided by compound, but it is convenient to consider the processes within these categories because current understanding of inputs to the ground is based closely on the available monitoring information. For wet deposition many of the processes leading to the observed concentrations in precipitation remain poorly known, but the monitoring networks provide reasonably detailed spatial and temporal fields for both oxidised nitrogen as NO_3^- and reduced nitrogen as NH_4^+. By contrast, dry deposition of gaseous nitrogen compounds is not routinely monitored at a network of locations in any country. The input estimates are provided by a combination of measured concentration fields and models of the deposition process which are validated and, or, parameterised by field measurements. There are, therefore, very different approaches for the provision of wet and dry deposition inputs. There are also important complexities in the landscape which lead to large inputs, for example, the proximity of large point sources of NH_3, or mountains with very large annual wet deposition. In this section we briefly examine the deposition processes and the current extent of understanding, followed by a summary of the deposition budgets and their spatial distribution.

2.4.1 WET DEPOSITION

The fixed nitrogen compounds present in the atmosphere include the reactive gases NO, NO_2 , HONO, HNO_3 and NH_3 and the aerosols containing NO_3^- and NH_4^+(see Section 2.2). Some of the gases are soluble and may be removed from atmosphere by falling precipitation, a process commonly known as a washout, or incorporated in cloud or rain within cloud by processes collectively known as rainout. The distinction between washout and rainout while commonly used in early work is entirely arbitrary. The processes contributing to NO_3^- and NH_4^+ in precipitation do not make the same distinction and operate throughout the lifetime of individual droplets. An index of the efficiency of precipitation scavenging of air pollutants is provided by the scavenging ratio, which is given by the ratio of the concentration of substance per unit mass of rain to that in surface air. The scavenging ratios for NO_3^- and NH_4^+ lie typically in the range 500 to 800 in UK data.

The processes which combine to determine the concentrations of NO_3^- or NH_4^+ in rain include the impaction, diffusion, and phoretic processes by which aerosols are captured by cloud and rain droplets (Hales, 1978). More important however, is the nucleation scavenging of aerosols in which the aerosols containing NO_3^- and NH_4^+ act as condensation nuclei for cloud droplet formation. These aerosols are hygroscopic and are readily activated as they are lifted through cloud base by orographic processes or updraughts in clouds.

The relative contribution to observed concentrations in precipitation of different scavenging processes has been considered by Garland (1978) who concludes that nucleation scavenging contributes approximately 65% of SO_4^{2-} in precipitation. Nucleation scavenging also contributes a similar fraction of the observed NO_3^- and NH_4^+ to precipitation (Cawse, 1975, Fowler, 1980).

Contamination of wet deposition collectors by NH_3 dry deposition onto the collector surfaces may represent a significant source of NH_4^+ in wet deposition measurements especially in areas of high NH_3 concentration. The relative contributions of the different direct scavenging processes to be observed for NO_3^- in precipitation in UK conditions

Table 2.2: Processes contributing to nitrogen in rain (From Fowler, 1980).

PROCESS	NITROGEN	
	Range of concentrations in rain at ground level ppm NO_3^-	**Average contribution wet-deposited nitrogen %**
Diffusiophoresis	10^{-3} – 10^{-2}	2.5
Brownian Diffusion	10^{-3} – 10^{-2}	2.5
Impaction and Interception	10^{-2} – 10^{-1}	10
Solution and Oxidation of Gaseous 'Species'	10^{-2} – 0.4	15 – 25 *(2)*
Cloud Condensation Nucleus Pathway *(4)*	10^{-1} – 5.0 *(3)*	60 – 70 *(1)*

(1) Considering rain with geometric mean NO_3^- concentration of 0.5 $\mu g\ g^{-1}$ (weighted for rain quantity)
(2) Uncertainty in this component necessarily leads to uncertainty in other components
(3) Lower limits of range deduced from average contribution and range of concentrations measured.
(4) This table considers the whole wet deposition pathway, no distinction between RAINOUT and WASHOUT.

are summarised in Table 2.2. Interesting as these processes are, the precise contributions made by the different aerosol scavenging processes to the observed concentrations in rain are not limiting current knowledge of wet deposition, its spatial distribution or links with emission process. The current estimates of wet deposition in the UK are limited mainly by the monitoring network for concentration and procedures to interpolate wet deposition in the high rainfall areas. The network comprises 32 weekly and daily collectors which are used to map the regional concentration fields for the major ions in precipitation. The volume weighted annual mean concentrations are interpolated using a kriging technique (Webster *et al.,* 1991) to provide an estimate of the spatial variability in concentration throughout the UK. The annual wet deposition is, in principle, obtained as the product of the mapped concentration and the annual precipitation. For individual sites at which the precipitation amount and concentration are monitored the procedure is straightforward. However, the wet deposition monitoring network is, for good practical reasons, restricted to accessible sites on low ground. The mountains, which are the areas of the country with the largest precipitation, are very difficult to monitor. If concentrations of the major ions did not change with altitude in these upland areas, the estimates of precipitation could still be used to calculate wet deposition. However, the concentrations of major ions do change with altitude in the uplands. Furthermore, the relationship between the composition of precipitation and altitude is highly variable across the upland regions. To obtain wet deposition throughout the country including the high elevation, high precipitation areas, methods for quantifying the orographic enhancement in deposition have been developed and incorporated into the wet deposition maps (RGAR, 1998).

Orographic enhancement of rainfall in the uplands occurs largely as a consequence of the washout of low level orographic clouds. The process is commonly referred to as seeder-feeder scavenging and is illustrated schematically in Figure 2.3.

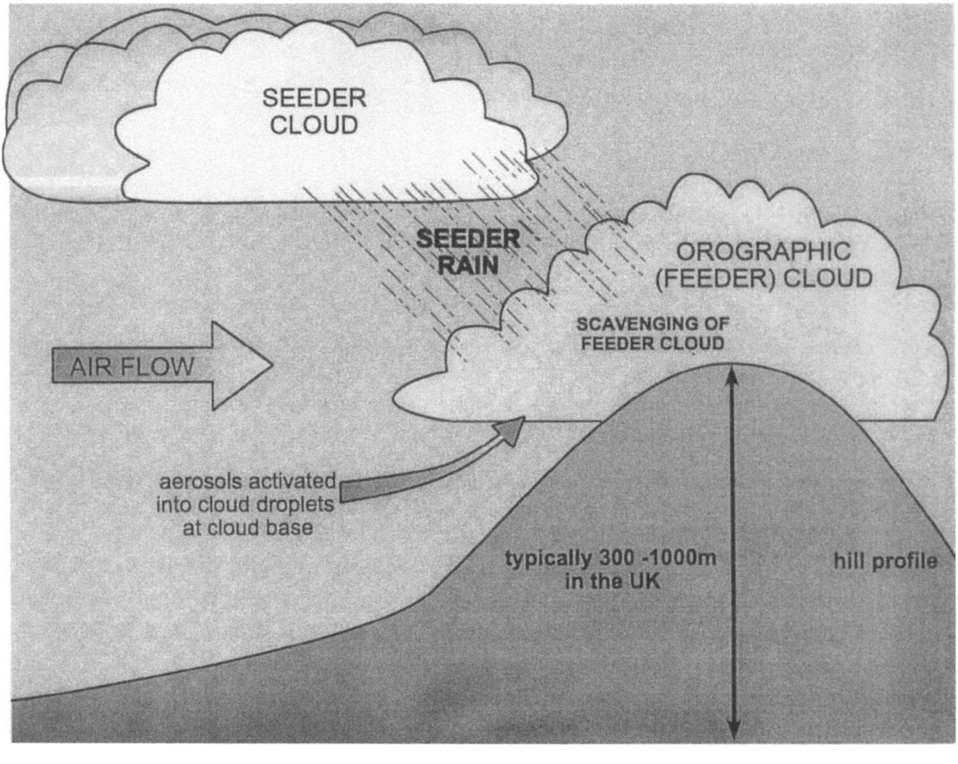

Figure 2.3: A schematic of the seeder-feeder wet deposition scavenging process.

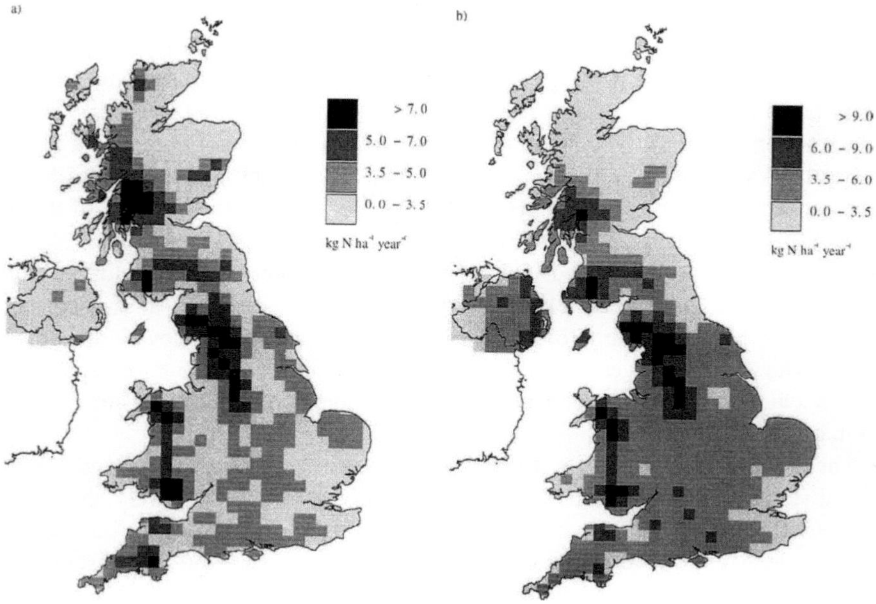

Figure 2.4: Mean annual wet deposition for the UK during 1992–94 of: (a) NO_3^- (kg N ha^{-1} yr^{-1}) and (b) NH_4^+ (kg N ha^{-1} yr^{-1}) (RGAR and CLAG, 1998).

The mechanism of seeder-feeder enhancement of precipitation modifies the composition of precipitation in the uplands because the feeder cloud incorporates soluble ions from the relatively polluted boundary layer. In fact the NO_3^- and NH_4^+ aerosols are very readily activated into cloud droplets at cloud base as air is lifted up hillsides and the concentration of the major ions in aerosol phase per unit volume of air is very similar to that in cloud water. The solute concentration declines with the increase in cloud liquid water content but in the absence of any additional entrained aerosol or soluble gases (NH_3 or HNO_3) the concentration per unit volume of air remains constant.

The seeder-feeder processes and their effects on wet deposition in the UK uplands have been extensively studied in field experiments (Fowler *et al*., 1988; Choularton *et al.,* 1988). The modelling of processes includes simulation of the underlying meteorology and precipitation scavenging at individual hills (Inglis *et al.,* 1995) and over extended areas of complex terrain (Dore *et al.,* 1990). These studies have been used to develop a simple method of extending the orographic enhancement to the UK to provide the basis for mapping wet deposition (RGAR, 1998). Several problems arise in the simplification. First, in average conditions with wind drift of the falling droplets, the washout of feeder cloud (Figure 2.3) leads to the point of maximum deposition on the ground being displaced downwind of the mountain summit, and the distance varies with windspeed. Second, the extent of orographic enhancement is strongly influenced by upwind topography and lastly, in the full complexity of a range of mountains, patterns of wet deposition are very difficult to simulate and even more difficult to validate with measurements.

These complications restrict the current spatial resolution of the wet deposition maps to 20 km x 20 km, and finer scale deposition estimates cannot be provided using this approach. The maps of wet deposition at this scale, and the underlying methods have been subject to a range of validation experiments using long-term catchment budgets of SO_4^{2-} and Cl^- (Reynolds *et al.*, 1997) and ^{210}Pb inventories in soil (Fowler *et al.*, 1995). The wet deposition of NO_3^- and NH_4^+ for the UK is shown in Figures 2.4a and b. The maximum values are found in the south Pennine hills, Cumbria and in Snowdonia. The areas of high wet deposition of NH_4^+ and NO_3- are very similar and each of the ions in these regions contributes approximately 10 kg N ha^{-1} for NO_3^- and NH_4^+. The cumulative total wet deposited NO_3^--N in the UK amounts to 110 k tonnes annually (averaged over the period 1992–1994). The value includes a small (national) contribution from cloud droplet deposition which is important at elevations in excess of 600 m. The equivalent wet deposition for NH_4^+-N in the UK is 120 k tonnes N annually. The spatial distribution in wet deposition follows the high rainfall areas of the west and north of the country.

The national networks for wet deposition are quite recent and only provide adequate spatial resolution for the UK maps since the mid 1980s. The relatively short series of data reveal no clear trends in NO_3^- or NH_4^+ with time. By contrast SO_4^{2-} concentrations and wet deposition have shown decreases during this period, but by substantially less than the decline in UK sulphur emissions (RGAR, 1998). It is not possible, therefore, to examine historical trends in NO_3^- or NH_4^+ throughout the country. The few data available for the second half of the last century, from the work at Rothamsted, suggest a wet deposition of nitrogen of 1–2 kg N ha^{-1} annually for NO_3^- and NH_4^+ (Lawes *et al.*, 1861). However these early data were estimated using very different methods and are not strictly comparable with the data from the last two decades. The remote arctic regions of Europe receive a wet deposited nitrogen input of about typically 1 kg ha^{-1} annually (EMEP, 1996).

2.4.2 DRY DEPOSITION

The range of gaseous nitrogen compounds differ in their reactivity with natural surfaces and solubility in water. Thus the rates at which these gases deposit range from the very reactive nitric acid HNO_3, which is deposited on to all natural surfaces at the rate at which turbulence in the atmosphere can transport the gas to the surfaces, to nitric oxide, which deposits at a very limited rate on to vegetation, soil or water. The other two gases NH_3 and NO_2 are the dominant components of the reduced and oxidised pollutant nitrogen respectively, and both are deposited onto natural surfaces at significant rates.

To summarise current knowledge of the dry deposition of nitrogen compounds it is convenient to outline the physical steps in the deposition pathway which are common to all gases. The different controlling steps for each gas will then be presented. The gases in the atmospheric boundary layer are advected by wind and transported to terrestrial surfaces by turbulence. As the turbulence is generated by frictional drag at the surface on the airflow above, surfaces which differ in their aerodynamic roughness lead to different rates of turbulent exchange with the atmosphere. Forests for example exert much greater drag on the airflow than grassland. Close to terrestrial surfaces, turbulence is suppressed by the viscosity of the air and very close to absorbing surfaces transfer relies on molecular diffusion to reach the leaf or soil surfaces. Finally, at the leaf or soil surfaces, the gases may react with external surfaces or diffuse further into the foliage via stomata or into soils.

These steps in the deposition pathway are commonly quantified using a simple resistance analogy in which the transfer is treated as an analogue of electrical current flowing through a network of resistances (Figure 2.5).

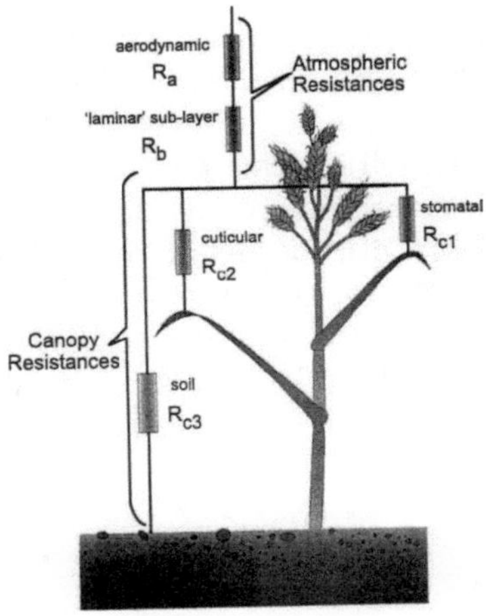

Figure 2.5: A resistance analogue of the dry deposition processes for pollutants onto vegetation and soil.

NO and NO_2 deposition to soil and vegetation: In controlled environment studies, the deposition of NO_2 onto vegetation has been shown to be regulated by stomatal uptake (Rogers *et al.*, 1979). The residual NO_2 uptake onto external, cuticular surfaces is detectable but very small, contributing to a few percent of the overall flux (Hargreaves *et al.*, 1992). Deposition rate is commonly expressed as a velocity, termed deposition velocity (Vg) which normalises the flux for ambient concentration, thus

$$Vg_{(z)} = \text{flux/concentration}_{(z)}$$

The subscript $_{(z)}$ denoting height is required since both the concentration and the deposition velocity vary with height. The deposition velocity for any of the pollutant gases is also the reciprocal of the total resistance to transfer between the reference height in the atmosphere and the absorbing surface (Figure 2.5). Rates of deposition to natural surfaces range from 1 mm s^{-1} to a few cm s^{-1} for most pollutant gases. The upper rate limit of the process is determined by the turbulent transport above the surface, which in turn is largely governed by wind velocity and the aerodynamic roughness of vegetation. Above the roughest surfaces, forests, the maximum rates of deposition are in the range of 5 to 10 cm s^{-1} and apply for gases which are absorbed by the forest canopy at the upper limit of turbulent transport. Nitric acid vapour and HCl fit into this category. For NO_2, the stomatal uptake limits the maximum rate of deposition to canopies of vegetation, the rates ranging with tree species and leaf area index but this limits maximum NO_2 deposition rates to the range 5 to 8 mm s^{-1}. In general forests have smaller maximum stomatal conductances than short agricultural crops (grasses and cereals) (Jarvis, 1976), largely to protect forests against the potentially very large transpiration loss of water. Short vegetation (<1 m) is protected to some degree by the larger aerodynamic resistances which limit deposition rates and crop conductances to values smaller than 2 cm s^{-1}.

A complication in the NO_2 deposition process, widely measured in earlier work (Duyzer and Fowler, 1994), is the reaction of NO emissions from soil with O_3 to form NO_2 in the crop canopy (Figure 2.6). This process leads effectively to a compensation point for NO_2 exchange such that ambient NO_2 will not deposit to vegetation wherever the internal canopy production from soil NO emissions exceeds the above canopy NO_2 concentrations. For the majority of the polluted regions of the UK the ambient NO_2 concentrations are substantially larger than the in-canopy NO_2 concentrations generated by soil NO emissions. However, the strongly positive temperature response of NO emission from soil (Skiba *et al.,* 1997) and the co-incidence of the high soil temperatures and stomatal uptake lead to the 'internal' cycle of NO production and uptake within crop canopies significantly reducing the net deposition of NO_2 from the atmosphere.

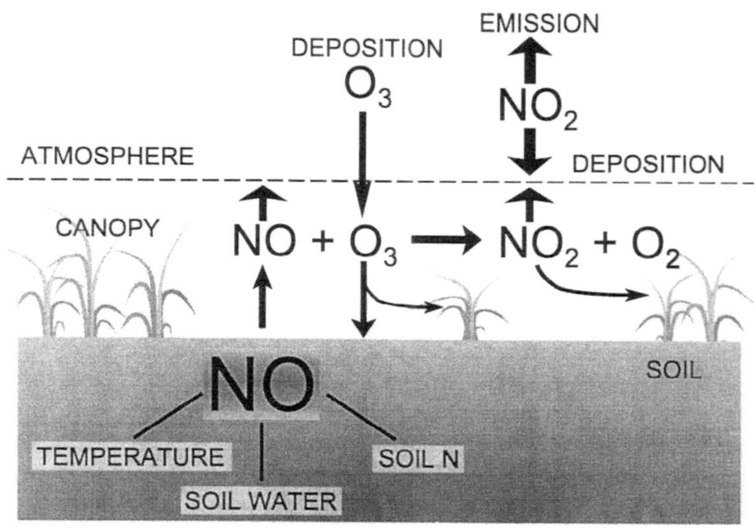

Figure 2.6: A schematic showing principal surface–atmosphere exchange processes for NO_2 (From RAGAR 1988)

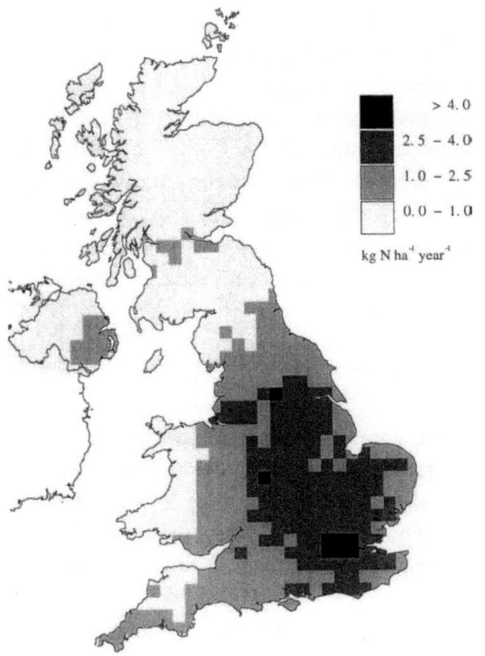

Figure 2.7: Mean annual dry deposition of NO_2 (kg N ha^{-1} yr^{-1}) for the UK during 1992–94. (CLAG, 1997)

Models are currently being developed to quantify the NO_2 exchange within canopies from soil emissions but their effect on the net UK NO_x budget has yet to be quantified.

The NO_2 deposition to the UK countryside is calculated using a combination of measured concentration fields and a dry deposition model which simulates the NO_2 uptake by stomata (RGAR, 1998). The stomatal resistance is calculated for the main classes of vegetation present (grassland, crops, moorland and forest) using solar radiation, temperature and wind velocity as the main meteorological inputs.

A consequence of the stomatal NO_2 uptake is that the majority of deposited NO_2 occurs during the spring and summer months and winter NO_2 deposition is small. The small winter deposition values also lead to a longer atmospheric lifetime for NO_2 in the winter. The longer atmospheric lifetime, combined with reduced boundary layer depth and larger winter emissions, lead to a marked increase in ambient NO_2 concentrations in the winter months. The NO_2 dry deposition to the UK ranges from 0.5 to 5 kg N ha^{-1} annually and is at a maximum in the polluted regions of central and southern England (Figure 2.7). It amounts to approximately 40k tonnes N annually and represents only 5% of UK emissions.

Nitric acid (HNO_3): The dry deposition input of nitric acid is limited only by atmospheric transfer and rates of deposition expressed as a deposition velocity exceed those for NO_2 by typically an order of magnitude. However, HNO_3 concentrations are small and are measured at very few locations. The data from a year of measurements in southern Scotland showed an average concentration of 80 ng m^{-3} HNO_3, while further south at High Muffles close to the coast in north-east England the concentration averages 500 ng m^{-3} HNO_3, with summer peak values of 2 µg HNO_3 m^{-3}.

With such limited spatial coverage of HNO_3 concentration quantifying its contribution to total oxidized nitrogen deposition in the UK is very speculative, but these concentrations can be used to scale the likely range of deposition in the UK. The range from the south Scotland and High Muffles data would lead to inputs of HNO_3 in the range 0.5 to 5 kg N ha^{-1} annually and a possible contribution to the total budget of oxidised nitrogen of the same order as NO_2 (40 k tonnes N ± 30). However, current knowledge of ambient HNO_3 is too uncertain to add this term to the atmospheric budget of oxidized nitrogen over the UK.

2.4.3 AMMONIA DEPOSITION

The emissions, atmospheric concentrations and deposition of ammonia are subject to considerable uncertainty, due partly to difficulties in the measurement of ammonia, but also to the wide range of sources and variability in their characteristics (Sutton *et al.*, 1995a). The problems of ammonia in the atmosphere have become a focus of scientific and political interest quite recently and our understanding of the processes is developing quickly.

The patchiness of ammonia emissions and the lack of a satisfactory monitoring network to capture this spatial heterogeneity precludes preparation of NH_3 concentration maps directly from measurements (RGAR, 1998). The approach taken to provide estimates of concentration and deposition has been to model NH_3 emissions, transport and deposition using a long range transport model (FRAME) to provide a high resolution map of surface NH_3 concentrations and to validate these concentrations with measurements of ambient NH_3 at approximately 70 measurement stations throughout the UK including transects in two of the major source areas in East Anglia and Cheshire (RGAR, 1998).

The application of deposition rates to the concentration field for NH_3 is more complex than that for NO_2. The exchange of NH_3 with vegetation is bi-directional with the potential for emission from vegetation whenever ambient air concentration is smaller than the canopy compensation point.

Field studies during the last two decades have shown that for semi-natural vegetation, NH_3 is deposited rapidly, especially onto wet canopies when deposition velocity approaches Vmax. However the long-term average deposition rates are smaller than Vmax and median canopy resistances for these surfaces are in the range 15–30 m s^{-1}. For agricultural crops, emission fluxes are not uncommon, especially following application of nitrogen containing fertilizer and during senescence (Sutton *et al.*, 1993). The presence of surface water was also shown to have a marked effect on canopy resistance.

To simulate the bi-directional exchange, and to quantify the cuticular and stomatal exchange of NH_3, a two leg canopy compensation point model of atmosphere–surface exchange has been developed (Sutton *et al.*, 1995a). The underlying equations are

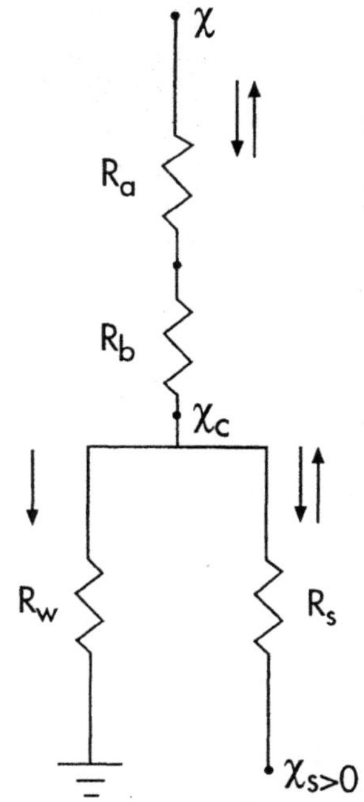

Figure 2.8: Schematic of ammonia 'canopy compensation point' model. The net flux is considered as a result of exchange between air concentrations (χ) and the 'canopy compensation point' (χ_c), which derives from competition between exchange to the leaf surface and transfer through stoma (R_s) with a stomatal compensation point (χ_s). R_a, R_b, and R_w are the resistances for transfer through the turbulent atmosphere, the quasi-laminar boundary layer and to the leaf cuticle surface, respectively. (From RGAR 1998).

described by Sutton *et al.* (1995a) and a schematic of the approach is provided in Figure 2.8. The important development here is the provision of methodology to quantify the combination of cuticular and stomatal fluxes simultaneously. Several important variables within the model are very site and time dependent. The model has, however, been tested against flux measurements over semi-natural vegetation and over fertilized crops and has been shown to simulate the bi-direction exchange of NH_3 satisfactorily (Sutton *et al.*, 1995b).

A version of the two leg model has been used to calculate NH_3 deposition in the UK by quantifying inputs to a range of vegetation classes (cropland, moorland, forest etc). When scaled for land use, this approach provides a map of the net inputs of NH_3 to the UK (Figure 2.9). The sum of dry deposition totals 110 k tonnes NH_3- N annually or 42% of the UK emissions.

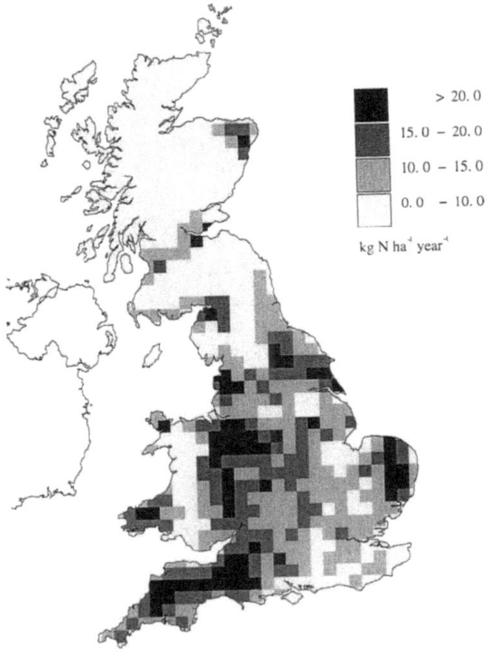

Figure 2.9: Mean annual dry deposition of NH_3 (kg N ha^{-1} yr^{-1}) to moorlands in Great Britain during 1992–94. (RGAR and CLAG, 1998)

2.4.4 TOTAL NITROGEN DEPOSITION IN THE UK

The wet deposition of NO_3- and NH_4^+ (Figures 2.4a and b) provides 110 k tonnes NO_3^- - N and 120 kt NH_4^+- N annually, the bulk of which is deposited in the high rainfall areas of the north and west of the country. The dry deposition inputs of 40 k tonnes NO_2 - N and 110 k tonnes NH_3 - N (Figures 2.7 and 2.9) provide the bulk of their input to the east and south of the UK.

These inputs of fixed nitrogen suggest quite different atmospheric lifetimes for reduced and oxidised nitrogen. For example, if the atmospheric budget for oxidized nitrogen (Figure 2.10a) is examined the total deposition represents only 20% of emissions, and the vast majority of UK emissions are advected out of the country by the wind. By contrast, the budget for reduced nitrogen (Figure 2.10b) implies that most of the UK emissions of NH_3 are returned to the UK mainland before the air reaches the coast, with the sum of wet and dry deposition representing 80% of the emissions plus imports. The implication is that reduced nitrogen has a much shorter atmospheric lifetime due to the much faster dry deposition and rapid incorporation of gaseous ammonia into acidic aerosol or cloud water.

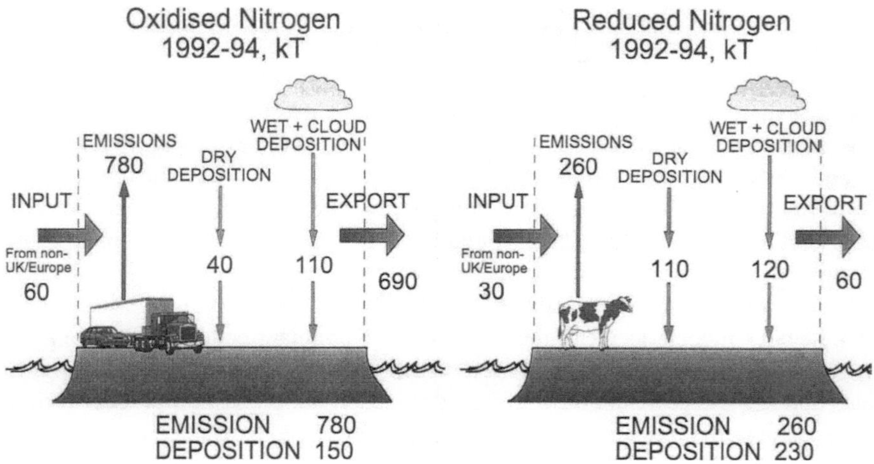

Figure 2.10: Schematic representation of the Great Britain (a) oxidised nitrogen budget 1992–94 and (b) reduced nitrogen budget 1992–94.(From RGAR 1998)

The budget values, however, demand further comment as they imply a satisfactory balance between the emission and deposition terms. In fact the export of reduced nitrogen at 60 k tonnes NH_3 - N is deduced as the residual term. In practice, measured aerosol NH_4^+ concentrations and the wind speed at the coast imply advection losses of aerosol NH_4^+ substantially larger than 60 k tonnes NH_4^+- N, requiring either the deposition terms to be reduced or the emission to be increased. Recent progress in quantifying both terms in the deposition budget suggest that error in emissions is more likely, and that UK NH_3 emission are of the order 350 k tonnes NH_3 - N. With emissions of this order, the UK budget for reduced nitrogen would balance reasonably well and international comparisons of livestock NH_3 emissions would also be more consistent than is presently the case.

2.5 Modelling Nitrogen Deposition

2.5.1. INTRODUCTION

Modelling the transport, transformation and deposition of atmospheric nitrogen poses a number of significant challenges. Early models of the long-range transport of air pollutants dealt largely with sulphur alone (e.g. Fisher, 1975; Eliassen, 1978). Compared with nitrogen, emissions of sulphur are more easily quantified and in the UK, and much of north-west Europe and North America, these emissions come mainly from large power stations, often with tall stacks where simplifying assumptions concerning mixing through the boundary layer are more likely to be tenable. In addition, compared with nitrogen, sulphur chemistry is relatively simple and until recently much better understood. The chemistry described in Section 2.2 and the removal processes described in Section 2.4 are difficult to represent in anything but the most complex model. The limitations of our understanding as well as on processing power, and the desire for relatively simple tools with which to explore present day and possible future deposition, means that many simplifying assumptions have to be made and processes represented by parameterisations. Models which included nitrogen species were not widely developed before the mid-1980s (e.g. Derwent and Nodop, 1986) and in many cases these did not include the chemical coupling known to take place between the various sulphur and nitrogen compounds. The approaches taken in a number of national scale deposition models currently in use in the UK have recently been reviewed (RGAR, 1998). The exercise highlighted the problems of handling low level emissions in single layer trajectory models to yield good estimates of pollutant concentrations (e.g. NO_2) close to source areas. Only those models which included a range of oxidised nitrogen species and some representation of the seeder-feeder wet deposition process were able to come close to the measured wet N deposition values. None of the models apparently performed well with respect to reduced nitrogen, but some of this may reflect the many uncertainties which have been referred to in previous sections. At the European scale, the EMEP model has become the primary tool in policy development and hence, is described in detail below.

2.5.2 THE EMEP MODEL

Under the terms of the 1979 LRTAP convention estimates of trans-boundary fluxes and depositions of pollutants have to be made for the member countries. This requirement is fulfilled by the EMEP model run at MSC–W (Barrett *et al.*, 1995). The original EMEP model was for sulphur only (e.g. Eliassen and Saltbones, 1983), but from 1987 oxidised and reduced nitrogen have been included in a coupled chemical scheme. The current standard model is a receptor orientated Lagrangian model with 150 km resolution, although a 50 km Lagrangian model and an Eulerian model have also been developed. The 150 km model has been used extensively to develop and assess abatement strategies within the UNECE (see Section 2.6). Up-to-date descriptions of the EMEP model are given in annual reports, a recent example of which is Barrett and Berge (1996). The model calculates concentrations and depositions of ten acidifying species and their precursors. Only a brief description will be given here, with the focus on nitrogen compounds. The model has a single layer which represents the mixed boundary layer over Europe and the north-east Atlantic. Initial concentrations for the calculated components are provided by a 3–D hemispheric model and some measurement data for forms of sulphate and nitrate; initial values of O_3, hydroxyl and peroxyacetyl radicals are prescribed. Exchange between the boundary layer and the free troposphere is represented. The model uses the 150 km x 150 km emissions data described in Section 2.3.1. Seasonal variability in emissions is generated, using country specific monthly scalings from GENEMIS (Lenhart and Friedrich, 1995) for SO_2 and NO_2 and a sine function for NH_3, indicative of the far greater uncertainty relating to this pollutant. Most of the chemistry in the EMEP model is linear (see Barrett and Berge, 1996 for details), although non-linear equations describe the formation of ammonium sulphate and ammonium nitrate. The chemical evolution of the air parcel is calculated with a time step of 15 minutes. Until January 1996 the EMEP model used meteorological input data derived from the Norwegian Weather Prediction model, but has moved on to the HIRLAM model (High Resolution Limited Area Model) which covers a larger area and has more vertical layers. Meteorological data (e.g. wind speed, temperature, rate of precipitation) are archived every six hours. The EMEP 150 km model also makes use of observed precipitation over land and radiosonde data for the height of the planetary boundary layer.

In common with most single layer models, the EMEP model tends to underestimate near surface concentrations and hence dry deposition fluxes. As a result, a local deposition correction factor is applied, 0.04 for dry deposition of NO_2 and 0.19 for dry deposition of NH_3. For NH_3, a wet deposition correction of up to 0.19 is also applied. The representation of dry deposition within the EMEP model depends upon the pollutant. For NO_2, HNO_3 and NH_3 a resistance analogy is used, where the grid square average reflects the different land uses in the square (land use data are supplied by RIVM in the Netherlands). For PAN, values based on maximum possible deposition velocities, adjusted for season and land use, are used. Constant values are prescribed for nitrate aerosols. The dry deposition velocities for gases are calculated for 1 m above the ground and then transformed to give 50 m values more typical of surface layer depth. Wet removal of pollutants is represented through scavenging ratios, with a probability function to describe the likelihood of encountering precipitation at any given point in a grid square. Maps and tables of modelled concentrations and depositions across the EMEP area are published annually. Modelled deposition fields for oxidised and reduced nitrogen for 1992 are shown in Figure 2.11. The oxidised N field (Figure 2.11a) shows a clear area of high deposition extending from central England eastwards in to Poland, while the reduced N field (Figure 2.11b) shows high levels of N deposition over much of western and central Europe. It is notable that for 1994, over much of the EMEP area, combined nitrogen deposition contributes more to total deposited acidity than does S. Comparison of the EMEP emissions and deposition maps (Figures 2.1 and 2.11).indicates the very different behaviour of oxidised and reduced N. A very high proportion (> 80% in almost all cases) of NO_x emitted by any country is exported, while for reduced N export levels are much lower (>40%). EMEP publishes import–export budgets for the member countries. Over the period 1985–1994, for example, it is estimated that Norway exported 90% of its NO_x emissions, but imported 91% of its oxidised N deposition; in contrast the percentages for the UK over the same period were 90% and 39% respectively. For reduced N the percentages are very different (47% and 67% for Norway, 50% and 20% for the UK) although the countries' relative positions as a net importer (Norway) and net exporter (UK) of pollutants remains.

Although EMEP are also responsible for a network of daily monitoring sites and model output is compared with the data from these, little use of made of the measured data to calculate deposition. Data from the EMEP sites, as well as other national monitoring programmes, were used by Van Leeuwen *et al.* (1996) to map wet deposition. Although the general patterns are similar to those generated by the EMEP model it is interesting to make a more detailed comparison.

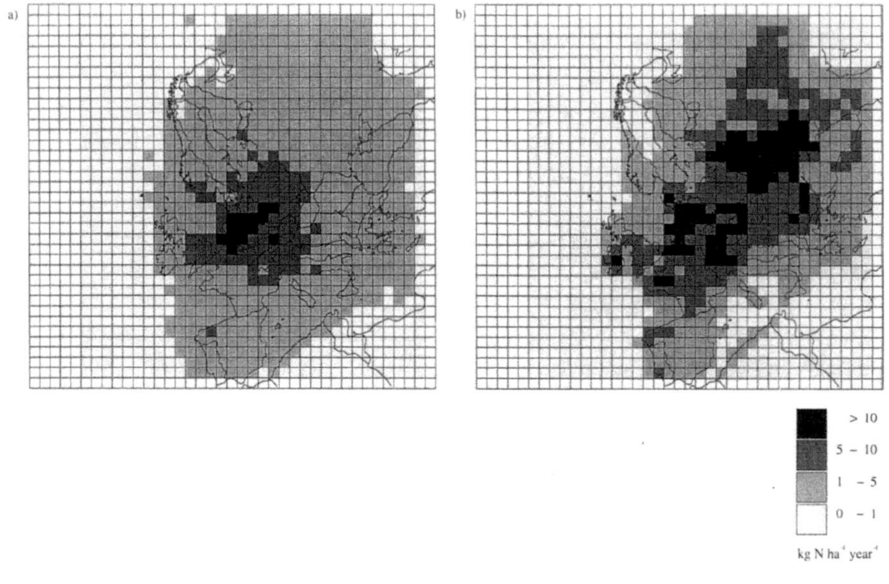

Fig. 2.11: (a): EMEP area modelled total oxidised N deposition for 1992 and ((b): EMEP area modelled total reduced N deposition for 1992

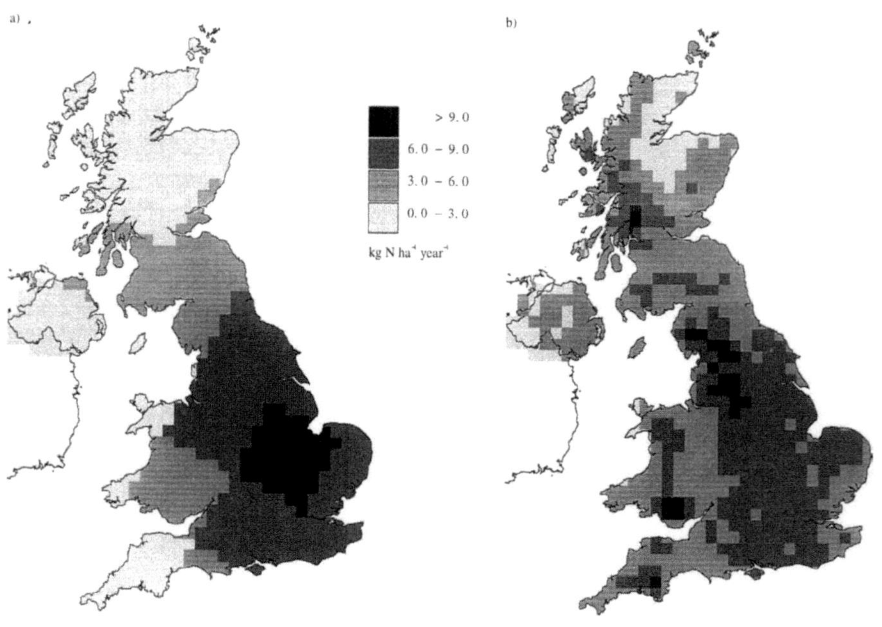

Figure 2.12: (a) EMEP modelled total oxidised N deposition to the UK for 1992 and ((b) RGAR (1998) total oxidised N deposition to the UK for 1992–94.

The pattern of EMEP modelled oxidised N deposition across the UK for 1992 is compared with the UK's own estimates for 1992–1994 (RGAR, 1998) in Figure 2.12. The EMEP total compares favourably with that from RGAR: NO_y-N EMEP total 124 k tonnes, RGAR total 150 k tonnes. The spatial distribution of the deposition is, however, quite different, with the EMEP model indicating the highest deposition in central England, while the data based on measurements show the highest deposition in the uplands of western and central Britain as well as near the major source area of London. ased on analyses carried out for the UK Review Group on Acid Rain (RGAR, 1998) it appears that the main discrepancy arises from the EMEP model's underprediction of wet deposition. This may be explained by the absence of any representation of the seeder-feeder process (see Section 2.4.1) from the EMEP model. The inability of the EMEP model to provide a good representation of the distribution of deposition across the UK, or of the relative importance of wet and dry deposition processes, has important implications both for the calculation of present day critical loads exceedances (and therefore ecosystem damage) and for the development of future policy based on the EMEP transfer matrices.

2.6 The Policy Context

Within the arena of international policy to control acidification, there has been more emphasis on deposited S than on N, the most recent piece of legislation within the UNECE being the signing of the second S protocol in Oslo in 1994. A protocol on oxides of nitrogen was agreed in Sofia in 1988, but this only aimed to freeze emissions at 1987 levels by 1994 (Murley, 1996). Within the EU there is policy in place to reduce NO_x emissions through the 1988 Large Combustion Plant Directive (LCPD) and its amendment. Under the terms of the LCPD, NO_x emissions are to be reduced by 15% by 1993 and 30% by 1998 relative to the 1980 base year. The European Commission's 5th Action Programme (1993) proposed further targets of stabilising NO_x emissions at 1990 levels by 1994 with a reduction of 30% by 2000. There are currently no internationally agreed targets for reducing emissions of NH_3, although individual countries, most notably the Netherlands, have set targets. Some of the difficulties in reducing ammonia emissions and in including these in integrated assessment models are discussed in ApSimon *et al.* (1995).

Both the UNECE and the EU are currently considering further legislation to limit environmental damage, in a cost effective way, from acidification, eutrophication and photochemical oxidants within a multi-pollutant, multi-effect framework (Amann *et al.*, 1996). The agreement of a second NO_x (now multi-pollutant) protocol, has slipped from its original timetable, but is likely to be 1998. EU strategies for acidification and ozone are likely to be tabled in the same year. Some of these issues are discussed in more detail in Chapters 1 and 7.

One of the effects of pollution control policies and industrial restructuring (particularly in electricity generation) has been a marked change in the balance between deposited S and N. Across the EMEP area, emissions of SO_2 (all source categories) fell by 30% between 1980 and 1990, while emissions of NO_x rose by 7% (although the trend is now downward – see 2.4.1). Emissions of NH_3 have apparently fallen by 3% over the same period although uncertainty in NH_3 emission is much too large to have confidence in such small changes. Current reduction plans (CRP) (Asmann *et al.*, 1996) would reduce SO_2 by 60%, reduce emissions of NO_x by 12% and emissions of NH_3 by 26% relative to 1980. Within the 15 members of the EU, the ratio of SO_2:NO_2 emission was > 1 in all countries except the Netherlands in 1980, by 1990 only 6 countries had a ratio of >1 and under CRP only Eire and Greece will emit more SO_2 than NO_x. As described in section 2.5.2 even using 1994 emissions, the EMEP model indicates that nitrogen contributes more H^+ than sulphur over much of Europe, with reduced N being more important than oxidised N except in the far north. Given the likely trends in emissions, this dominance of acidic deposition by nitrogen will increase. Some examples of modelled future S and N deposition in the UK and their effects on CL exceedance are given in Section 2.7

2.7 Modelling Future N Deposition – A UK Case Study

Within the UK, the Hull Acid Rain Model (HARM) has been used to model annual depositions of sulphur and nitrogen at a scale of 20 km x 20 km. The model is a receptor orientated Lagrangian statistical model which follows the coupled behaviour of SO_2, NO_x, NH_3 and HCl in air parcels which travel across the EMEP emissions grid and then the UK emissions grid to a series of designated receptor sites. The model has a single layer, but includes a number of oxidised N species and a parameterisation of the seeder-feeder process. Detailed descriptions of the model's treatment of sulphur and nitrogen are given in Metcalfe *et al.* (1995) and Metcalfe *et al.* (1998). Comparisons of model output with data from the national monitoring networks, with our best estimates of deposition and with output from other models (RGAR, 1998) have shown that HARM performs well, particularly for S and oxidised N.

Here we have used HARM to model N deposition using 1994 emissions and compare this output with modelled deposition under a possible emissions scenario for 2010. HARM modelled N deposition for 1994 is shown in Figure 2.13. Modelled NO_y-N deposition is 143 k tonnes for the UK which compares very well with the RGAR total for 1992–94 of 150 k tonnes. HARM NH_x-N is well below the RGAR total (74 k tonnes compared with 230 k tonnes), but this mainly reflects the model's inability to reproduce the dry NH_3-N field, itself a product of model output (see Section 2.4.3). HARM modelled total N deposition (oxidised + reduced) is 217 k tonnes, with the highest depositions across England, Wales and parts of southwest Scotland.

Under the 2010 scenario UK emissions of NO_x were reduced from 2102 k tonnes to 1186 k tonnes, SO_2 emissions from 2726 k tonnes to 980 k tonnes and NH_3 emissions were held constant. EMEP area emissions of NO_2, SO_2 and NH_3 were all reduced in line with published current emissions reduction plans (CRP) (Amann *et al.*, 1996). Modelled N deposition in 2010 is shown in Figure 2.14. Under this scenario, oxidised **N** deposition is reduced to 67 k tonnes, 47% of the 1994 total. The spatial pattern of deposition persists, but the amounts are considerably reduced, with no grid cell receiving more than 6 kg N ha yr^{-1}.

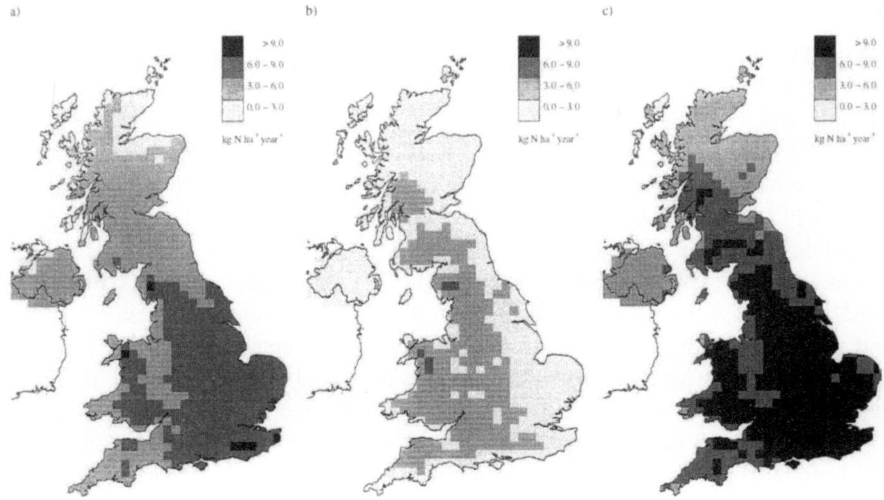

Figure 2.13: HARM modelled N deposition a) oxidised, b) reduced, c) total for 1994

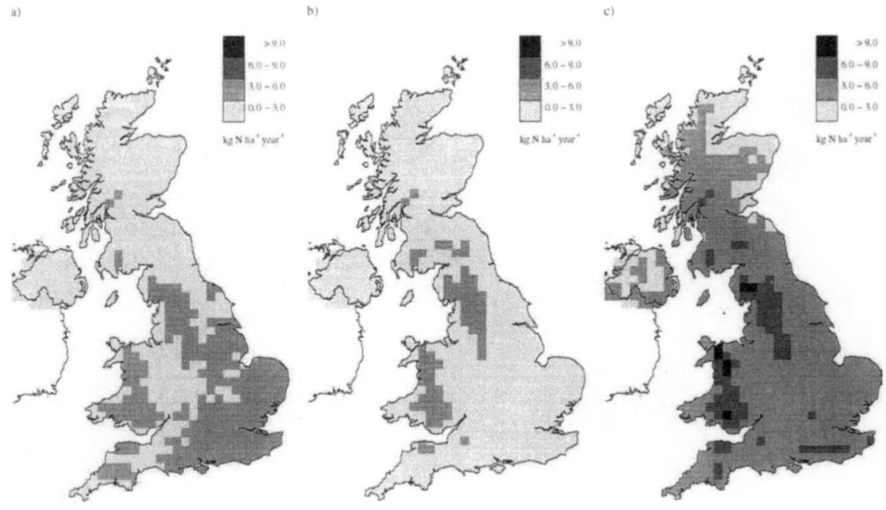

Figure 2.14: HARM modelled N deposition a) oxidised, b) reduced, c) total for 2010

Reduced N deposition declines to 50 k tonnes (77% of 1994) and reflects not only the direct impact of cutting NH_3 emissions across the EMEP area outside the UK, but also the benefits of lower SO_2 emissions.The significance of the coupling between sulphur and reduced nitrogen through the formation of ammonium aerosols (see Section 2.2.2), which cuts wet deposition of NH_4 as SO_2 emissions are reduced, has been discussed in more detail in Metcalfe *et al.*(1997). Modelled total N deposition in 2010 is 117 k tonnes. Areas of high total N deposition (> 9 kg N ha yr^{-1}) persist in the mountains of Wales, especially Snowdonia and in Cumbria.

One way of exploring the environmental effects of sulphur and nitrogen deposition together is to use the critical loads function (CLF), which defines an acidity critical load and a limiting critical load of nitrogen for selected receptors (Bull, 1995). As well as indicating areas where the critical values are exceeded (ie unprotected), the CLF also gives an indication of which pollutants need to targetted for further emissions control. Figure 2.15 shows the distribution of CLF classes under deposition from the 1994 and 2010 model runs. CLF data from December 1996 were used in this exercise. The 1994 map (Figure 2.15a) shows most of England and Wales in CLF region 3, where reductions in both S and N deposition are required. Much of central and eastern Scotland, the western fringes of England and Wales and parts of eastern central England (especially near the large coal-fired power stations of the Trent and Ouse valleys) fall in to CLF region 2 where reductions in S are the first priority.

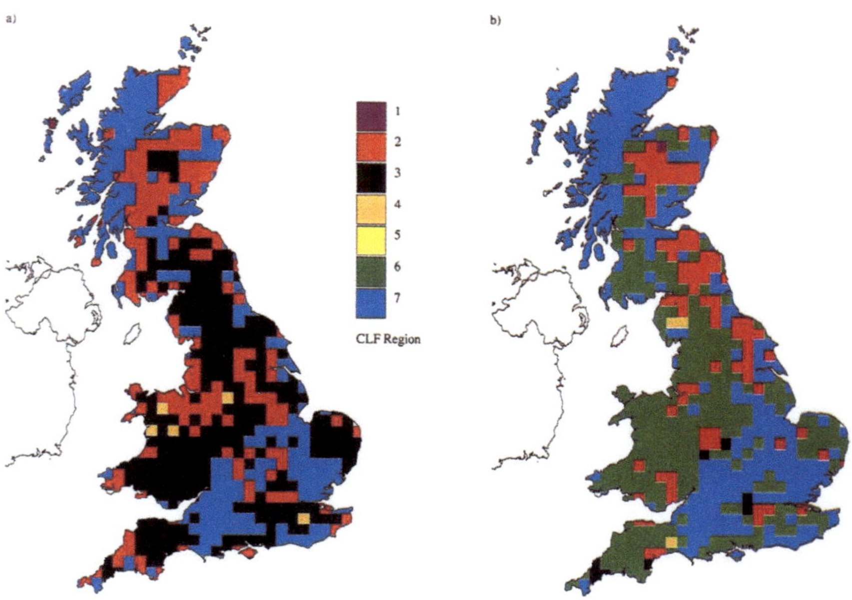

Figure 2.15: (a) Critical load function regions using HARM modelled total N deposition for 1994 and (b) Critical load function regions using HARM modelled total N deposition for 2010. Region 1: reduce S; Region 2: reduce S and then S or N; Region 3: reduce S and N; Region 4: reduce N and then S or N; Region 5: reduce N; Region 6: option to reduce S or N; Region 7: protected.

Only 5 grid cells, of which 3 are in north Wales, lie in region 4 where N deposition must be cut first. The 2010 CLF map (Figure 2.15b) shows a very different distribution; most of central and eastern England and western Britain fall in region 6 (S + N reduction). Along the western side of the country the impact of reductions in SO_2 emissions is clear, but to the east a band of region 2 (S reductions required) has emerged from Scotland to the Humber. nterestingly, the grids in region 4 have changed, with 2 in Cumbria and one in the southwest. It appears, from this exercise, that further reductions in both S and N deposition will be required beyond 2010. As HARM apparently underestimates reduced N deposition, it is also likely that it provides a minimum estimate of the reduction in N emissions required to achieve ecosystem protection.

2.8 Conclusions

Our understanding of the chemistry of atmospheric N compounds has improved considerably in recent years, but this has also highlighted its great complexity. Although much work clearly remains to be done, it has been suggested that our understanding is sufficient for modelling deposition at the regional scale (RGAR, 1998). The development of emissions inventories for the UK becomes increasingly sophisticated (see Salway *et al.*, 1996). The uncertainties in emissions estimates are, however, considerable (20–30% for UK NO_x) and become increasingly uncertain with increasing scale (Graedel *et al.*, 1995; Lee *et al.*, 1997). Estimates of ammonia emissions are particularly uncertain (see Section 2.3.2) and given that reduced N is likely to play an increasingly important role in the N loading of the atmosphere, this is a matter of some concern.

The developments in understanding emissions, deposition and effects of reduced nitrogen in Europe have been stimulated by observed terrestrial effects (Heij and Erisman, 1997). It is of considerable interest that in North America reduced nitrogen has attracted little attention to date. However, with growing interest in the role of regional nitrogen deposition on the sequestration of atmospheric carbon by forests, as well as local effects of eutrophication of land and freshwaters this will change in the near future.

Better understanding of N chemistry and of the processes which remove N from the atmosphere have allowed improved estimates of deposition to be made, although reliable estimates of dry deposition (particularly of NH_3) remain problematic. The methods currently used in the UK to derive national maps of wet and dry deposited oxidised and reduced N have been discussed in Section 2.4. The quality of any deposition estimate is ultimately limited by the spatial and temporal resolution of monitoring networks and by the range of pollutants they measure. It is clear that reliable measured concentration fields for NH_3 and HNO_3 would be particularly valuable. The UK budgets for oxidised and reduced N (Figure 2.10) indicate rather different behaviour, reflecting the generally longer atmospheric lifetime, and hence transport distance, of oxidised N (Barrett and Berge, 1996). This difference in transport distance becomes of particular interest when we attempt to model N deposition and consider the effects of emissions controls both nationally and internationally.

Although the EMEP model is that used by both the UNECE and the EU, it is clear that it does not provide a particularly satisfactory representation of N deposition across the UK, especially in its underestimation of wet N deposition in many areas of upland Britain where ecosystems are often particularly sensitive. The spatial resolution of the current EMEP model is not appropriate for mapping CL exceedance within the UK. Application of a UK scale model assists policy makers in quantifying the environmental benefits of proposed emissions reduction strategies. It is clear, however, that setting future emissions targets for NO_x will be a complex process, which will have to take in to account a range of environmental and health issues beyond considerations of acidification and eutrophication (Grennfelt *et al.*, 1994). Any control of NH_3 emissions is likely to be relatively limited given the practical difficulties involved. The assessment of the environmental impacts of N deposition either in the present, or in the future, must take in to account the uncertainties in the deposition estimates being used and consider whether the spatial and temporal scale of those estimates is appropriate to the receptor ecosystem of interest.

Acknowledgements

Funding for the work described here was largely provided under DETR contract EPG 1/3/98 (SEM and JDW), EPG 1/3/93 (RGD) and EPG 1/3/28 and EPG 1/3/31 (DF, MAS, RIS).

References

Amann, M., Bertok, I., Cofala, J., Gyarfas, F., Heyes, C., Klimont, Z. and Schöpp, W. (1996) Cost effective control of acidification and ground-level ozone. Second Interim Report to the European Commission, DG–XI. December 1996.

Ambio (1997) Nitrogen: a present and future threat to the environment. Special Issue. *Ambio* **26** (5).

ApSimon, H., Couling, S., Cowell, D. and Warren, R.F. (1995) Reducing the contribution of ammonia to nitrogen deposition across Europe. *Water, Air and Soil Pollution,* **85**, 1891–1896.

Atkinson, R. (1994) Gas-phase tropospheric chemistry of organic compounds. *Journal of Physical Chemistry Reference Data Monographs,* **2**, 1–216.

Barrett, K., Seland, Ø., Foss, A., Mylona, S., Sandnes, H., Styve, H. and Tarrasón, L. (1995) *European Transboundary Acidifying Air Pollution.* EMEP/MSC–W Report 1/95.

Barrett, K. and Berge, E. eds. (1996) *Transboundary Air Pollution in Europe.* EMEP MSC–W Report 1/96.

Berge, E. (ed) (1997) Transboundary Air Pollution in Europe. EMEP MSC–W Report 1/97.

Berge, E., Styve, H. and Simpson, D. (1995) *Status of emissions data at MSC-W.* EMEP/MSC–W Report 2/95.

Bower, J.S., Broughton, G.F.J., Stedman, J.R. and Williams, M.L. (1994) A winter NO_2 smog episode in the UK. *Atmospheric Environment,* **28**, 461–475.

Brimblecombe, P. (1986) *Air composition and chemistry.* Cambridge Environmental Chemistry Series 6. CUP. First edition.

Bull, K.R. (1995) Critical loads – Possibilities and constraints. *Water, Air and Soil Pollution,* **85**, 201–212.

Cawse, P.A. (1974) *A survey of atmospheric trace elements in the UK (1972–73).* AERE Harwell, Oxfordshire.

Choularton, T.W., Gay, M.J., Jones, A., Fowler, D., Cape, J.N. and Leith, I.D. (1988) The influence of altitude on wet deposition. Comparison between field meausurements at Great Dun Fell and the predictions of a seeder-feeder model. *Atmospheric Environment* **22**, 1363–1371.

CLAG (1997) *Deposition Fluxes of Acidifying Compounds in the United Kingdom.* Critical Loads Advisory Group Report Prepared for the UK Department of the Environment.

Dentener, F.J. and Crutzen, P.J. (1994). A three-dimensional model of the global ammonia cycle. *Journal of Atmospheric Chemistry,* **19**, 331–369.

Derwent, R.G. and Nodop, K. (1986) Long-range transport and deposition of acidic nitrogen species in north-west Europe. *Nature* **324**, 356–358.

Derwent, R.G., Middleton, D.R., Field, R.A., Goldstone, M.E., Lester, J.N. and Perry, R. (1995) Analysis and interpretation of air quality data from an urban roadside location in central London over the period from July 1991 to July 1992. *Atmospheric Environment* **29**, 923–946.

DoE (1995) Official estimates of ammonia emissions for the United Kingdom. Submitted to the UNECE/EMEP. Air Quality Division, Department of the Environment. London.

Dore, A.J., Choularton, T.W., Fowler, D. and Storeton-West, R. (1990) Field measurements of wet deposition in an extended region of complex topography. *Quarterly Journal of the Royal Meteorological Society* **116**, 1193–1212.

Dragosits, U., Sutton, M.A. and Place, C.J. (1996) The spatial distribution of ammonia emissions in Great Britain for 1969 and 1988 assessed using GIS techniques. *Atmospheric Ammonia: emission, deposition and environmental impacts. Poster Proceedings.* M.A. Sutton *et al.* eds.

Duyzer, J. and Fowler, D. (1994) Modelling land atmosphere exchange of gaseous oxides of nitrogen in Europe. *Tellus*, **46B**, 353–372.

Eliassen, A. (1978) The OECD study of long-range transport of air pollutants: long range transport modelling. *Atmospheric Environment,* **12**, 476–487.

Eliassen, A. and Saltbones, J. (1983) Modelling of long-range transport of sulphur over Europe: a two-year model run and some model experiments. *Atmospheric Environment,* **17**, 1457–1473.

Erisman, J.W., Vermetten, A.W.M., Asman, W.A.H., Wijers-IJpelaan, A. and Slanina, J. (1988) Vertical distribution of gases and aerosols: the behaviour of ammonia and related compounds in the lower troposphere. *Atmospheric Environment* **22**, 1153–1160.

Fay, J.A., Golomb, D. and Kumar, S. (1986) Modeling of the 1900–1980 trend of precipitation acidity at Hubbard Brook, New Hampshire. *Atmospheric Environment* **20**, 1825–1828.

Finlayson-Pitts, B.J. and Pitts, J.N. (1986) *Atmospheric chemistry: fundamentals and experimental techniques.* J. Wiley & Sons.

Fisher, B.E.A. (1975) The long-range transport of sulphur dioxide. *Atmospheric Environment,* **9**, 1063–1070.

Fowler, D. (1980) Removal of sulphur and nitrogen compounds from the atmosphere in rain and by dry deposition. In: *Ecological impact of acid precipitation*, eds. D. Drablos and A. Tollan, 22–32. Oslo–As: SNSF Project.

Fowler, D., Cape, J.N., Leith, I.D., Choularton, T.W., Gay, M.J. and Jones, A. (1988) The influence of altitude on rainfall composition at Great Dun Fell. *Atmospheric Environment,* **22**, 1355–1362.

Fowler, D., Mourne, R. and Branford, D. (1995) The application of ^{210}Pb inventories in soil to measure long-term average wet deposition of pollutants in complex terrain. *Water Air and Soil Pollution,* **85**, 2113–2118.

Galloway, J.N. and Roche, H. (1991). Regional atmospheric budgets of S and N fluxes: how well can they be quantified. *Proceedings of the Royal Society of Edinburgh,* **97B**, 61–80.

Galloway, J.N. (1995) Acid deposition: perspectives in time and space. *Water, Air and Soil Pollution,* **85**, 15–24.

Garland, J.A. (1978) Dry and wet removal of sulphur from the atmosphere. In: *Sulphur in the Atmosphere* (eds. R.B. Husar, J.P.Lodge, Jr., D.J. Moore) 349–362. Pergamon Press

Graedel, T.E., Benkovitz, C.M., Keene, W.C., Lee, D.S. and Marland, G. (1995) Global emissions inventories of acid-related compounds. *Water, Air and Soil Pollution,* **85**, 25–36.

Grennfelt, P., Hov, Ø. and Derwent, R.G. (1994) Second generation abatement strategies for NO_x, NH_3, SO_2 and VOCs. *Ambio,* **23**, 425–433.

Hales, J.M. (1978) Wet removal of sulphur compounds from the atmosphere. In: *Sulphur in the Atmosphere* (eds. R.B. Husar, J.P.Lodge, Jr., D.J. Moore) 389–401. Pergamon Press.

Hargreaves, K.J., Fowler, D., Storeton-West, R.L. and Duyzer, J.H. (1992) The exchange of nitric oxide, nitrogen dioxide and ozone between pasture and the atmosphere. *Environmental Pollution,* **75**, 53–59.

Heij, G. J. and Erisman, J. W. (1997). Acid atmospheric deposition and its effects on terrestrial ecosytems in the Netherlands. Studies in Environmental Science 69. Elsevier.

INDITE (1994) *Impacts of Nitrogen Deposition in Terrestrial Ecosystems.* Report of the United Kingdom Review Group on the Impacts of Atmospheric Nitrogen.

Inglis, D.W.F., Choularton, T.W., Wicks, A.J., Fowler, D., Leith, I.D., Werkman, B. and Binnie, J. (1995) Orographic enhancement of wet deposition in the United Kingdom: case studies and modelling. *Water Air and Soil Pollution,* **85**, 2119–2124.

IPPC (1990) *Climate change 1990.* CUP.

IPCC (1995) *Climate change 1994. Radiative forcing of climate change and an evaluation of the IPCC IS92 emission scenarios.* CUP.

IPCC (1996) *Climate change 1995. The Science of Climate Change.* CUP.

Jarvis, P.G., James, G.B. and Landsberg, J.J. (1976) Coniferous forest. In: *Vegetation and the Atmosphere, Vol. 2: Case Studies* (J.L. Monteith, ed). Academic Press, London.

Lawes, J.B., Gilbert, J.H. and Pugh, E. (1861) On the sources of nitrogen for vegetation. *Philosophical Transactions of the Royal Society, II* 431–577.

Lee, D. S., Köhler, I., Grobler, E., Roher, F., Sausen, R., Gallardo-Klenner, L., Olivier, J.G., Dentener, F.J. and Bouwman, A.F. (1997) Estimations of global NO_x emissions and their uncertainties. *Atmospheric Environment* **31**, 1735–1749.

Leighton, P.A. (1961). *Photochemistry of air pollution.* Academic Press.

Lenhart, L. and Friedrich, R. (1995) European emission data with high temporal and spatial resolution. *Water, Air and Soil Pollution* **85**, 1897–1902.

Lindley, S.J., Longhurst, J.W., Watson, A.F. and Conlan, D.E. (1996) Procedures for the estimation of regional scale atmospheric emissions – an example from the north west region of England. *Atmospheric Environment,* **30**, 3079–3091.

Mayewski, P.A., Lyons, W.B., Spencer, M.J., Twickler, M.S., Buck, C.F. and Whitlow, S. (1990) An ice-core record of atmospheric response to anthropogenic sulphate and nitrate. *Nature,* **346**, 554–556.

Metcalfe, S.E., Whyatt, J.D. and Derwent, R.G. (1995) A comparison of model and observed network estimates of sulphur deposition across Great Britain for 1990 and its likely source attribution. *Quarterly Journal of the Royal Meteorological Society* **121**, 1387–1411.

Murley, L. (ed.) (1996) *NSCA Pollution Handbook*, Bryton UK p. 509

Metcalfe, S.E., Derwent, R.G., Whyatt, J.D. and Dyke, H. (in press) Nitrogen deposition and strategies for the control of acidification and eutrophication across Great Britain. *Water, Air and Soil Pollution* **107**, 121–145

Reynolds, B., Fowler, D., Smith, R.I. and Hall, J.R. (1997) Atmospheric inputs and catchmemnt solute fluxes for major ions in five Welsh upland catchments. *Journal of Hydrology* **194**, 305–329

RGAR (1998) *Acid Deposition in the United Kingdom 1986–1995.* 4th Report of the Review Group on Acid Rain. Critical Loads Advisory Group Report Prepared for the UK Department of the Environment.

Rogers, H.H., Jeffries, H.E. and Witherspoon, A.M. (1979) Measuring air pollutant uptake by plants: nitrogen dioxide. *Journal of Environmental Quality,* **8,** 55–557.

Salway, A.G., Goodwin, J.W. and Eggleston, H.S. (1996) *UK Emissions of Air Pollutants 1970–1993.* AEA/RAMP/16419127/R/001/ISSUE1.

Skiba, U., Fowler, D. and Smith, K. (1994) Emission of NO and N_2O from soils. *Environmental Monitoring and Assessment,* **31**, 153–158.

Skiba, U., Fowler, D. and Smith, K.A. (1997) Nitric oxide emissions from agricultural soils in temperate and tropical climates: Sources, control and mitigation options. *Nutrient Cycling in Agroecosystems,* **48**, 139–153.

Sutton, M.A., Fowler, D., Moncrieff, J.B. and Storeton-West, R.L. (1993) The exchange of atmospheric ammonia with vegetated surfaces. II: Fertilized vegetation. *Quarterly Journal of the Royal Meteorological Society,* **119**, 1047–1070.

Sutton, M.A., Place, C.J., Eager, M., Fowler, D. and Smith, R.I. (1995a) Assessment of the magnitude of ammonia emissions in the United Kingdom. *Atmospheric Environment* **29**, 1393–1411.

Sutton, M.A., Schjorring, J., and Wyers, G.P. (1995b) Plant-atmosphere exchange of ammonia. *Philosophical Transactions of the Royal Society of London Series A* **351**, 261–278.

The Norwegian State Pollution Control Authority (1987) 1000 lake survey 1986 Norway. Report 283/87.

Trenberth, K.E. and Guillemot, C.J. (1994) The total mass of the atmosphere. *Journal of Geophysical Research,* **99**, 079–23,088.

United Nations (1997) Present status of emission data and emission database. EB.AIR/GE.1/1997/3. Geneva.

Van Leeuwen, E.P., Draaijers, G.P. and Erisman, J.W. (1996) Mapping wet deposition of acidifying components and base cations over Europe using measurements. *Atmospheric Environment* **30**, 2495–2511.

Wayne, R.P., Barnes, I., Biggs, P., Burrows, J.P., Canosa-Mas, C.E., Hjorth, J., LeBras, G., Moortgat, G.K., Perner, D., Poulet, G., Restelli, G. and Sidebottom, H. (1991) The nitrate radical: physics, chemistry and the atmosphere. *Atmospheric Environment,* **25A**, 1–206.

Webster, R., Campbell, G.W. and Irwin, J.G. (1991) Spatial analysis and mapping the annual mean concentrations of acidity and major ions in precipitation over the United Kingdom in 1986. *Environmental Monitoring and Assessment,* **16**, 1–17.

THE ROLE OF PLANT AND SOIL PROCESSES IN DETERMINING THE FATE OF ATMOSPHERIC NITROGEN

B. L. WILLIAMS AND H. A. ANDERSON
Macaulay Land Use Research Institute, Craigiebuckler, Aberdeen, AB15 8QH, UK

3.1 Introduction

Currently, in Europe and North America, as S emissions decrease and N deposition increases (Galloway, 1995), NH_4 and NO_3 will replace sulphate, as the major soil-acidifying mobile ions and supply nutrient N in excess. This applies especially to those terrestrial ecosystems which historically have been N-limited and are susceptible to vegetation change (Bråckenhielm and Qinghong, 1995). This deposition is now perceived as one of the major threats to the structure and functioning of natural and semi-natural ecosystems (Bobbink and Roelofs, 1995).

Areas of natural and semi-natural vegetation seldom receive additions of fertilizer N and the underlying soils have a wide range of N contents ranging from 700 kg ha^{-1} in the surface 10 cm of peaty gleysols (Williams, 1992a) to over 10,000 kg ha^{-1} in the surface 30 cm of *Agrostis–Festuca* grassland (Perkins, 1978). The cycling of this N and its availability to plants in peaty, acid soils is frequently restricted by conditions limiting microbial activity and organic matter decomposition such as acidity and waterlogging. For extreme examples of mineral soils, such as ecosystems on wind-blown sand, the small quantity of total- and available-N can be limiting for plant growth (Miller *et al.*, 1979). However, in upland ecosystems, inorganic inputs from wet and dry deposition are likely to be significant in relation to the quantities of N cycling through the plant-microbial-soil system irrespective of the total N capital. As a result, atmospheric deposition could alter the balance between soil microbial activity, the availability of nutrients such as P, soil acidity, and carbon turnover. Such changes will affect the composition of vegetation communities, plant growth, water quality, and gaseous emissions of carbon dioxide and nitrogen oxides. For peatlands, which worldwide store a significant proportion of the global C reserves, there is a genuine concern that enrichment with N could change their function from sinks to sources of C (Francez, 1991). Thus the capacity for changes in soil function and structure, affecting both hydrology and hydrochemistry, will be governed not only by direct physicochemical effects driven by proton loadings and enhanced NH_4 and NO_3 ion concentrations, but also by biologically-mediated processes altered by pollutant-N inputs.

Soils beneath natural and semi-natural vegetation differ from their agricultural counterparts mainly through the influence of land management. Thus acidification in arable and grazed soils is counteracted by the application of lime. Large areas of soils in the developed regions are cultivated for food production and are frequently bordered by their uncultivated cousins, and influence them, especially by gaseous N-emissions, eg. by spreading manures and slurries. In northern Europe, altitude and aspect can often dictate that large areas are not viable for agricultural production and remain as forest or treeless moorland. Thus the soils vulnerable to increased nitrogen loading may vary from largely-mineral, with a stratified organic topsoil which can form over centuries, to deep peats with layers developed from vegetation sequences emerging over millennia. Pollutant inputs in precipitation can thus face very different hydrological pathways through rapidly-draining coarse-grained mountain soils, layered podzolic profiles with perched water-tables, imperfectly- and poorly-drained peaty gleys and other gleysols, and peats which may form deposits several metres thick.

In this chapter we review the extent of current knowledge on the ways in which plant and soil processes transform inorganic nitrogen derived from atmospheric deposition and the impact that this has on carbon turnover and soil acidification. The influence of land use management and soil conditions (particularly organic soils) on the fate of nitrogen deposition are discussed and the measured fluxes of inorganic nitrogen in field studies on a plot and catchment scale presented.

3.2 Processes Involving N Transformations

Inorganic N is generally transformed to organic forms at different stages in the passage through the plant canopy to the soil horizons. Assimilation by plants is a major process which is influenced by soil and environmental factors, whereas in the soil, micro-organisms and abiotic processes can be involved.

3.2.1 ASSIMILATION BY VEGETATION

Vegetation forms a barrier to direct entry of inorganic N into the soil, and atmospheric N enters only after interaction with the plant foliage. In forest ecosystems, this interaction is readily measured and foliar uptake of inorganic N by N-deficient species is well known (Miller *et al.*, 1979). Throughfall and stemflow waters beneath forest canopies contain dissolved organic matter that contains N (Qualls and Haines, 1991). Passage of this material to the forest floor and underlying soil is likely have a profound effect on microbial activity.

In peatland ecosystems, *Sphagnum* mosses are compact plants that have adapted to grow on the surface of raised bogs utilizing the NO_3 in rainwater (Woodin *et al.*, 1985) and low ambient levels of NH_4 ions (Meade, 1984). Calculations have shown that the N accrued in bogs over long periods of time account for varying amounts of the total input (Damman, 1988), which indicates controls on the assimilation of atmospheric N by bog vegetation, such as P availability (Aerts *et al.*, 1992). Provided N is assimilated and there is concomitant growth and fixation of C by the plant, the system remains in balance but N assimilation without growth will result in plant tissues with decreased C:N ratios and a greater tendency to decompose at senescence, with remineralization of N in the long term (Aerts *et al.*, 1992). Williams and Silcock (1997) showed that *Sphagnum magellanicum* assimilated NH_4NO_3 -N in amounts well in excess of that falling locally as wet deposition, up to 3.5 g N m^{-2} during the growing season compared to 0.3 g N m^{-2} in the rainfall. However, when this experiment was repeated at four other sites in Europe the uptake of N at the largest application rate varied considerably from 0 to almost 100% (Williams *et al.*, 1998). While low P availability explained the poor uptake of N at one site in Finland, sites in Estonia and France were relatively rich in this nutrient. At both French and Estonian sites the experiment was located in tussocks of the moss well above (> 22 cm) the water table, and this appeared to be another factor limiting N-uptake. Using ^{15}N-labelling, Williams *et al.*, (1998b) showed a recovery of between 10 and 100% during the year of a regular 2-weekly *in situ* application to the mosses *S. capillifolium* and *S. recurvum* of NH_4NO_3 equivalent to 30 kg N ha^{-1} yr^{-1}. Complete recovery occurred during the growing season and values were least for the pool colonising species *S. recurvum* when the water table was high and close to the moss surface. Under these conditions the inorganic N probably diffused into the underlying peat.

Under controlled environmental conditions, there was no preference for NH_4 over NO_3 by either species (Silcock and Williams, 1995), but mosses treated with inorganic-N released dissolved organic-N (DON) into the waters retained by the moss. Similar increases in extractable DON were measured by Williams and Silcock (1997) when *Sphagnum magellanicum* was treated every two weeks with the equivalent of 100 kg N ha^{-1} yr^{-1} during the growing season. The chemical nature of this DON was not established, but Baxter *et al* (1992) reported increases in the amino acid N in the tissues of mosses treated with inorganic-N. Similarly in Finland, Karsisto *et al.* (1996) reported large increases in the free amino acids, in particular, arginine, glutamic acid and aspartic acid in *S. fuscum* and *S. capillifolium* treated with NH_4NO_3. Yesmin *et al.* (1995) studied the drainage waters from peat turves taken from different sites along a N pollution gradient where the predominant vegetation is *Calluna vulgaris* L. At sites where the N pollution load was largest the leachates contained greater concentrations of DON. The origin and nature of this DON was not investigated, but it indicates the increasing need to characterise DON in order to determine its bioavailability and thus its fate. Similar responses by trees to additions of inorganic-N have been reported and it is suggested that these increases in the amino acid contents of foliage are a response to stress and a detoxifying mechanism for the removal of NH_4-N in plant tissues (Nashholm *et al.*,1994). As a result, DON appears to be released into the surrounding soil and waters with important consequences for the cycling of N.

Thus when atmospheric N reaches the plant canopy there are several possibilities:

a) It is assimilated by the plant and returned to the soil in litter at senescence, and in the absence of a growth response, the litter has a reduced C:N ratio.

b) N is assimilated and the growth response of the vegetation results in the transfer of organic matter to the soil in root exudates which in turn influence soil microbial activity and chemistry.

c) The assimilated N accumulates in the plant because of nutrient imbalances and leaks out of the tissues as DON
d) Inorganic-N is not assimilated by the vegetation and passes into the soil directly.

Which of these options predominates will vary between sites and in time. Between-site variation is caused by many factors and P content or availability is one that is important for peatland and heathland ecosystems which are inherently low in this element (Williams, 1996). In peatlands, water table depth and the availability of other nutrients such as K could also have some influence.

3.2.2 N MINERALISATION AND IMMOBILISATION

Inorganic-N penetrating the vegetation canopy and reaching the litter and soil horizons can be expected to enter the mineralisation–immobilisation cycle and influence the size and activity of the soil microbial biomass. In N-deficient soils with high C contents, additional N inputs are likely to be immobilised in the microbial biomass. In an experiment involving the addition of ^{15}N labelled $(NH_4)_2SO_4$ to litters from *Calluna*, *Molinia* and *Deschampsia*, (Van Vuuren and Van der Erden, 1992) all showed retention of the added NH_4-N. In *Molinia* litter the immobilisation of N predominated, whereas *Calluna* and *Deschampsia* both retained and released N during an 11 month incubation period. Yesmin *et al.* (1996) reported greater retention of NH_4- than NO_3-N in peat and its associated vegetation of *Calluna* or grass when treated for 12 months with $(NH_4)_2SO_4$ and HNO_3. Kristensen and McCarty (1996) measured rates of gross mineralisation and immobilisation of N in the mor humus from untreated control plots beneath *Calluna* and reported rapid assimilation of ^{15}N during the first 24 hr period. In contrast, net mineralisation occurred in humus from fertilized plots treated with 15 kg N ha^{-1} yr^{-1}, which was greater than that from plots treated with 35 kg ha^{-1} yr^{-1}. There was no nitrification in these very acidic samples (pH_{H2O} 3.3–4.0).

The availability of C substrates and therefore, the rate of decomposition, is the principal factor determining N immobilisation. In soils with high C:N ratios, typical of forest and peaty soils, immobilisation of inorganic-N by assimilation and incorporation into microbial material is an important process. The term immobilisation, however, is used loosely because the process involves inorganic-N entering pools in organic forms which have slower rates of turnover than the inorganic pool and are not directly available for uptake by plants. Nevertheless the microbial biomass is considered to be a pool with a relatively rapid turnover rate and can be considered as a source of readily available plant nutrients (Jenkinson and Ladd, 1981). Methods of measuring rates of gross mineralisation and immobilisation of N in soils are based on ^{15}N isotope dilution methods. The rate calculations depend on two main assumptions, that there is complete mixing of the added $^{15}NH_4$ with the soil pool, and that the time for the immobilisation and remineralisation of ^{15}N by the microbial biomass is long relative to the time taken to carry out the measurements (Schimel,1996). Wessel and Tietema (1992) used a model that incorporated the recycling effect that showed that there are clear differences in rates of gross mineralisation in different litter types. An alternative method, directly measures the incorporation of ^{15}N into the microbial biomass, using fumigation with chloroform to release labelled N from the biomass (Davidson *et al.*, 1991).

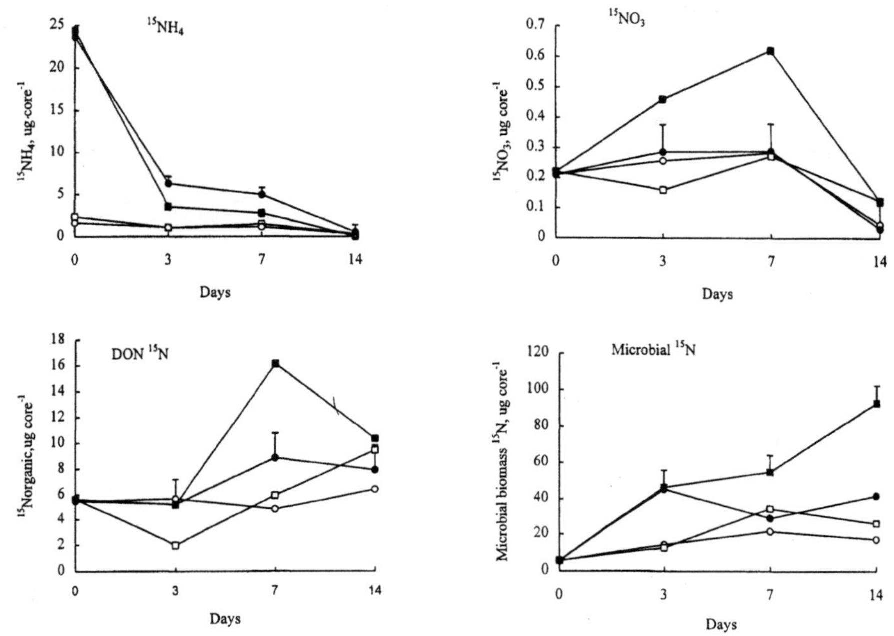

Figure 3.1: ^{15}N content, μg core^{-1}, of nitrogen fractions in the surface 5 cm of peat cores treated with 22 μg $^{15}NH_4$ at 99.8 atom % and incubated at 10°C. Open symbols no ^{15}N, closed symbols + $^{15}NH_4$. Circles – no added N in the field, squares – NH_4NO_3 added every two weeks, equivalent to 30 kg N ha^{-1} yr^{-1}. Error bars show standard error of difference between labelled and unlabelled samples.

On a raised bog, direct additions of $^{15}NH_4$ to peat cores which had been treated *in situ* with different amounts of unlabelled NH_4NO_3 for 12 months, resulted in the disappearance of most the NH_4-N within 3 days, little labelling of the NO_3-N pool and rapid labelling of the microbial-N pool (Figure 3.1). The rate of ^{15}N labelling of the microbial biomass appeared to be marginally greater in peat which had not been amended previously with NH_4NO_3, but the effect was small ($P < 0.05$). The rates of gross mineralisation and immobilisation calculated from the rates of ^{15}N dilution were not significantly different in the N30 compared with the N0 treatment (Table 3.1). Immobilisation exceeded the gross mineralisation rate during each time interval, but the rates decreased rapidly with time. Labelling of the microbial material released after chloroform fumigation yielded values of 0.79 and 0.70 atom% for ^{15}N incorporation into the microbial biomass in N0 and N30, respectively. The microbial biomass N continued to increase in size after the first 3 days and appeared to be greater in the N30 cores amended with NH_4NO_3 in the field compared with N0 (Table 3.2). Combining the biomass values with the ^{15}N abundance gave values for the rate of assimilation of label during the first 3 days of 27.1 and 13.8 mg ^{15}N day^{-1} m^{-2} in N0 and N30, respectively. The lower values obtained by this method compared with those obtained by the isotope dilution method have been attributed to the uneven labelling of the microbial biomass and the extraction of only half of the total microbial material after fumigation (Davidson *et al.*, 1991)

Table 3.1: Calculated rates (kg ha^{-1} day^{-1}) of gross mineralisation and immobilisation of N in the 0–5 cm layer of peat beneath *Sphagnum capillifolium* treated *in situ* every two weeks for 12 months with NH_4NO_3 at 0 (N0) and 30 (N30) kg N ha^{-1} yr^{-1}. Standard error of difference between N treatments in parentheses.

	Mineralisation		Immobilisation	
Interval	N0	N30	N0	N30
0–3 days	1.2	1.3 (0.3)	1.6	1.5 (0.3)
3–7 days	0.04	0.16 (0.06)	0.05	0.20 (0.08)
7–14 days	0.02	0.06 (0.03)	0.09	0.12 (0.005)

Table 3.2: Microbial biomass-N, kg ha^{-1} in the 0–5 cm layer, beneath *Sphagnum capillifolium* treated *in situ* every two weeks for 12 months with NH_4NO_3 at zero (N0) and 30 (N30) kg N ha^{-1} yr^{-1}.

Day	NO	N30	sed
0	11.2	20.2	
3	15.5	16.5	3.9
7	20.1	36.4	8.3
14	22.9	30.8	7.7

Accumulation of NH_4-N in soils is the net effect of gross mineralisation and immobilisation, and judging by the seasonal patterns of net mineralisation in different soils, the relative rates of these two processes vary. Increased inputs of atmospheric N could be expected to lead to increased net mineralisation of N and, indeed, Morecroft *et al.* (1994) reported that experimental additions of NH_4NO_3 between 35 and 140 kg N ha^{-1} yr^{-1} to semi-natural grassland on acidic and calcareous soils stimulated mineralisation in both. Morecroft *et al.* (1992) reported increasing rates of N mineralisation in soils along two altitudinal gradients in the Scottish Highlands; contrary to the authors' expectations, based on the decline in temperatures with increasing altitude, the rates of net mineralisation were either the same or greater at higher altitudes than those measured below. The sites selected for this work were comparable in terms of vegetation and no altitude trends in common soil characteristics were detected. Atmospheric deposition at the different altitudes was not measured, but Fowler *et al.* (1988) reported higher concentrations of NH_4-N in cloudwater compared with rainwater, and variations in N deposition rates with altitude have been postulated (INDITE,1995). The interactions between climate, soils and N deposition are difficult to investigate *in situ*, and there is a clear need for such studies to separate out the effects of the different factors.

Black *et al.* (1993) reported elevated NO_3 concentrations in streamwaters draining very acid hill peats in eastern Scotland and concluded that it was derived directly from NO_3 in rainwater, rather than from the activities of nitrifiers in the soil. This implies that there are circumstances in which N immobilisation is not active because of a combination of acidity and low temperatures.

3.2.3 NITRIFICATION AND DENITRIFICATION

Nitrification is a key process in relation to acidification and N movement out of soils and into surface waters. Predicting which soils actively nitrify is difficult, particularly for acid forest and moorland soils, and the characteristics of nitrifying organisms prevent them from being easily cultured in the laboratory. Oxidation and nitrification by autotrophic organisms of the NH_4-N reaching the soil beneath natural vegetation is not considered an active process in acid mineral soils (Killham, 1990) or in ombrotrophic bogs (Rosswall and Granhall, 1980). Yet, stagnopodzols in Wales, beneath Sitka spruce, and on clearfelling, nitrify freely (Stevens and Hornung, 1988) despite having acid pH values around 4.1. In a study of acidic and calcareous soils beneath grassland, Morecroft *et al.* (1994) reported increased nitrification rates in the calcareous soil (pH 6.1) in reponse to N additions whereas the acid soil (pH 3.9) nitrified during spring and summer in plots treated with 70 and 140 kg N ha^{-1} yr^{-1}. Litter from Sitka spruce growing on a brown forest soil released NO_3 during incubation *in situ* after fertilization with NH_4NO_3, rock phosphate and potassium chloride which also had the effect of raising pH_{CaCl2} from 3.6 to 4.4 (Williams, 1983). Nitrification in the underlying humus occurred after a lag phase, but only in closed samples which allowed NH_4 accumulation to occur. Application of molecular biological techniques have begun to distinguish which groups of bacteria are involved in the oxidation of ammonia in soils and sediments (Kowalchuk *et al.*, 1997) and this approach will help advance the understanding of nitrification in acid soils.

Nitrification by heterotrophic organisms that utilize organic substrates has been identified in some acid soils and slow growing nitrifiers utilizing urea have been isolated and cultured (Allison and Prosser, 1991). Adams (1986a) reported a positive correlation between organic N extracted from acid litters and humus from beneath larch, spruce, pine and heather and net nitrification during incubation. In this study, nitrification was greater in litter than in humus and was more active beneath larch and spruce than under pine and heather. Using acetylene to block autotrophic nitrification and sodium chlorate to inhibit chemautotrophic oxidation, Adams (1986b) showed that a heterotrophic pathway was involved. Nitrification activity in a range of different acid peat types increased with increasing mineral components suggesting that in an organic soil a mineral surface could provide a microhabitat for nitrification to occur (Williams, 1984).

In N-limited natural and semi-natural ecosystems, there is normally considerable capacity for competition for inputs of mineral N. In the northern temperate zone, temperature is a major factor limiting microbial activity and mineralisation processes in the soil. In most soils which release NO_3 to drainage waters, nitrification occurs more rapidly than N-uptake, and in warm dry years, an acidifying pulse of NO_3 can appear in drainage waters at the end of a drought. One such event was noted in 1982 at the Solling forest in Germany (Matzner and Thoma, 1983), where soil NO_3 concentrations increased dramatically and were accompanied by Ca and Al release, the cation ratio being related to the base (Ca) saturation of the different sites. Nitrification can be performed by both autotrophic and heterotrophic nitrifiers, with the autotrophic populations, previously thought to be acid-sensitive, having the potential to become acid-adapted in ecosystems where the background pH has been gradually lowered (Killham, 1990). The pH dependence of nitrification also leads to a situation where attempted amelioration of acidic ecosystems by the application of lime can lead to accelerated nitrification (Kaila, 1954; Viro, 1962), which could result in a worsening situation, with the possibility of downstream eutrophication.

The effect of soil temperature in nitrification processes is important, but does not preclude NO_3 production in winter. Stottlemyer and Rutkowski (1990) found little relationship between snowmelt NO_3 and peak stream NO_3 discharge, and concluded that this independence suggested additional factors, such as winter mineralisation or nitrification, contributed to the stream ouputs. The catchment under investigation showed little freezing of the top 5 cm of the soil profile under the snowpack, with the snowmelt loss of NO_3 only appearing in the equivalent mean cumulative stream discharge of 12% of the total. This contrasted sharply with results from a site with frozen soils (Cadle *et al.*, 1984), where some 30% of the snowpack NO_3 was lost in stream discharge at melt. More NO_3 remains in the snowpack up to final melt over frozen soils, conditions likely to lead to reduced nitrification rates, a reduced biological uptake and possibly a reduced residence time of the NO_3 in the ecosystem on release, resulting in larger acidic pulses in the output stream. Cadle *et al.* (1984) also concluded that 50–70% of the combined acidic species (mainly sulphate and NO_3) were released during the first 20% of snowmelt. Similar flow-related NO_3 pulses occurring at snowmelt have been seen by Mitchell *et al.* (1996), in the Arbutus Watershed in the Adirondacks. Again this site exhibited a positive correlation between annual drainage water flux and inorganic-N flux from the catchment, a significant temperature-dependent output of NO_3, but also an overall 85% retention of input N.

Nitrate entering the soil will be competed for by plant uptake and denitrifiers although little denitrification activity is considered possible in acid soils (Klemmedtsson *et al.*, 1977). Urban *et al.* (1988) concluded that the moss vegetation was the predominant sink for NO_3-N on bogs in Minnesota and Ontario and that rates of denitrification as measured by nitrous oxide emission were small, less than 2.5 μg N m^{-2} h^{-1}. Measurements of nitrous oxide fluxes on bogs in central Finland showed that highest values were of the order of 5 μg N m^{-2} h^{-1} in the surface 10 cm of bogs (Regina *et al.*, 1996). These values correlated positively with the total contents of nitrogen, phosphorus and calcium in the peat and negatively with the depth of the water table, ie. values were greater in minerotrophic than in ombrotrophic bogs. In the acid horizons of forest and bog soils, Müller *et al.* (1980) reported little denitrification potential, which increased in the less acid underlying mineral soil horizons. Martikainen and De Boer (1993) measured rates of denitrification of the order of 0.5 to 1 kg N ha^{-1} yr^{-1} in the acid F horizon (pH 3.7) beneath Douglas fir and concluded that the production of nitrous oxide was occurring during the nitrification process. Dutch and Ineson (1990) obtained a slightly greater rate of N loss in the range 1–3 kg ha^{-1} yr^{-1} beneath Sitka spruce on an upland peaty gley soil which increased to 10–40 kg ha^{-1} yr^{-1} on clearfelling, indicating the capacity of soil microorganisms to denitrify excess NO_3, if C availability and the soil moisture conditions are favourable. The denitrification rates measured corresponded to the range of inputs of N in forest ecosystems (see Wilson and Emmett, Chapter 5). Skiba *et al.* (1997) reported rates of N_2O emissions from a range of soil and vegetation types in Northern Britain. Greatest emissions of 0.9 kg N ha^{-1} yr^{-1} were obtained for mown grassland, 0.5 kg N_2O-N for deciduous woodland and 0.4 kg N ha^{-1} yr^{-1} for coniferous forest. In the same study, the emission of N_2O was sensitive to inputs and decreased as the distance from the source, a poultry farm, increased. Mean annual emission rates from organic soils in upland areas ranged between 0.1 and 0.3 kg N_2O-N ha^{-1} yr^{-1} and increased with altitude and greater atmospheric deposition, up to 40 kg N ha^{-1} yr^{-1}. These emissions of N_2O were also influenced by temperature and soil moisture.

3.2.4 PRODUCTION OF DISSOLVED ORGANIC-N (DON)

Dissolved organic N (DON) has been reported in throughfall waters reaching the forest floor (Qualls and Haines, 1991), and is increasingly released from peats in response to additions of pollutant-N and -S (Yesmin *et al.*,1995), and similar findings have been reported during catchment manipulation experiments (Emmett *et al.*,1995; see below). In the N budget for Thoreau's Bog, Hemond (1983) described the DON fraction as a low concentration pool of organic-N which was the source of NH_4 and which was replenished by decomposition of peat organic-N. At this site runoff waters accounted for 1 kg N ha^{-1} yr^{-1} which compares with approximately 4 kg ha^{-1} calculated for a reseeded, limed and urea fertilized blanket bog in the north of Scotland (Williams and Young, 1994).

In the surface 5 cm of an undisturbed raised bog in NE Scotland the DON component of the waters surrounding the moss ranged between 0.05 and 0.2 kg ha^{-1} over a period of 15 months (Williams and Silcock, unpublished results). Addition of NH_4NO_3 every 2 weeks at a rate equivalent to 30 kg N ha^{-1} yr^{-1} (N30) significantly ($P < 0.05$) increased the DON pool on one sampling occasion only, with ^{15}N-labelled inorganic-N being incorporated into the DON fraction mainly between October and March, the recovery of added ^{15}N being < 0.5 %. DON extracted with 0.5 M K_2SO_4 from the underlying peat at 5–10 cm depth showed a marked seasonal pattern, being greatest during summer and least during winter. The extreme values of this DON pool were enhanced significantly by the N30 treatment and also differed significantly in the profiles beneath two contrasting moss species, *S. capillifolium,* a hummock former and *S. recurvum*, a pool colonizer. In cores labelled with ^{15}N, the heavy isotope in this N pool in the peat did not increase above natural abundance, indicating that changes in its size were not caused by inorganic N directly, but most probably indirectly via the mosses at the surface. In the absence of the moss, direct injection of $^{15}NH_4$ into peat resulted in a small increase in ^{15}N labelled DON after 7 days incubation (Figure 3.1)

Dissolved organic N (DON) has been recognised as an important component of stream waters draining upland catchments (Edwards *et al.*,1996). However, clear links between any impact of increased atmospheric N deposition and the movement of DON out of soils and into streamwaters have not been established. One reason for this is the procedural definition of DON and its complex nature (Williams and Edwards, 1993). Currently, DON is defined as the difference between inorganic- and total-N determinations in waters and soil extracts. These solutions have usually been filtered through 0.45 μm pore diameter membranes, and a major proportion of the "dissolved" N may be associated with smaller particulate material. This may be particularly true where clays and other mineral particles are present in extracts or river waters, and are capable either of fixing NH_4-N in the lattice structure (Hillier, 1996) or of adsorbing proteins to active surfaces (Delanoue *et al.,* 1997).

3.3 Impact Of N Deposition On Carbon Cycling

The fertilizing effect of nitrogen deposition on plant production is a major influence on the input of carbon to the decomposition cycle (see Aerts and Bobbink, Chapter 4).

3.3.1 LITTER DECOMPOSITION AND MICROBIAL ACTIVITY

Plant litter is the principal carbon resource reaching the soil surface and its nitrogen content and C:N ratio has been used for many years as a guide to its quality and rate of decomposition (Heal *et al*., 1997) along with the content of carbohydrate, lignin and polyphenolic compounds (Fog, 1988). In semi-natural and forest ecosystems, litter decomposition and subsequent release of nutrients is a vital part of the cycles of C and nutrients essential for sustaining plant growth.

From measurements of increased N concentrations and reduced C:N ratios in the tissues of mosses on bogs receiving high and low inputs of atmospheric N in Sweden it has been predicted that the decomposition of these materials will be enhanced (Aerts *et al*., 1992). The C:N ratio of litter has been used as a general indicator of litter quality and its rate of decomposition and net mineralisation of N, but the chemical nature of the carbon, eg. carbohydrate, lignin or protein, will also have an effect on the rate of carbon loss (Fog, 1988). Threshold values of C:N ratios at which gross mineralisation of organic N exceeds the rate of immobilisation can be expected to vary with the composition or quality of decomposing litter (Heal *et al*., 1997).

Verhoeven *et al.* (1990) compared cellulose decomposition rates using cotton strips in fens and bogs in the Netherlands, but found little difference between rates measured at sites having a wide range of atmospheric N inputs. The implication of this is that additional inputs of inorganic N have no direct impact on microbial activity. Williams and Silcock (1997) measured microbial biomass carbon in peat beneath *Sphagnum magellanicum* treated at four rates of NH_4NO_3, corresponding to 0, 10, 30 and 100 kg N ha^{-1} yr^{-1}. Increases in microbial carbon occurred where the moss had shown an increase in growth, in response to the lowest application of N of 10 kg N ha^{-1} yr^{-1}, but not at higher rates of N addition. Rates of CO_2 evolution from the peat were decreased beneath the N-treated areas leading to significantly decreased specific rates of respiration (Table 3.3), indicating a less-stressed microbial population (Wardle and Ghani, 1995). These results were confined to specific depths beneath the moss and indicate influences of N mediated by the living plant, i.e. the equivalent in mosses of a rhizosphere effect. Johnson *et al.* (1996) compared the impact of long-term N additions to acidic heathland and grassland or calcareous grassland, and reported increases in the microbial biomass with N treatment in the heathland and a decrease in the acidic grassland. This effect was observed during the summer, but not in samples taken in winter, and no changes in microbial biomass were found in the calcareous grassland soil during either season. A larger microbial biomass may be interpreted as a mobilisation of C, since microbial material is regarded as a more active pool than soil organic matter.

Table 3.3: Mean specific respiration rates (g CO_2-C h^{-1} mg^{-1} microbial-C) calculated from the basal rates of respiration at 10 and 22°C and microbial biomass-C measured using the substrate induced respiration (SIR) method in peat cores treated for 4 months with NH_4NO_3 every 2 weeks at rates equivalent to 0(N0), 10(N10), 30(N30) and 100 (N100) kg N ha^{-1} yr^{-1}

Depth, cm	N0	N10	N30	N100	†sed
5–10	8.7	2.2	2.0	3.5	1.00
10–15	3.3	3.8	3.7	1.5	1.17
20–25	2.2	2.4	2.6	3.5	1.53
40–45	5.5	4.9	4.8	9.8	2.62

†sed = standard error of difference between N treatment means at each depth.

The impacts of direct additions of inorganic N on the decomposition of organic matter have been comprehensively reviewed (Fog, 1988). Negative or zero effects of N additions on rates of decomposition are a widespread phenomenon for substrates having high C:N ratios such as wood and forest humus. Satisfactory explanations have not been forthcoming; Van Vuuren and Van der Erden (1992) reported no effects of additional inputs of $(NH_4)_2SO_4$ at rates of 5, 20 and 80 kg N ha^{-1} during 11 months on the decomposition of litters of *Calluna vulgaris* L. and *Molinia caerulea* (L.) Moench, but there was a temporary increase in the rate for *Deschampsia flexuosa* (L.) Trin., which decomposed fastest of the three litters. At first sight, reductions in rates of carbon dioxide evolution appear advantageous in relation to global climatic change, but increases in soluble organic compounds or microbial biomass C as a result of N additions may merely delay the release of C as CO_2 and divert it via alternative pathways.

In forest ecosystems, the impact of N addition has been studied mainly in N fertilized sites where N additions have exceeded the quantities in atmospheric deposition. Nevertheless, additions of NH_4NO_3 at rates equivalent to 150 kg N ha^{-1} as NH_4NO_3 altered the community structure of fungi colonizing the litter of Scots pine in Swedish forests (Arnebrant *et al.*, 1990). In the same plots, inorganic N decreased the rate of respiration and CO_2 evolution from decomposing coniferous litter (Bääth *et al.*, 1981). This effect which was still present for 3–5 years after fertilization, has been reported widely in a range of soils,. and cannot be explained by the increased acidity that follows uptake of the fertilizer N (Nohrstedt *et al.*, 1989). Increased acidity does reduce microbial activity and biochemical activity in decomposing plant litter (Brown, 1985), but the effects of N can vary and have been discussed in detail by Fog (1988).

3.3.2 METHANE EMISSIONS

Methanotrophs are methane-oxidizing bacteria, present in aerobic soils, which are also capable of utilizing NH_4 ion as a substrate and oxidizing it to nitrite and NO_3 (Verstraete, 1980); competition between NH_4 and methane leading to the inhibition of methane oxidation is therefore possible if NH_4 inputs to soils increase. Gulledge *et al.* (1997) reported inhibition of methane oxidation and consumption in taiga soils fertilized with NH_4. In aerobic soils, methanotrophs utilize atmospheric methane and inhibition occurs where there is sufficient NH_4 present depending on the activity of NH_4 oxidation (Conrad, 1996). In peatland and wetland areas where methanogenesis is active at lower depths, methane oxidation is more likely to occur in the surface horizons above the water table, which are those layers most likely to be impacted by anthropogenic ammonium-N inputs. Thus, one consequence of increased NH_4 in N deposition would be the inhibition of methane oxidation particularly in peat and wetland areas.

It is clear that for increased N supply to influence the activity of methanogens at lower depths in peats it is necessary for the N to reach the deeper layers. This would require reduced uptake by both microbes and plants in the upper horizons, which would require in turn, severely limiting factors such as extremely poor P availability (Williams *et al.*, 1998b).

3.4 Influence Of Soil Conditions

3.4.1 WATER TABLE DEPTH

The soils under discussion are largely undrained unless they have been afforested or improved for grazing. Consequently, the water table depth fluctuates throughout the season and there is little information about how this affects the plants' assimilation of N and the fate of N in the soil. In raised bogs, the growth of *Sphagnum* moss, the principal peat forming plant, is reduced at high water table levels the extent of the decrease depending on the species (Grosvernier *et al.*, 1997). Williams *et al.* (1998b) reported that at five European mire sites receiving N in wet deposition in the range 7–13 kg N ha^{-1} yr^{-1}, *S. magellanicum* growth was smaller where the mean water table level was lower and that this accounted for the greater concentrations of total N in the moss tissues. Woodin and Lee (1987) reported that previous exposure of *S. fuscum* to N in atmospheric deposition reduced its capacity to retain NO_3-N which would suggest that increased N concentrations in the plant would have the same effect. It follows from this argument that drainage of peatlands impairs the growth of mosses and their capacity to retain atmospheric nitrogen. This does not apply to woody plants, which benefit from improved root growth when the water table level falls (Ingram, 1992) and which will assimilate any additional N in atmospheric deposition. However, long term changes in the vegetation and organic matter dynamics of drained bogs could be expected to lead to reduced carbon storage (Vompersky *et al.*, 1992) a process that would be enhanced by nitrogen deposition.

Changes in water table will also influence the N transformation processes in soil. Denitrification of atmospherically derived NO_3 will be reduced on drained mires compared with wetter and more nutrient rich soils (Regina *et al.*, 1996) whereas the improved aeration will encourage organic matter decomposition and immobilisation of N.

3.4.2 SOIL PHOSPHORUS STATUS

Aerts *et al.* (1992) have implicated P status and availability as factors in the assimilation of atmospheric N by the vegetation in peatlands where soil P levels are intrinsically low. Williams and Silcock (1997) reported increased concentrations of molybdate reactive P in the peat beneath *S. magellanicum* receiving N additions, and attributed this to the plant's response to increased P supply through increased activity of phosphatase enzymes. Recent results by Johnson *et al.* (1998) on the characterisation of microbial communities in soil using BIOLOG® have indicated increased organic P utilisation in heathland soil treated with N, but this was not the case in a fertilized grassland. Additions of potassium orthophosphate to peats increased N uptake by mosses as shown by the use of ^{15}N labelled NH_4NO_3 (Williams *et al.*, 1998c) although retention is highly efficient (Table 3.4). It is assumed that P is the main limiting nutrient, but effects of K which is also inherently low in peaty soils cannot be ruled out. However, with this in mind, applications of P fertilizers to semi-natural vegetation to increase N retention may not be a solution because increased P concentrations in peats can lead to increased organic matter decomposition rates (Williams, 1992b).

3.4.3 LAND USE CHANGE

Land use change to areas of natural or semi-natural vegetation frequently involves changes in N inputs in the form of atmospheric deposition, fertilizer applications or returns from grazing animals. On the other hand, the changes in soil chemistry which accompany afforestation and improvement of upland soils for grazing can potentially alter the various N transformation processes, especially nitrification, and have a less obvious impact on the capacity of soils to retain N. As the result of drainage, ploughing and fertilization with P and K, to establish tree growth on blanket bog and deep peat, the surface horizons become more acidic through loss of base cations (Williams *et al.*, 1978). Microbial biomass-C in these soils is reduced (Williams and Sparling, 1988) and net mineralisation is increased (Williams *et al.*, 1979). The extent to which the peat dries out depends on the tree species and the thickness of the peat. Lodgepole pine (*Pinus contorta*) can tolerate waterlogged conditions and after canopy closure, peats undergo progressive drying and shrinking, developing wide cracks (Pyatt and Craven, 1979). Topography influences peat thickness and the reversible drying which occurs in shallow organic soils results in greater acidity in the absence of tree growth (Table 3.4). On the thicker deposits the acidity of the unplanted peat was less and increased to a greater extent by tree growth than on the shallower peats. None of the studies on changes in the dynamics of N in peat on afforestation have showed increased rates of nitrification which is consistent with the downward trend in pH.

Table 3.4: pH_{CaCl2} and base saturation percent in peat from the surface 15 cm layer of blanket bog planted with lodgepole pine (LP) and adjacent unplanted areas (0) where the peat thickness varied from 0.6 to 4m.

Peat thickness	0.6 m		1 m		1.5 m		4 m	
	0	LP	0	LP	0	LP	0	LP
pH	3.20	3.01	2.99	2.97	3.37	3.09	3.54	3.12
Base sat %	13.4	12.0	10.0	10.0	16.1	10.5	19.1	10.5

Techniques to improve the grazing value of upland soils such as blanket bog have included liming, fertilization with N, P and K and reseeding with appropriate seed mixtures. Liming highly organic acid soils results in decreased availability of N, attributed to increased immobilisation by the microbial biomass especially during Spring (Williams and Wheatley, 1992). At the same time, potential denitrification rates are increased compared to the unimproved peat (Wheatley and Williams, 1989) and the recovery of NH_4NO_3 fertilizer in herbage was only 15% of that applied (Williams and Wheatley, 1996). Nitrification was not activated by the application of ground limestone at this site although peatland limed with shell sand and reseeded showed active nitrification compared to peat from adjacent unimproved areas (Table 3.5).

Analysis of the runoff draining plots of reseeded blanket bog showed that the area was behaving as a sink for N in wet deposition (Williams and Young, 1994) though over a period of years the gain in N concentration in the surface10 cm suggested that losses of carbon and decreases in the C:N ratio of the peat had occurred (Williams and Wheatley, 1996).

Table 3.5: Net ammonification and nitrification (as mg N kg^{-1} dry wt), after incubatiom for 30 days at 30°C, in peat from adjacent semi-natural and improved sites at Port Voller, Lewis.

	NH_4-N			NO_3-N		
Depth cm	Semi-natural	Improved	sed	Semi-natural	Improved	sed
0–5	6.0	367	21.4	<2	138	66.5
5–10	34.0	190	56.2	<2	16	15.8
10–15	81.8	16	18.6	<2	7	5.2

While N immobilisation is a predominant process in organic soils, especially if they have been limed (Williams and Wheatley, 1989), there is less information about the processes and soil N dynamics in a range of hill and upland mineral soils including podzols and peaty gleys beneath extensively grazed grasslands. Floate (1970) compared rates of mineralisation of N returns in faeces and plant litters in *Agrostis–Festuca* and

Nardus grasslands concentrating on the net mineralisation of N. Thomas *et al.*(1988) suggested that substantial amounts of urine-N were lost from the soil-plant system and large amounts of NO_3-N have been measured in soil solutions collected from beneath grazed swards (Cuttle *et al.*, 1996). Sheep urine can comprise single additions of large quantities of N as urea, equivalent to 500 kg N ha^{-1} yr^{-1}, but the impact of grazing animals on soil N transformations has been studied mainly beneath grass–clover plots in intensively managed grassland rather than in grazed indigenous vegetation, where these animal returns represent large N additions when compared with atmospheric inputs. However, there is a possibility that the combined effects of defoliation of vegetation and N returns by grazing animals will have some influence on the ultimate fate of N in atmospheric deposition to these sites.

3.5 Acidification Of Soils By Enhanced N-Deposition

Acidification of soils and surface waters is considered to be the most damaging consequence of S and N deposition in the atmosphere. Increased acidity is a threat to herbaceous vegetation (Houdijk *et al.*, 1993), and leads to decreased microbial activity (Härtl and Cerny, 1981). With declining emissions of S, N is gaining more importance as the N emissions are showing few signs of abating (Hornung and Langan, Chapter 1).

Studies into the acidification of soils by anthropogenic S inputs largely focus on the ability of the relatively mobile sulphate anion to remove alkaline and alkaline–earth cations from soil exchangeable cation pools and the gradual depletion of parent material sources supplied by mineral weathering. This process leads to the replacement of the base cation soil buffering capacity by one dependant on iron and aluminium phases (Ulrich, 1991). Sorption–desorption processes can lead to significant acidification impacts (Anderson *et al*, 1995), but the NO_3 ion is more mobile than sulphate, being relatively unaffected by adsorption processes and, once past the rooting zone it is capable of leaching base cations in a manner similar to sulphate (Nihlgard, 1985).

Soil nitrogen transformation processes involve either the consumption or release of protons (Van Breemen *et al.*, 1983). Ammonification of organic nitrogen consumes protons and provided the NH_4 is taken up by plant roots there is no net effect on acidity. Increasingly, attention has been drawn to N-inputs arising from enhanced emissions of ammonia from intensive agricultural activities (Sommer and Hutchings, 1997), since nitrification of these in the soil gives rise to two protons for each mole of NO_3 produced, accelerating the acidification process. Nitrification by autotrophic microorganisms releases protons which are neutralised if plant uptake involving the exchange of hydroxyl ions follows. At one forested site in North Wales, Stevens *et al.* (1994) have estimated that the annual rate of soil acidification due to these N transformations amounts to 1.01 keq ha^{-1} yr^{-1}, substantially more than that due to anthropogenic S-deposition (0.37 keq) or base cation uptake by the tree crop (0.51 keq).

Denitrification of NO_3 through to N_2 gas consumes ten H^+ for each atom of N (Nommik, 1956) and the process, which is sensitive to low pH, can markedly increase in activity in acid soils treated with lime (Klemmedtsson *et al.*,1977). Thus the attempted amelioration of acid-impacted sites by the application of limestone, while producing a temporary increase in surface water pH (Miller *et al.*, 1995), can lead to decreased outputs of NO_3.

Soil acidification attributed to N deposition is largely caused by NO_3 leaching and concomitant removal of base cations though cation displacement by NH_4^+ ions can also reduce the base status of granitic soils (White and Cresser, 1996). Mobilisation and leaching of aluminium by excess NO_3 is potentially a very damaging process for surface waters and Rawlins *et al.* (1996) showed experimentally that an increase in temperature from 4 to 15.4 °C was sufficient to increase the NO_3 concentration in column leachates from 15–20 to 120 mg l^{-1} which was accompanied by an increase in monomeric aluminium from 1–3 to 12 mg l^{-1}. Ecosystems where NO_3 leaching occurs have been described as being N saturated (Aber *et al.*, 1989). This concept applied on a catchment scale implies that atmospheric N enters the system and exits unmodified whereas there are many intervening processes which may be involved between input and outputs. It follows from the concept of N saturation that soils, vegetation and the ecosystems they comprise, have a N uptake capacity which is determined by a number of limiting factors. By considering the individual soil processes it may be seen that in reality the system is more complex than the concept allows.

3.6 Nitrogen Flux Measurements In The Field

An important consequence of increased N inputs is the potential change and magnitude of N fluxes in the biogeochemical cycle. Such changes have been monitored and detected on a mass balance basis where outputs are compared to inputs within various segments of the ecosystem and values reported in the literature.

3.6.1 FOREST CANOPIES AND ORGANIC SOIL HORIZONS

Responses to increasing N-deposition in mineral soils are very similar to those in highly organic systems, especially in natural or semi-natural sites where topsoils are mainly organic, and plant growth may significantly influence the chemical and biological responses in the underlying mineral layers. In the Netherlands, large N inputs of 40–60 kg N ha^{-1} yr^{-1}, mainly as ammonium sulphate, can result from mutually enhanced dry deposition to vegetation surfaces of NH_3 (volatilised from intensive animal farming) and of SO_2 (from fossil fuels) (Van Breemen *et al.,* 1982). Van Breemen *et al.* (1987) investigated the effects of such inputs under oak–birch woodland, in a series of four soils developed on sandy to loamy Pleistocene deposits. Three of the soils were highly acidic Inceptisols and Entisols (pH 3–4), while the fourth was a calcareous Mollisol (pH 6.5–7.5), and substantial nitrification in all of them resulted in an annual budget of 3–9 kmol H^+ ha^{-1}, which formed the major part of the total soil acidification taking place on the site. In the Inceptisols, nitrification amounted to 4–14 kmol H^+ ha^{-1} annually in the well-developed organic layer forming the forest floor, with NO_3 formation invariably exceeding NH_4 removal, indicating that part of this NO_3 was derived from organic-N in leaf litter. It was estimated that on this site, the mean annual increase in the organic N pool was at least 40 kg ha^{-1} between 1960 and 1984, a value which was equal to or somewhat lower than the 1984 atmospheric N input. The wetter Inceptisol and the Mollisol both had NO_3 leaching at 90 cm depth, in excess of the inputs (Table 3.6). Van Breemen *et al.* (1987) argued that the differences in N retention between the four soil systems could be explained by:

i) differences in N inputs over many years causing some of the systems to become N-saturated (as noted earlier, N-inputs derived from agriculture may be locally intense). Thus the lesser inputs to the Udipsamment may be the major reason for its biota still utilizing part of the incoming N.

ii) differences in biomass production and thus in capacity to assimilate N (the site had been coppiced up to 1939; note the presence of alder, capable of N-fixation).

In addition to the net NO_3 production in the Haplaquept and Haplaquoll, they also had the greatest H^+ production (7.5–8.6 kmol ha^{-1}), indicating that the finer soil textures may influence the degree of nitrification, whereas the coarser textures and better aeration probably led to better N-retention in organic matter.

Where sufficient soil hydrochemical budget data have been reported, it is possible to carry out a direct comparison between regions and under different vegetation communities, although the majority of suitable publications deal with forested sites. In this comparison, the hypothesis of N saturation can be tested by assuming that all data describe situations in which either NH_4 or NO_3 saturation (or both) already occur, i.e. the incoming N-deposition, whether at the bulk deposition or canopy throughfall stages, is in excess to ecosystem requirements and the same amount of N will exit from the site. Various assumptions are made:

(a) chloride ion is assumed to be conservative after it contacts the soil surface and is derived *in total* from marine inputs

(b) the soil hydrology is assumed to be downward within the profile, although lateral water movement on hillslopes is acceptable

Table 3.6: Annual inorganic-N budgets for throughfall and drainage at 90 cm depth under mixed oak–birch forest in the Netherlands (Van Breemen et al., 1987).

Soil	Vegetation	Input	Output	Retained (+denitrified)
	N flux kg ha^{-1} yr^{-1}			
Aeric Haplaquept	Oak & birch	54.5 ± 4.6	78.5 ± 18.6	4.0 ± 23.4
Umbric Dystrochrept	Oak	56.2 ± 5.1	28.1 ± 12.4	28.1 ± 11.6
Aquic Udipsamment	Oak	44.6 ± 2.8	22.5 ± 10.7	22.1 ± 9.7
Typic Haplaquoll	Oak & birch (+alder & poplar)	62.8 ± 6.8	87.6 ± 26.5	-25.1 ± 21.8

The vegetation–soil–stream continuum is treated as a series of boxes draining into the next in the following order :– atmospheric deposition (BD), canopy throughfall (TF), litter drainage (S), soil horizons (O E B BC C) and stream (Str). In Table 3.7, a summary is shown of several site studies which report the output of inorganic-N i.e. N-saturation according to Aber *et al.* (1989); using this treatment allows the visualisation of N-retention and -release at various points within the different systems.

In Table 3.7, there are three datasets from Beddgelert, a site where NO_3 in the output streams is present in larger concentrations than those seen elsewhere in Welsh forested sites (Stevens *et al.,* 1989). The general pattern of Stevens' data was repeated in Adamson *et al.* (1993), with NO_3 release being dominant in the upper organic and the E horizons, with partial retention in the B horizons being insufficient to prevent NO_3 leaching. The inter-site comparison of the 1993 publication also revealed that although the soil water NO_3 concentrations were largest at Beddgelert, the Cumbrian site at Dodger Wood was actually leaching the greater proportion of NO_3 in the B horizon. The Northbrae site in NE Scotland differed markedly in showing net retention of NO_3 in both the O and B horizons. Under hybrid larch, there was greater utilisation of NO_3, with only the Dodger Wood site showing NO_3 leaching from the organic topsoil. In later publications (Stevens *et al.,* 1994; Hughes *et al.*, 1994), throughfall and soil drainage waters were compared between open moorland control sites with those under Sitka spruce stands of increasing age class. This comprehensive study included Beddgelert along with another four forests in Wales, in all comprising 25 study sites, where soils were dominantly stagnopodzols underlain by Ordovician or Silurian shale, slate or mudstone. The data show throughfall NO_3 retention in surface runoff, but NO_3 leaching under the organic topsoils at all sites, with partial retention in the B horizons gradually decreasing as the forest aged, with extensive NO_3 leaching under the oldest stands, in which nitrification was also very active, resulting in significant soil acidification.

Interestingly, the 1989 Beddgelert data (Stevens *et al.,* 1989) show net uptake of bulk deposition input-N, both as NO_3 and NH_4 (Table 3.7), whereas in the pollution-impacted Loch Ard sites, Miller *et al.* (1990) found both N-forms being enhanced in the throughfall under Sitka spruce at Kelty Water and lesser amounts as NO_3 under Norway spruce at Loch Chon. Both sites showed extensive NO_3 retention by the soils (peaty gleys at Kelty, humus–iron podzols at Chon, with large NH_4 release in the litter being counteracted by uptake within the soil profile. This pattern of NH_4 release from organic upper layers and retention within the mineral soils is widely seen in the Table 3.6 data, the exception being the most easterly Northbrae data (Adamson *et al,* 1993), which may reflect the greater rainfall at the western sites.

3.6.2 TRANSPLANTED SOIL CORES

In the EC funded CORE project, six forest sites with specific physical and chemical characteristics (Raubuch *et al.*, 1995) were selected along a transect of increasing N and S deposition (Table 3.8). Buffer ranges (Ulrich, 1991) of the soils ranged from 'cation exchange' ranges ($4.2 < pH < 5.0$) at Kilkenny and Haldon to 'aluminium' ranges ($3.0 < pH < 4.2$) at Fontainebleau and Solling, with Grizedale and Wekerom soils being borderline between the two ranges, but with their exchange sites dominated by Al. The aim of the project was to reciprocally transplant soil cores between each site to separate the influence of microclimate from deposition of N and S (Berg *et al.*, 1994; Raubuch *et al.*, 1995).

Table 3.7: Retention and release of NO_3 and NH_4 under conifer forest, assuming N saturation; [1] Stevens et al., (1989); [2] Adamson et al., (1993). Str^1 = stream chemistry compared with that calculated from Surface+L+O+E output; Str^2 = stream compared with B+C output.

			Observed (mg l^{-1})			Calculated (mg l^{-1})		% Retention	
Site	Vegetation	Sample	Cl	NO_3-N	NH_4-N	NO_3-N	NH_4-N	NO_3-N	NH_4-N
Beddgelert	50–yr	BD	5.10	0.19	0.24	0.19	0.24	0	0
[1]	Sitka	TF	16.00	0.56	0.40	0.60	0.75	6	47
	spruce	L	15.70	0.84	0.39	0.55	0.74	–59	–84
		O	16.80	1.41	0.42	0.59	0.79	–90	–1
		E	14.70	1.55		0.51		–41	
		B	19.90	0.98		0.70		108	
		C	15.80	0.81		0.55		5	
		Str^1	11.10	0.72		0.39		181	
		Str^2	11.10	0.72		0.39		–19	
Beddgelert	Hybrid	TF	18.30	0.39	0.20	0.39	0.20	0	0
[2]	larch	O	20.60	0.28	0.31	0.44	0.23	36	–38
		B	22.80	0.24	0.03	0.49	0.25	18	122
	Sitka	TF	27.80	0.62	0.48	0.62	0.48	0	0
	spruce	O	30.20	3.06	0.46	0.67	0.52	–354	12
		B	37.70	2.22	0.06	0.84	0.65	120	81
Dodger	Hybrid	TF	18.90	0.67	0.58	0.67	0.58	0	0
Wood	larch	O	21.50	0.82	0.22	0.76	0.66	–8	67
[2]		B	24.20	0.52	0.01	0.86	0.74	46	39
	Sitka	TF	16.30	0.57	0.57	0.57	0.57	0	0
	spruce	O	17.30	0.65	0.75	0.60	0.60	–7	-24
		B	23.70	1.51	0.04	0.83	0.83	–77	113
Northbrae	Hybrid	TF	10.20	0.49	0.64	0.49	0.64	0	0
[2]	larch	O	14.20	0.11	0.31	0.68	0.89	84	65
		B	17.90	0.11	0.01	0.86	1.12	21	47
	Sitka	TF	9.60	0.56	1.10	0.56	1.10	0	0
	spruce	O	16.00	0.21	0.87	0.93	1.83	78	53
		B	32.50	1.07	0.02	1.90	3.72	5	74

Table 3.7: (contd): Retention /release of NO_3 and NH_4 under conifer forest, assuming NO_3- and NH_4-N saturation; [3] Stevens et al., (1994) and Hughes et al., (1994)–D* = surface runoff, NH_4-N* = no data reported (see text); [4] Miller et al., 1990.

			Observed (mg l^{-1})			Calculated (mg l^{-1})		% Retention	
Site	Vegetation	Sample	Cl	NO_3-N	NH_4-N*	NO_3-N	NH_4-N	NO_3-N	NH_4-N
Beddgelert,	Moorland	TF	8.08	0.25		0.25		0	
Dyfi,	control	D*	7.77	0.11		0.24		55	
Dyfnant,		O	7.52	0.18		0.24		-32	
Hafren,		B	8.47	0.18		0.27		10	
& Tywi	Sitka spruce	TF	9.36	0.42		0.42		0	
[3]	0–15 yrs	D*	8.90	0.07		0.40		82	
		O	8.51	0.20		0.38		-40	
		B	9.61	0.03		0.43		52	
	Sitka spruce	TF	14.08	0.96		0.96		0	
	30–45 yrs	D*	12.23	0.46		0.83		45	
		O	16.17	1.23		1.10		-46	
		B	16.74	1.01		1.14		23	
	Sitka spruce	TF	17.37	0.90		0.90		0	
	45–60 yrs	D*	13.44	0.56		0.70		19	
		O	18.44	1.15		0.96		-34	
		B	16.24	1.91		0.84		–104	
Loch Ard	Norway	BD	4.63	0.16	0.23	0.16	0.23	0	0
(Chon)	spruce	TF	7.68	0.27	0.25	0.27	0.38	-9	3
[4]	(35 yrs)	L	7.06	0.17	0.68	0.25	0.23	35	–246
		BC	9.62	0.06	0.05	0.34	0.32	59	224
		Stream	7.96	0.04	0.10	0.28	0.26	–15	-40
(Kelty)	Sitka	BD	5.54	0.15	0.35	0.15	0.35	0	0
	spruce	TF	9.66	0.46	0.40	0.27	0.61	-92	-49
	(35 yrs)	L	6.84	0.31	0.41	0.33	0.28	65	-116
		BC	6.70	0.11	0.08	0.32	0.28	59	113
		Stream	7.68	0.14	0.18	0.37	0.32	4	-19

Table 3.8: Site characteristics from the CORE project; NS – Norway spruce, SP – Scots pine. The elemental values are annual mean kg ha^{-1} in canopy throughfall.

Site	Rainfall mm	Vegetation	Age yrs	NO_3-N	NH_4-N	SO_4-S	H^+
Kilkenny (IRL)	826	NS	20	3.3	5.9	9.0	0.01
Haldon (UK)	1006	NS	45	6.6	9.3	39.7	0.04
Grizedale (UK)	1851	NS	40	6.2	7.5	33.2	0.88
Fontainebleau (F)	697	SP	30	7.0	5.0	10.6	0.06
Wekerom (NL)	750	SP	35	11.7	38.0	28.6	0.08
Solling (D)	1088	NS	109	19.5	19.5	45.5	0.99

Throughfall and soil solution chemistries were monitored, the latter being contrasted against 'control' soil solutions produced from 'control' lysimeters of the Haldon soil at each site, watered *pro rata* with the appropriate volume of distilled water.

Unsurprisingly, the control lysimeters produced NO_3 in the leachates, the relative amounts being due solely to climatic factors at the host sites, with Wekerom (212 kg N ha^{-1} yr^{-1}) leaching the largest annual amounts. The comparisons within hosts for the transplanted soils showed that the leaching of NH_4 and NO_3 appeared to be pH-related, with the 'cation exchange' buffered soils giving small yields of NH_4 and large outputs of NO_3 (up to 120 mg l^{-1}). In a consideration of temperature dynamics, nitrification rates and NO_3 leaching, it was found that the host site characteristics specified nutrient dynamics, i.e. NO_3 leaching with time was comparable between sources, indicating independence of soil textural class. A relationship was noted between bulk soil pH and NO_3 leached, with the high pH calcareous soils showing greater losses (see above) and the low pH acidic soils smaller values, up to 55 mg l^{-1}. The amounts of NH_4 leached were negligible with all available being efficiently converted to NO_3, which Berg *et al.*, (1994) construed as pH setting limits to the extent of ammonification, which in turn limits the rate of nitrification.

In the transplanted soils of the CORE project, sulphate behaved as a conservative inert anion (Raubuch *et al.*, 1995), with no relationship with the leaching of calcium and aluminium cations. NO_3 dominated the anion and cation leaching, a strong Ca–NO_3 relationship being evident in the control, Kilkenny and Haldon soils, and an equally dominant Al–NO_3 relationship in the Grisedale, Werkerum and Solling soils. As concluded by Berg *et al.*, (1994), the depletion of base cations by NO_3 leaching will eventually lead to a transfer of the 'cation exchange' buffered soils into the 'aluminium' buffered range, with concomitant increases in soil solution Al contents. Subsequent plant root damage and increased availability of substrates for N-mineralisation will eventually lead to enhanced NO_3 leaching and severe soil acidification due to nitrification.

3.6.3 CATCHMENT STUDIES

Plot studies such as those reviewed above can provide detail which is very useful in bridging the gap between process and catchment studies, which in turn form a bridge towards the understanding of regional responses to N and S deposition. In the United States, the Integrated Forest Study (and related work) demonstrated that high-elevation southern Appalachian spruce–fir (SASF) ecosystems received the largest annual atmospheric loadings of N and S in North America, amounting to ca. 1.9 Keq N ha^{-1} yr^{-1} and 2.2 Keq S ha^{-1} yr^{-1} (Johnson and Lindberg, 1992). In a study of one sub-region of SASF, the Great Smoky Mountains, Flum and Nodvin (1995) concluded that:

i) most high elevation catchments in the area were N saturated, with NO_3 being the dominant strong acid anion in the output streams, especially those draining old growth forests.

ii) continued large atmospheric N loadings would spread N saturation downslope into catchments where second growth forests were now maturing.

iii) sulphate outputs were smaller than expected and this phenomenon might be related to the N dynamics of the systems.

Within the Smoky Mountain project, a detailed study of the Nolan Divide catchment by Nodvin *et al.* (1995) showed that in the two streams draining the basin, NO_3 concentrations ranged between 0.6–1.0 mg NO_3-N l^{-1}, approximately similar to the combined sulphate and chloride outputs.

3.6.4 ECOSYSTEM MANIPULATION

"Nitrogen saturation" of ecosystems, whose cited examples more-correctly indicate "NO_3 saturation", has been visualised as occurring in several stages by Stoddard (1994), who developed a ranking system based on annual N budgets and seasonal streamwater trends. Thus a Stage 0 system is N-deficient, showing little or no NO_3 loss regardless of the magnitude of input N. In upland areas unimpacted systems only exhibit significant concentrations of NO_3 in runoff during snowmelt (Wright and Tietema, 1995). (Impaction as defined here includes severe physical disturbance of the system, such as clear-felling of forests (Pardo *et al.*, 1995) and reclamation of natural and semi-natural landcovers for agriculture). The next (Stage 1 in NO_3 saturation) is manifested by significant NO_3 export during the winter months, when the vegetation is dormant, as seen at Hubbard Brook (Likens *et al.*, 1977), and this is followed by Stage 2 in which N retention declines and a significant loss of NO_3 also occurs during the summer months, when vegetation uptake is maximal. This stage of NO_3 saturation was induced following three years of ammonium sulphate addition at an annual rate of 25 kg N ha^{-1} to a forested catchment at Bear Brook in Maine (Kahl *et al.*, 1993). At Stage 3, the terrestrial system becomes a net source of N (as NO_3) to the aquatic system, with outputs exceeding inputs, and biological control on surface water NO_3 concentration is essentially absent (Aber *et al.*, 1989).

In Bear Brook, powdered ammonium sulphate was applied bi-monthly to the 10.2 ha West Bear Brook catchment, with the adjacent 10.9 ha East Bear Brook acting as an untreated control. Both catchments have shallow spodosols developed on 1–5m of till overlying metapelites with granite intrusion; the vegetation cover is 40–60 year old mixed hardwood forest dominated by beech (*F. grandiflora*), with red spruce (*P. rubens*) and balsam fir (*A. balsamea*). The experimental treatment is small relative to the total estimated annual vegetative requirements of 80–120 kg ha^{-1}, adding approximately 10% annually to the estimated total N pool of 2500 kg N ha^{-1} in upper soil horizons. The experiment elevated the total N deposition of the catchment to ca. 25% greater than some of the high-elevation large N-deposition catchments of North America. Prior to the treatment, the two catchments varied in parallel, with NH_4 being undetectable in the outputs and NO_3 showing a marked seasonality, with virtually zero output during the growing season rising to a maximum of ca. 0.2 mg NO_3-N in the spring. The site was close to N saturation before the experiment, as judged by a relatively small retention of 75% of the N in wet and estimated dry deposition (50% of wet). Factors judged by Kahl *et al.* (1993) to be contributing to the potential for N saturation included a smaller C:N ratio and a larger forest floor N concentration, compared to lower-elevation sites in Maine.

After three years' treatment, NO_3 outputs more than doubled in the treated catchment. Retention of the experimental N additions within the catchment decreased each year, from 94% in the first to 74% in the third, indicating that although the system continued to accumulate N, this was happening at an increasingly slower rate over the period. The treatment resulted in a fundamental shift in the seasonal discharge/NO_3 concentration relationship. Prior to treatment, regression of the NO_3 concentration as a function of discharge had an approximately-zero/zero discharge/NO_3 intercept, which remained after six months of the treatment, even though the NO_3 concentration had increased relative to discharge. By the third year of treatment, NO_3 concentrations were chronically elevated for most of the summer, with summer baseflow values as large as 0.94 mg NO_3-N l^{-1}. Kahl *et al* (1993) interpreted these findings as indicating the leaching of N into the deeper hydrological pathways which sustained the summer baseflows. Kahl *et al.* (1993) concluded that, before the experiment began, Bear Brook was transitionally into the first N-saturation stage of chronic or near-chronic NO_3 export during the nongrowing season. The treatment drove the system to the second stage of chronic loss of NO_3 to surface waters even during summer baseflow.

In Europe, the EXMAN project was established in 1989 to observe inputs and outputs in conifer stands in Northwest Europe. The deposition range, from only moderately-polluted Atlantic areas through decreasing maritime influences and increasing anthropogenic N and S effects, ranged from southern Ireland to Germany (Table 3.9). In intensive agricultural regions, NH_4 inputs were sufficiently large to cause N saturation, with nitrification-driven acidification a major soil process. Another European Community funded programme, NITREX included a group of European experiments in which N deposition was drastically changed to whole catchments (or large forest stands) at 8 sites (Table 3.9) spanning an N deposition gradient (Wright and Van Breemen, 1995). In these sites, intensive efforts were made to quantify inputs and throughputs, in the form of both canopy throughfall and soil drainage outputs (Tietema and Beier, 1995). In the NITREX programme, these data were recovered from control sites. In Table 3.9, the behaviour of both NH_4 and NO_3 at all of the EXMAN/NITREX–

Site (Country)	Project	Vegetation	Veg age yrs	Soil	Inputs[1]		Data divergence from saturation model[2]			
							NH_4-N		NO_3-N	
					NH_4-N	NO_3-N	TF	Output	TF	Output
Höglwald (D)	Exman	NS	39	Typic Haplumbrept	6.0	4.6	-0.04	0.99	0.38	–2.46
Ballyhooley (IRL)	Exman	NS	52	Typic Haplorthod	3.4	1.4	0.44	0.92	0.44	–0.36
Solling (D)	Exman/Nitrex	NS	58	Aquic Dystrochrept	7.2	6.1	0.16	0.98	0.05	–0.32
Ysselsteyn (NL)	Nitrex	SP	45	Humic Haplorthod	21.0	11.0	0.33	0.95	0.62	–0.31
Speuld (NL)	Nitrex	DF	31	Orthic Podzol	15.0	8.0	0.24	0.95	0.36	–0.22
Kootwijk (NL)	Exman	DF	39	Plaggic Dystrochrept	11.1	5.6	0.20	1.00	0.37	–0.08
Harderwijk (NL)	Exman	SP	80	Typic Udipsamment	10.3	4.6	0.09	1.00	0.25	0.53
Aber (UK)	Nitrex	SS	33	Ferric Stagnopodzol	4.6	6.3	0.54	0.87	0.61	0.66
Sogndal (N)	Nitrex	Alpine		Lithic Haplumbrept	1.4	1.3	1.00	0.98	1.00	0.98
Gårdsjön (S)	Nitrex	NS	84	Orthic Humic Podzol	4.6	5.6	0.72	1.00	0.57	0.99
Klosterhede (DK)	Exman/Nitrex	NS	72	Typic Haplorthod	3.3	4.4	–0.50	0.99	0.26	1.00
Alptal (SCH)	Nitrex	NS	185	Umbric Gleysol	6.6	8.3	0.64	NP	0.54	NP

Table 3.9: Site characteristics for the EXMAN and NITREX field-sites. [1] **Inputs are quoted as annual means during 1989-94 (Exman) and 1991-93 (NITREX);** [2] **The saturation model assumes *total* exclusion of both ammonium and nitrate ions from the system and the values quoted use the Gårdsjön volumetric throughput data as a driver:- canopy interception (0.7 x BD) and soil evapotranspiration (runoff = 0.5 x Throughfall). Data divergence is expressed as proportionate retention (+ve) or release (-ve).**

control sites have been compared using enhancement factors based upon methods similar to those used in Table 3.8:

a) Output data is compared with modelled values in which total nitrogen saturation is assumed i.e. both NH_4 and NO_3 will pass unaltered through the plant /soil system and appear in soil drainage;

b) Sites can be compared by applying known (site-specific) canopy interception and total evapotranspiration losses (those favoured are derived from sites with shallow soils lying on impervious bedrock.

Table 3.9 has throughfall and soil water data modified by applying the Gårdsjön canopy interception and evapotranspiration factors. These determinands are the one extreme of the dataset, the other uses the Aber factors (Canopy interception 0.4 x BD, Runoff 0.8 x Throughfall), in which all of the soil drainage outputs are increased proportionately. The data in Table 3.9 have been ranked according to NO_3 outputs relative to throughfall inputs and, as such, require some comment. All of the sites retain NH_4 ion regardless of the interception/evapotranspiration factors; with the exception of Aber, the maximum NH_4 loss to soil drainage is 3%, interception/transpiration drivers used; with the exception of Aber, the maximum NH_4 loss to soil drainage which lies within the standard error for the estimate. The Höglwald output of NO_3 (246% of the throughfall input to the forest floor) is very large when compared to the inputs (relative to those from the Dutch sites). One very understandable reason for the size of the output is that the 39-year old stand of Norway spruce used in EXMAN had been planted on a site previously occupied by native hardwoods (mainly *F. sylvatica*). The deciduous litter and derived organic topsoils will have exerted a prolonged and major influence on the carbon and nitrogen dynamics of the site (Tietema and Beier, 1995) and the outputs will be biased by these factors. When the inputs are compared with the rankings, the three most polluted sites appear to be anomalous. Höglwald being explicable, Solling can be understood as a very-impacted site, given its soil parent materials and its geographical position relative to the major European acidifying outputs of 1950–90. The data from Ballyhooley is puzzling, the site being in a relatively-pristine area of Europe, with small agricultural inputs of NH_4 and even-smaller amounts of precipitated NO_3. Otherwise, the dataset is very understandable, with the large-input sites releasing enhanced amounts of NO_3 in comparison to the more-pristine areas. There is also a tendency for the more-poorly drained soils to lose NO_3 at greater rates than their better-drained counterparts.

However, the major objective of NITREX lay in altering the inputs of N to the eight sites (Wright and Van Breemen, 1995), accomplished by a series of input exclusions ("roof" experiments) and enhancements ("spraying" experiments). In the roof experiments, natural inputs were deflected from entire forest catchments by under-canopy roofs for chemical modification, and subsequent application, by spraying in a "pristine" form to the forest floor. At the other sites, nitrogen inputs were enhanced by spraying NH_4 and NO_3 salts on to forest floors. The outcome was largely in agreement with the conclusions of Kahl *et al.* (1993) in Bear Brook, namely there was a rapid surface water response which did not appear to be immediately related to any vegetation or soil factors. At annual inputs of <10 kg ha^{-1}, the major part of the N was retained in the system, whereas an annual input of >25 kg ha^{-1} led to substantial NO_3 outputs. Changes were relatively slow at sites which were regarded as non-saturated (e.g.

"sprayed" Gårdsjön and Klosterhede – see Table 3.9), with annual NO_3 output increasing to ca. 3 kg ha^{-1} over two years, but fast large responses occurred at saturated sites ("roofed" Speuld and Ysselsteyn), where annual N output decreased eight-fold (80 to 10 kg ha^{-1}).

The initial lack of response at unimpacted sites was very evident in the soil solution studies at Gårdsjön (Stuanes *et al.,* 1995). At the treatment catchment G2 and its control F1, soil solutions were recovered by a mixture of tray and tension lysimetry in a transect which ran through skeletal soils at the top of the catchment, through increasingly-wetter soils towards the catchment outlet (Figure 3.2). During the first two years of the treatment (the annual application of 35 kg-N ha^{-1} NH_4NO_3), soil solution response was confined to organic topsoils, possibly strongly-influenced by the timing of sampling vs application. There appeared to be larger effects occurring within the control soils, rather than in the treatment.

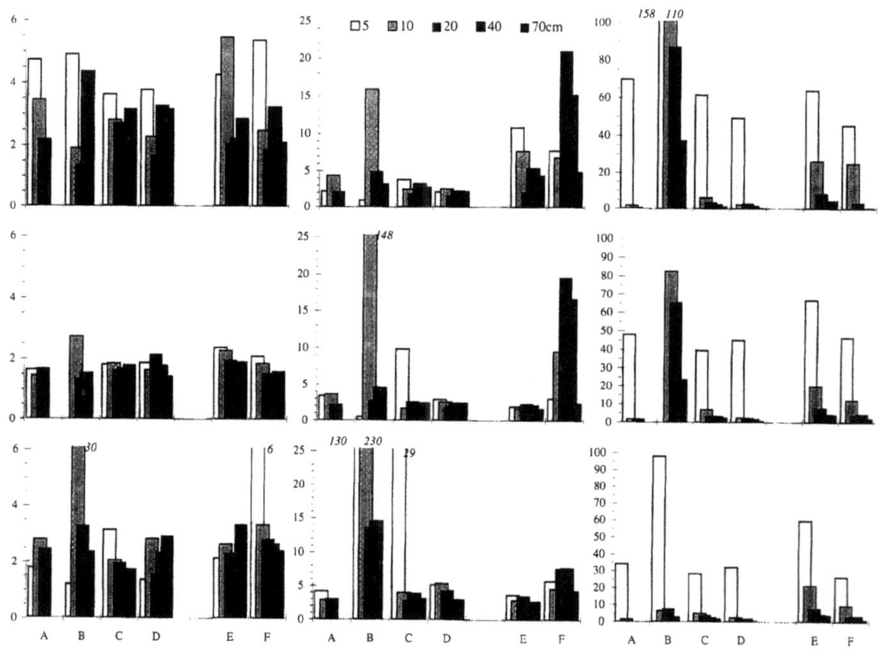

Figure 3.2: Volume-weighted concentrations of NH_4-, NO_2- and organic -N, μmol l^{-1} in the Lake Gårdsjön NITREX soil water data (after Stuanes *et al.*, 1995); A–D are podzolic soils within the treatment catchment G2, ranging from the dry upper parts of the catchment (A, B) through the mid levels (C), to the wetter parts around the catchment outlet (D); E, F are the corresponding mid-point and wet soils from the control catchment (see text). Water samples were recovered by tension lysimetry at the depths indicated.

The delayed soil responses seen in NITREX have led to the development of the current concept of a *cascade of response* to increased N inputs (Tietema *et al.,* 1995). It is argued that since the external impact affects the aqueous phase, and water is the principal transport medium, the first response will be seen in the runoff. Depending on hydrological routing, i.e. the physical nature of catchment soils, different soils will react with throughflow water and affect hydrochemistry at their independent rates. The short-term effect on outputs will be small, but will increase with time, during which the strongly-retained fraction of soil solution has time to react with biological and abiotic compartments. Biological factors such as microbial immobilisation and adsorption of N act rapidly within the major soil N-pool, and will therefore buffer any impaction by increased inputs, delaying any response until the pools themselves have been significantly altered. This change will then influence vegetation and only then, due to the interaction between plant roots and soil solution, will dramatic effects be seen in soil hydrochemistry. The cascade system can now be regarded as a series of feedback loops with water, soil, faunal and vegetation components thoroughly enmeshed. Based on the fast response of the vegetation and soil responses in saturated systems (e.g. Ysselsteyn), compared to the delayed response at Gårdsjön, Tietema *et al.* (1995) concluded that the soil solution-dependent pools were "one-way" in their response to changed N concentrations. In non-saturated systems, they are capable of removing large amounts of N from the increasing quantities emerging in soil solution, whereas in saturated systems they largely retain their N capital, rather than release large quantities of mineralised-N.

3.7 Conclusions

On soils with sufficient nutrient supply, the additional N from atmospheric deposition will be assimilated by the vegetation and C and N will be incorporated in greater amounts but in balance. However, many soils beneath semi natural and natural vegetation in upland areas are inherently nutrient poor in essential nutrients such as P and K with the result that their low concentrations will limit the growth response of vegetation to N deposition. Where the inorganic N is assimilated by the vegetation without additional growth, the C:N ratio of the litter will be reduced and subsequent increases in carbon and N turnover result. The plant can also transmit the impact of N inputs to the soil through the roots and processes such as root exudation. In the past, only economically important cropping systems have been studied in this degree of detail, but this approach is now required for the extensively managed vegetation in hill and upland areas.

Studies on *Sphagnum* mosses have shown how these can assimilate inorganic N without concomitant growth and transform it to organic forms, probably amino acids, which can be released into the surrounding waters as DON. Production of DON in forest ecosystems treated with inorganic-N has also been reported. The factors governing the production of this, N its characteristics and fate in pristine conditions as well as those of high N inputs are not clear at present and these need elucidating if the effects of increased N deposition are to be fully understood.

The nutrient status of the soils, as in some peats, are such that the vegetation behaves as though it were N saturated and inorganic N could penetrate the canopy and pass directly into the soil. The review of flux measurements in forests suggests that this already occurs in some locations at present and nitrate is released into surface waters. However, it is not clear which factors control nitrification in forest soils. The activity of this process varies considerably between sites and it is far from clear if the supply of N is the only controlling factor (see Wilson and Emmett, Chapter 5). The occurrence of nitrification combined with nitrate leaching has a strongly acidifying effect on the soil with removal of base cations from the soil surface horizons. This process potentially leads to the most damaging consequences of nitrogen deposition on ecosystems.

The results of ecosystem manipulation experiments have to be interpreted with caution, in particular, exclusion experiments, which have unintended confounding factors in the design (Gundersen, 1994). In acid grassland soils and peatland, nitrification is less active and nitrate in wet deposition may only penetrate into the profile if the depth of the water table is low. At high water table levels NO_3 will be lost in runoff, or more likely lost through denitrification. Ammonium ions reaching the deeper layers could reduce the methane oxidation activity.

In future, newer agricultural techniques for the disposal of animal manures will reduce the emissions of NH_3 into the atmosphere and this will result in a greater proportion of NO_3 in wet deposition. By changing from spreading slurries on land to injection into the soil, the excess nitrogen load will be partially transferred from the atmosphere to the soil and drainage waters. The gaseous nitrous oxide subsequently released from the soil is a greenhouse gas and the continued acquisition of information to obtain a better understanding of the soil processes leading to less complex, but effective, models has been identified an essential requirement (Denmead, 1997).

References

Aber J. D., Nadelhoffer, K.J., Steudler, P.A., and Melillo, J.M. (1989) Nitrogen saturation in northern forest ecosystems. *Bioscience,* **39**, 378–386.

Adams, J.A. (1986a) Nitrification and ammonification in acid forest litter and humus as affected by peptone and ammonium-N amendment. *Soil Biology and Biochemistry* **18**, 45–51.

Adams, J.A. (1986b) Identification of heterotrophic nitrification in strongly acid larch humus. *Soil Biology and Biochemistry,* **18**, 339–341.

Adamson, J.K., Hornung, M., Kennedy, V.H., Norris, D.A., Paterson, I.S. and Stevens, P.A. (1993) Soil solution chemistry and throughfall under adjacent stands of Japanese larch and Sitka spruce at three contrasting sites in Britain. *Forestry,* **66**, 51–68.

Aerts, R., Wallén, B. and Malmer, N. (1992) Growth-limiting nutrients in *Sphagnum* dominated bogs subject to low and high atmospheric nitrogen supply. *Journal of Ecology,* **80**, 131–140.

Allison, S.M. and Prosser, J.I. (1991) Urease activity in neutrophilic autotrophic ammonia-oxidising bacteria isolated from acid soils. *Soil Biology and Biochemistry,* **23**, 45–52.

Anderson H.A., Peacock S., Berg A. and Ferrier R.C. (1995) Interactions between anthropogenic sulphate and marine salts in the Bs horizons of acidic soils in Scotland. *Water, Air and Soil Pollution,* **85**, 1083–1088.

Arnebrant, K., Bååth, E. and Söderström, B. (1990) Changes in microfungal community structure after fertilization of Scots pine forest soil with ammonium nitrate or urea. *Soil Biology and Biochemistry,* **22**, 309–312.

Bååth, E., Lundgren, B. and Söderström,B. (1981) Effects of nitrogen fertilization on the activity and biomass of fungi and bacteria in a podzolic soil. *Zentralblatt für Bakteriologie Mikrobiologie und Hygiene Originale 1 Abt.*, C 1981 2, 90–98.

Baxter, R., Emes, M.J., and Lee, J.A. (1992) Effects of experimentally applied increase in ammonium on growth and amino acid metabolism of *Sphagnum cuspidatum* Ehrh. ex. Hoffm. from differently polluted areas. *New Phytologist,* **120**, 265–274.

Berg, M.P, Anderson, J.M., Beese, F., Bolger, T., Couteaux, M.M., Henderson, R., Ineson, P., McCarthy, F., Palka, L., Raubuch, M., Splatt, P., Verhoef, H.A. and Willison, T. (1994) Effects of temperature and air pollutants on mineral-N dynamics in reciprocally transplanted forest soils. In *Ecosystem Manipulation Experiments: scientific approaches, experimental design and relevant results.* (Eds A. Jenkins, R.C. Ferrier and C. Kirby). Ecosystems Research Report 20, European Community, Brussels. Pp29–37.

Black, K.E., Lowe, J.A., Billett, M.F. and Cresser, M.S. (1993) Observations on the changes in nitrate concentrations along streams in seven upland moorland catchments in northeast Scotland. *Water Research,* **27**, 1195–1199.

Bobbink R. and Roelofs J.G.M. (1995) Nitrogen critical loads for natural and semi–natural ecosystems: the empirical approach. *Water, Air and Soil Pollution,* **85**, 2413–2418.

Bråckenhielm S. and Qinghong L. (1995.) Impact of sulfur and nitrogen deposition on plant species assemblages in natural vegetation. *Water, Air and Soil Pollution,* **85**, 1581–1586.

Brown, K.A. (1985) Acid deposition: effects of sulphuric acid at pH 3 on chemical and biochemical properties of bracken litter. *Soil Biology and Biochemistry,* **17**, 31–38.

Cadle S.H., Dasch J.M. and Grossnickle N.E. (1984) Retention and release of chemical species by a Northern Michigan snowpack. *Water, Air and Soil Pollution,* **22**, 303–319.

Conrad, R. (1996) Soil microorganisms as controllers of atmospheric trace gases (H_2, CO, CH_4, N_2O and NO). *Microbiological Reviews* **60**, 609–640.

Cuttle, S.P., Hallard, M., Gill, E.K. and Scurlock, R.V. (1996) Nitrate leaching from sheep-grazed upland pastures in Wales. *Journal of Agricultural Science,* **127**, 365–375.

Damman, A.W.H. (1988) Regulation of nitrogen removal and retention in *Sphagnum* bogs and other peatlands. *Oikos,* **51**, 291–305.

Davidson, E.A., Hart, S.C., Shanks, C.A. and Firestone, M.K. (1991) Measuring gross mineralization, immobilization, and nitrification by ^{15}N isotopic pool dilution in intact soil cores. *Journal of Soil Science,* **42**, 335–349.

Delanoue, A., Holt, D.M., Woodward, C.A., McMath, S.M., Smith, S.E. and Anderson, H.A. (1997) Effect of pipe materials on biofilm growth and deposit formation in water distribution systems. Proceedings of WQTC, Denver. AWWA, Denver, P–48, 399–409.

Denmead, O.T. (1997) Progress and challanges in measuring and modelling gaseous nitrogen emissions from grasslands: an oeverview. In Gaseous Nitrogen Emissions from Grasslands (Eds S.C. Jarvis and B.F. Pain) CAB International,Wallingford, UK. p 423–438.

Dutch, J. and Ineson, P. (1990) Denitrification of an upland forest site. *Forestry,* **63**, 363–377.

E.E.A. (1995) Air emissions of SO2, NOx, N_2O and NH_3 in Europe. *Corinair 1990 Summary Table (24 August 1995).* European Environment Agency, Copenhagen, Denmark.

Edwards, A.C., Ron Vaz, M.D., Porter, S. and Ebbs, L. (1996) Nutrient cycling in upland catchment areas: the significance of organic forms of N and P. In *Advances in Hillslope Processes* (Eds M.G. Anderson and S.M. Brooks) Volume I pp 253–262. Wiley and Sons Ltd 1996.

Emmett, B.A., Brittain, S.A., Hughes, S., Kennedy, V., Norris, D., Reynolds, B., Silgram, M. and Stevens, P.A. (1995) Effects of enhanced nitrogen deposition in a Sitka spruce stand in upland Wales. In *Ecosystem Manipulation Experiments: scientific approaches, experimental design and relevant results.* (Eds A. Jenkins, R.C. Ferrier and C. Kirby). Ecosystems Research Report 20,. European Community, Brussels. Pp 1–5.

Floate, M.J. (1970) Mineralization of nitrogen and phosphorus from organic materials of plant and animal origin and its significance in the nutrient cycle in grazed upland and hill soils. *Journal of the British Grassland Society,* **25**, 295–302.

Flum, T. and Nodvin, S.C. (1995) Factors affecting streamwater chemistry in the Great Smoky Mountains, USA. *Water, Air and Soil Pollution,* **85**, 1707–1712.

Fog, K.(1988) The effect of added nitrogen on the rate of decomposition of organic matter. Biological Reviews **63**, 433–462.

Fowler, D., Cape, J.N., Leith, I.D., Choularton, T.W., Gay, M.J. and Jones, A. (1988) The influence of altitude on rainfall composition. *Atmosphere Environment* **22**, 1355–62.

Francez, A-J. (1991) *Production primaire et accumulation de matière organique dans les tourbières à les Sphaignes des Monts du Forez (Puy-de Dôme).* Ph. D. thesis, University of Paris, France.

Galloway, J.N. (1995) Acid deposition: perspectives in time and space. *Water, Air and Soil Pollution,* **85**, 15–24.

Grosvernier, P., Matthey, Y. and Buttler, A. (1997) Growth potential of three Sphagnum species in relation to water table level and peat properties with implications for their restoration in cut-over bogs. *Journal of Applied Ecology,* **34**, 471–483.

Gulledge, J., Doyle, A.P. and Schimel, J.P. (1997) Different NH_4^+- inhibition patterns of soil CH_4 consumption: a result of distinct CH_4-oxidizer populations across sites. *Soil Biology and Biochemistry,* **29**, 13–22.

Gundersen, P. (1994) Unintended differences between natural and manipulated forest plots. In *Ecosystem Manipulation Experiments: scientific approaches, experimental design and relevant results.* (Eds A. Jenkins, R.C. Ferrier and C. Kirby). Ecosystems Research Report 20,. European Community, Brussels. Pp 335–343.

Härtl O. and Cerny M. (1981) Veränderungen in Fichtenwaldböden durch Langzeitwirkung von Schwefeldioxid. Mitt. FBVA Wien 137/II, 233–240.

Heal, O.W., Anderson, J.M. and Swift, M.J. (1997) Plant litter quality and decomposition: an historical overview. In *Driven by Nature: plant litter quality and decomposition.* Ed. by G. Cadisch and KE Giller. pp 3–30. CAB International, Wallingford, UK.

Hemond, H.F. (1983) The nitrogen budget of Thoreau's bog. *Ecology,* **64**, 99–109.

Hillier, S. (1996) Preliminary observations on the composition and origin of colloids and fine particles in the river DON, Scotland. Abstracts 33rd Annual Meeting Clay Minerals Society, Gatlinburg, Tennesee.

Houdijk, A.L.F.M., Verbeek, P.J.M., Vandijk, H.F.G. and Roelofs, J.G.M. (1993) Distribution and decline of endangered herbaceous heathland species in relation to the chemical-composition of the soil. *Plant and Soil,* **148**, 137–143.

Hughes, S., Norris, D.A., Stevens, P.A., Reynolds, B. and Williams, T.G. (1994) Effects of forest age on surface water and soil solution aluminium chemistry in stagnopodzols in Wales. *Water, Air and Soil Pollution,* **77**, 115–139.

Hultberg, H., Dise, N.B., Wright, R.F., Andersson, I. and Nyström U. (1993) Nitrogen saturation induced during winter by experimental ammonium nitrate addition to a forested catchment. *Environmental Pollution,* **84**, 145–147.

INDITE (1995) Impact of nitrogen deposition on terrestrial ecosystems. Report of the United Kingdom Review Group on Impacts of Atmospheric Nitrogen. Department of the Environment, London.

Ingram, H.A.P. (1992) Introduction to the ecohydrology of mires in the context of cultural perturbation. In *Peatland Ecosystems and Man: An Impact Assessment,* (Eds O.M. Bragg, P.D. Hulme, H.A.P. Ingram and R.A. Robertson), Department of Biological Sciences, University of Dundee, UK. p 67–93.

Jenkinson, D.S. and Ladd, J.N. (1981) Microbial biomass in soil: measurement and turnover. In *Soil Biochemistry* (Eds E.A. Paul and J.N. Ladd) pp 415–471, Volume 5. Dekker, New York.

Johnson, D., Leake, J., Lee, J. and Campbell, C.D. (1996) Changes in the activity and size of the soil microbial biomass in response to simulated pollutant nitrogen deposition. *Abstracts of the Nitrogen Deposition Workshop,* Macaulay Land Use Research Institute, Aberdeen., UK.

Johnson, D., Leake, J., Lee, J. and Campbell, C.D. (1998) Changes in soil microbial biomass and microbial activities in response to seven years simulated pollutant nitrogen deposition on a heathland and two grasslands. *Soil Biology and Biochemistry (submitted).*

Johnson, D.W. and Lindberg, S.E. (1992) *Atmospheric deposition and forest nutrient cycling.* Springer Verlag, New York.

Kahl, J.S., Norton, S.A., Fernandez, I.J., Nadelhoffer, K.J., Driscoll, C.T. and Aber, J.D. (1993) Experimental inducement of nitrogen saturation at the watershed scale. *Environmental Science and Technology,* **27**, 565–568.

Kaila A. (1954) Nitrification in decomposing organic matter. *Acta Agriculture Scandinavia* **4**, 17–32.

Karsisto, M., Kitunen, V., Jauhiainen, J. and Vasander, H. (1996) Effect of N deposition on free amino acid concentrations in two *Sphagnum* species. In *Northern Peatlands in Global Climatic Change* (Eds R. Laiho, J. Laine and H. Vasander). pp 23–29. Academy of Finland, Helsinki.

Killham, K. (1990) Nitrification in coniferous forest soils. *In Nitrogen Saturation in Forest Ecosystems.* (Eds O. Brandon and R.F. Hüttl), pp 31–44. Kluwer.

Klemmedtsson, L., Svensson, B.H., Lindberg, T. and Rosswall, T. (1977) The use of acetylene in quantifying denitrification in soils. *Swedish Journal of Agricultural Research*, **1**, 179–185.
Kowalchuk, G.A., Stephen, J.R., De Boer, W., Prosser, J.I., Embley, T.M. and Woldendorp, J.W. (1997) Analysis of proteobacteria ammonia-oxidising bacteria in coastal sand dunes using denatured gradient gel electrophoresis and sequencing of PCR amplified 16S rDNA fragments. *Applied and Environmental Microbiology*, **63**, 1489–1497.
Kristensen, H.L. and McCarty, G.W. (1996) Immobilization and mineralization of N in acid raw humus from a lowland heath. *Abstracts of the Nitrogen Deposition Workshop*, Macaulay Land Use Research Institute, Aberdeen. UK
Likens, G. E., Bormann, F.H., Pierce, R.S., Eaton, J.S. and Johnson, N.M. (1977) *Biogeochemistry of a Forested Ecosystem*. Springer–Verlag. New York.
Martikainen, P.J. and De Boer, W. (1993) Nitrous oxide production and nitrification in acidic soil from a Dutch coniferous forest. *Soil Biology and Biochemistry*, **25**, 343–347.
Matzner E. and Thoma E. (1983) Auswirkungen eines saisonalen Versauerungsschubes im Sommer/Herbst 1982 auf den chemischen Bodenzustandverscheidener Waldökosysteme. Allg. Forstzschr. **38**, 683–686.
Meade, R. (1984) Ammonia-assimilating enzymes in bryophytes. *Physiologia Planta*, **60**, 305–308.
Miller, H.G., Cooper, J.M., Miller, J.D. and Pauline, O.J. (1979) Nutrient cycles in pine and their adaptation to poor soils. *Canadian Journal of Forest Research* **9**, 19–26.
Miller, J.D., Anderson, H.A., Ferrier, R.C. and Walker, T.A.B. (1990) Hydrochemical fluxes in two forested catchments in central Scotland. *Forestry*, **63**, 311–331.
Miller, J.D., Anderson, H.A., Harriman, R. and Collen, P. (1995). The consequences of liming highly acidified catchments in central Scotland. *Water Air and Soil Pollution*, **85**, 1015–1020.
Mitchell M.J., Raynal D.J. and Driscoll C.T. (1996) Biogeochemistry of a forested watershed in the central Adirondack mountains: temporal changes and mass balances. *Water, Air and Soil Pollution*, **88**, 355–369.
Morecroft, M.D. Marrs, R.H. and Woodward, F.I. (1992) Altitudinal and seasonal trends in soil nitrogen mineralization rate in the Scottish Highlands. *Journal of Ecology*, **80**, 49–56.
Morecroft, M.D., Sellers, E.K. and Lee, J.A. (1994) An experimental investigation into the effects of atmospheric nitrogen deposition on two semi-natural grasslands. *Journal of Ecology*, **82**, 475–483.
Müller, M.M., Sundman, V. and Skujins, J. (1980) Denitrification in low pH spodosols and peats determined with the acetylene inhibition method. *Applied and Environmental Microbiology*, **40**, 235–239.
Nashholm, T., Edfast, A.B. Ericsson, A. and Norden, L.G. (1994) Accumulation of amino acids in some boreal forest plants in response to increased nitrogen availability. *New Phytologist*, **126**, 137–143.
Nihlgard B. (1985) The ammonium hypothesis – an additional explanation of the forest die-back in Europe. *Ambio*, **14**, 2–8.
Nodvin, S.C., Van Miegrot, H., Lindberg, S.E, Nicholas, N.S. and Johnson, D.W. (1995) Acidic deposition, ecosystem processes, and nitrogen saturation in a high altitude southern Appalachian watershed. *Water, Air and Soil Pollution*, **85**, 1647–1652.
Nômmik, H. (1956) Investigations on denitrification in soil. *Acta Agriculturae Scandinavica*, **6**, 195–228.
Nohrstedt, H.-Ö, Arnebrant, K., Bääth, E. and Söderström, B.(1989) Changes in carbon content, respiration rate, ATP content, and microbial biomass in nitrogen-fertilized pine forest soils in Sweden. *Canadian Journal of Forest Research*, **19**, 323–328.
Pardo, L.H., Driscoll, C.T. and Likens, G.E. (1995) Patterns of nitrate loss from a chronosequence of clear-cut watersheds. *Water, Air and Soil Pollution*, **85**, 1659–1664.
Perkins, D.F. (1978) The distribution and transfer of energy and nutrients in the *Agrostis–Festucca* grassland ecosystem. In *Production Ecology of British Moors and Montane Grasslands* (Eds O.W. Heal and D.F. Perkins) pp 375–393.Springer–Verlag, Berlin 1978.
Pyatt, D.G. and Craven, M. (1979) Soil changes under even-aged plantations. In The Ecology of Even-Aged Forest Plantations. Eds E.D. Ford, D.C. Malcolm and J. Atterson. Institute of Terrestrial Ecology, Cambridge. pp 369–386.
Qualls, R.G. and Haines, B.L. (1991) Geochemistry of dissolved organic nutrients in water percolating through a forest ecosystem. *Soil Science Society of America Journal*, **55**, 1112–1123.

Raubuch, M., Anderson, J.M., Beese, F., Berg, M.P, Bolger, T., Couteaux, M.M., Henderson, R., Ineson, P., McCarthy, F., Palka, L., Splatt, P., Verhoef, H.A. and Willison, T. (1995) Ionic balances of reciprocally transplanted forest soils. In *Ecosystem Manipulation Experiments: scientific approaches, experimental design and relevant results.* (Eds A. Jenkins, R.C. Ferrier and C. Kirby). Ecosystems Research Report 20, 1–5. European Community, Brussels. Pp 38–43.

Rawlins, B.G., Hornung, M. and Baird, A.J. (1996) Water flux and temperature controls on nitrate leaching and Al mobilisation in a coniferous forest soil. *Abstracts of the Nitrogen Deposition Workshop,* Macaulay Land Use Research Institute, Aberdeen. UK

Regina, K., Nykänen, H., Silvola, J. and Martikainen, P.J. (1996) Fluxes of nitrous oxide from boreal peatlands as affected by peatland type, water table level and nitrification capacity. *Biogeochemistry,* **35**, 401–418.

Rosswall, T. and Granhall, U. (1980) Nitrogen cycling in a subarctic mire. *Ecology of a Subarctic Mire* (ed. M. Sonesson). Ecological Bulletin (Stockholm) 30, pp 209–234.

Schimel, J. (1996) Assumptions and errors in the $^{15}NH_4^+$ pool dilution technique for measuring mineralization and immobilization. *Soil Biology and Biochemistry,* **28**, 827– 828.

Silcock, D.J. and Williams, B.L. (1995) The fate and effects of inorganic nitrogen inputs to raised bog vegetation. *Ecosystem Manipulation Experiments. Ecosystem Research Report **20*** (Eds A. Jenkins, R.C. Ferrier and C. Kirby) European Commission, Brussels. Pp 44–48.

Skiba,U., Sheppard, L., McDonald, J. and Fowler, D. (1997) Nitrous oxide emissions from forest and moorland soils in Northern Britain. Proceedings 7th International Workshop on Nitrous Oxide Emissions, Cologne, 1997(eds K.H. Becker and P. Wiesen), Bericht Nr. 41.Bergische Univeritaet Gesamthochschule, Wuppertal. p 247–253.

Sommer, S.G. and Hutchings, N.J. (1997) Components of ammonia volatilization from cattle and sheep production. In Gaseous Nitrogen Emissions from Grasslands (Eds S.C. Jarvis and B.F. Pain) CAB International,Wallingford, UK. p79–94.

Stevens, P.A. and Hornung, M. (1988) Nitrate leaching from a felled Sitka spruce plantation in Beddgelert Forest, North Wales. *Soil Use and Management,* **4**, 3–9

Stevens, P.A., Hornung, M. and Hughes, S. (1989) Solute concentrations, fluxes and major nutrient cycles in a mature Sitka-spruce forest, North Wales. *Forest Ecology and Management* **27**, 1–20.

Stevens, P.A., Norris, D.A., Sparks, T.H. and Hodgson, A.L. (1994) The impacts of atmospheric inputs on throughfall, soil and stream water interactions for different aged forest and moorland catchments in Wales. *Water, Air, and Soil Pollution,* **73**, 297–317.

Stoddard, J.L. (1994) Long-term changes in watershed retention of nitrogen: its causes and aquatic consequences. In *Environmental Chemistry of Lakes and Reservoirs,* (Ed. L.A. Baker). Advances in Chemistry Series No. 237, American Chemical Society, Washington. Pp223–284.

Stottlemyer R. and Rutkowski D. (1990) Multiyear trends in snowpack ion accumulation and loss, Northern Michigan. *Water Resources Research,* **26**, 721–737.

Stuanes, A.O., Kjønaas, O.J. and van Miegrot, H. (1995). Soil solution response to experimental addition of nitrogen to a forested catchment at Gårdsjön, Sweden. *Forest Ecology and Management,* **71**, 99–110.

Thomas, R.J., Logan, K.A.B., Ironside, A.D. and Bolton, G.R. (1988) Transformations and fate of sheep urine-N applied to an upland U.K. pasture at different times during the growing season. *Plant and Soil,* **107**, 173–181.

Tietema, A., Wright, R.F., Blanck, K., Boxman, A.W., Bredemeier, M., Emmett, B.A., Gundersen, P., Hultberg, H., Kjønaas, O.J., Moldan, F., Roelofs, J.G.M., Schleppi, P., Stuanes, A.O. and van Breemen, N. (1995) NITREX: the timing of response of coniferous ecosystems to experimentally-changed nitrogen deposition. *Water, Air and Soil Pollution,* **85**, 1623–1628.

Tietema, A. and Beier, C. (1995) A correlative evaluation of nitrogen cycling in the forest ecosystems of the EC projects NITREX and EXMAN. *Forest Ecology and Management,* **71**, 143–151.

Ulrich B. (1991) An ecosystem approach to soil acidification. In *Soil Acidity* (Eds. B. Ulrich and M.E. Sumner), Springer–Verlag, Berlin. Pp 28–79.

Urban, N.R., Eisenreich, S.J. and Bayley, S.E. (1988) The relative importance of denitrification and nitrate assimilation in midcontinental bogs. *Limnology and Oceanography,* **33**, 1611–1617.

van Breemen, N., Burrough, P.A., Velthorst, E.J., van Dobben, H.F., de Wit, T., Ridder, T.B. and Reinjnders, H.F.R. (1982) Soil acidification from atmospheric ammonium sulphate in forest canopy throughfall. *Nature* (London), **229**, 548–550.

van Breemen, N., Mulder, J. and Driscoll, C.T. (1983) Acidification and alkalinization of soils. *Plant and Soil,* **75**, 283–308.

van Breemen, N., Mulder, J. and van Grinsven, J.J.M. (1987) Impacts of acid atmospheric deposition on woodland soils in the Netherlands: II. Nitrogen transformations. *Soil Science Society of America Journal*, **51**, 1634–1640.

van Vuuren, M.M. and van der Erden, L.J. (1992) Effects of three rates of atmospheric nitrogen deposition enriched with ^{15}N on litter decomposition in a heathland. *Soil Biology and Biochemistry,* **24**, 527–532.

Verhoeven, J.T.A., Maltby, E. and Schmitz, M.B. (1990) Nitrogen and phosphorus in fens and bogs. *Journal of Ecology,* **78**, 713–726.

Verstraete, W. (1980) Nitrification. In 'Terrestrial Nitrogen Cycles' (Eds F.E. Clark and T. Rosswall). *Ecological Bulletins* (Stockholm) **33**, 303–314.

Viro P.J. (1962) Factorial experiments on forest hums decomposition. *Soil Science.*, **95**, 24–30.

Vompersky, S.E., Smagina, M.V., Ivanov, A.I. and Glukhova, T.V. (1992) The effect of forest drainage on the balance of organic matter in forest mires. In *Peatland Ecosystems and Man: An Impact Assessment,* (Eds O.M. Bragg, P.D. Hulme, H.A.P. Ingram and R.A. Robertson), Department of Biological Sciences, University of Dundee, UK. p17–22.

Wardle, D.A. and Ghani, A. (1995) A critique of the microbial metabolic quotient (qCO_2) as a bioindicator of disturbance and ecosystem development. *Soil Biology and Biochemistry,* **27**, 1601–1610.

Wessel, W.W. and Tietema, A. (1992) Calculating gross N transformations rates of ^{15}N pool dilution experiments with acid forest litter:analytical and numerical approaches. *Soil Biology and Biochemistry,* **24**, 931–942.

Wheatley, R.E. and Williams, B.L. (1989) Seasonal changes in rates of potential denitrification in poorly drained reseeded blanket peat. *Soil Biology and Biochemistry,* **21**, 355–360.

White, C.W. and Cresser, M.S. (1996) Effects of enhanced N deposition upon the base status of podzols derived from granitic parent materials : A laboratory simulation experiment. *Abstracts of the Nitrogen Deposition Workshop,* Macaulay Land Use Research Institute, Aberdeen. UK

Williams, B.L. (1983) Nitrogen transformations and decomposition in litter and humus from beneath closed-canopy Sitka spruce. *Forestry,* **56**, 17–32.

Williams, B.L. (1984) The influence of peatland type and the characteristics of peat on the content of readily mineralizable nitrogen. *Proceedings 7th International Peat Congress* 4, 410–418. Dublin 1984.

Williams, B.L. (1992a) Nitrogen dynamics in humus and soil beneath Sitka spruce (*Picea sitchensis* (Bong.) Carr.) planted in pure stands and in mixture with Scots pine (*Pinus sylvestris* L.). *Plant and Soil,* **144**, 77–84.

Williams, B.L. (1992b) Microbial nitrogen transformations in peatland disturbed by drainage and fertilizers. In *Peatland Ecosystems and Man: An Impact Assessment,* (Eds O.M. Bragg, P.D. Hulme, H.A.P. Ingram and R.A. Robertson), Department of Biological Sciences, University of Dundee, UK. p59–64.

Williams, B.L. (1996) Total, organic and extractable-P in humus and soil beneath Sitka spruce planted in pure stands and in mixture with Scots pine. *Plant Soil,* **182**, 177–183.

Williams, B.L., Buttler,A., Grosvernier, Ph., Francez, A-J, Gilbert, D., Ilomets, M., Jauhiainen, J., Matthey, Y., Silcock, D.J. and Vasander, H. (1998a) The fate of ammonium added to *Sphagnum magellanicum* carpets at five European mire sites. *Biogeochemistry* (in press)

Williams, B.L., Cooper, J.M. and Pyatt, D.G. (1978) Effects of afforestation with *Pinus contorta* on nutrient content, acidity and exchangeable cations in peat. *Forestry,* **51**, 29–35.

Williams, B.L., Cooper, J.M. and Pyatt, D.G. (1979) Some effects of afforestation with lodgepole pine on rates of nitrogen mineralization in peat. *Forestry,* **52**, 151–160.

Williams, B.L. and Edwards, A.C. (1993) Processes influencing dissolved organic nitrogen, phosphorus and sulphur in soils. *Chemistry and Ecology,* **8**, 203–215.

Williams, B.L. and Silcock, D.J. (1997) Nutrient and microbial changes in the peat profile beneath *Sphagnum magellanicum* in response to additions of ammonium nitrate. *Journal of Applied Ecology,* **34**, 961–970.

Williams, B.L., Silcock, D.J. and Young, M.Y. (1998b) Seasonal dynamics of N in two Sphagnum moss species and the underlying peat treated with $^{15}NH_4$ $^{15}NO_3$ *Biogeochemistry* (in press)

Williams, B.L., Silcock, D.J. and Young, M.Y. (1998c) Nitrogen–phosphorus interactions in peat. Submitted to *Écologie.*

Williams, B.L. and Sparling, G.P. (1988) Microbial biomass carbon and readily mineralized nitrogen in peat and forest humus. *Soil Biology and Biochemistry*, **20**, 579–561.

Williams, B.L. and Wheatley, R.E. (1989) Nitrogen transformations in poorly-drained reseeded blanket peat under different management systems. *International Peat Journal,* **3**, 97–106.

Williams, B.L. and Wheatley, R.E. (1992) Mineral nitrogen dynamics in poorly drained blanket peat. *Biology and Fertility of Soils,* **13**, 96–101.

Williams, B.L. and Wheatley, R.E. (1996) Seasonal responses of herbage to N fertilizer and changes in peat nutrient contents on poorly drained reseeded blanket bog limed and fertilized with P and K. *International Peat Journal,* **6**, 113–127.

Williams, B.L. and Young, M.E. (1994) Nutrient fluxes in runoff on reseeded blanket bog, limed and fertilized with urea, phosphorus and potassium. *Soil Use and Management,* **10**, 173–180.

Woodin, S. J. and Lee, J. A. (1987) The fate of some components of acidic deposition in ombrotrophic mires. *Environmental Pollution* **45**, 61–72

Woodin, S., Press, M.C. and Lee, J.A. (1985) Nitrate reductase activity in *Sphagnum fuscum* in relation to wet deposition of nitrate from the atmosphere. *New Phytologist,* **99**, 381–388.

Wright, R.F. and Tietema, A. (1995) Ecosystem response to 9 years of nitrogen addition at Sogndal, Norway. *Forest Ecology and Management,* **71**, 133–142.

Wright, R.F. and van Breemen, N. (1995) The NITREX project: an introduction. *Forest Ecology and Management,* **71**, 1–5.

Yesmin, L., Gammack, S.M., Sanger, L. and Cresser, M.S. (1995) Impact of atmospheric N deposition on inorganic- and organic-N outputs in water draining from peat. *The Science of the Total Environment,* **166**, 201–209.

Yesmin, L., Gammack, S. and Cresser, M.S. (1996) Changes in N concentrations of peat and its associated vegetation over 12 months in response to increased deposition of ammonium sulfate or nitric acid. *The Science of the Total Environment,* **177**, 281–290.

THE IMPACT OF ATMOSPHERIC NITROGEN DEPOSITION ON VEGETATION PROCESSES IN TERRESTRIAL, NON-FOREST ECOSYSTEMS

R. AERTS[1, 2] AND R. BOBBINK[2, 3]

[1]*Department of Systems Ecology, Vrije Universiteit, De Boelelaan 1087, NL–1081 HV Amsterdam, The Netherlands*
[2]*Department of Plant Ecology and Evolutionary Biology, Utrecht University, P.O. Box 800.84, 3508 TB Utrecht, The Netherlands*
[3] *Department of Ecology, Catholic University of Nijmegen, Toernooiveld, NL–6525 ED Nijmegen, The Netherlands*

4.1 Introduction

During the past few decades, the atmospheric deposition of nitrogen compounds has become a major threat to the species diversity of terrestrial ecosystems in north-west Europe. The two environmentally most damaging air-borne N compounds are nitrogen oxides (NO_x) and ammonia (NH_3). Nitrogen oxides originate mainly from stationary and non-stationary burning of fossil fuel. High levels of ammonia emissions especially occur in areas with high densities of animal husbandry (Buysman *et al.*, 1987). This is most apparent in the Netherlands, where the amount of deposited NH_x is significantly related to the distribution of intensive farming systems (Asman and Maas, 1986). The deposition of reduced and of oxidised N compounds results in eutrophication and acidification of nutrient-poor ecosystems (Heij and Schneider, 1991; Bobbink *et al.*, 1992a). Soil acidification as a result of the deposition of atmospheric pollutants, such as oxidised and reduced nitrogen compounds, is characterised by a wide variety of effects (Figure 4.1). First, there is a loss of buffer capacity (ANC) and a decrease in soil pH. Furthermore, it leads to increased leaching of base cations, especially Ca^{2+} and Mg^{2+}, and to increased availability of metals (Al^{3+} and Fe^{2+}). Moreover, inhibition of nitrification occurs due to lower pH, and causes an increase of ammonium to nitrate ratios. Finally, acidification may lead to a decrease in the decomposition rate of organic matter.

The occurrence and rate of these processes are mainly determined by the quantities of atmospheric deposition and the abiotic conditions, especially buffer capacity and nitrification potential, at the site. A decrease in pH is dependent on the buffering of the soil (e.g. Ulrich, 1983, 1990). Acidifying compounds, deposited on calcareous soils, will not cause a change in acidity. In these soils overlying calcareous bedrock, soluble $Ca(HCO_3)_2$

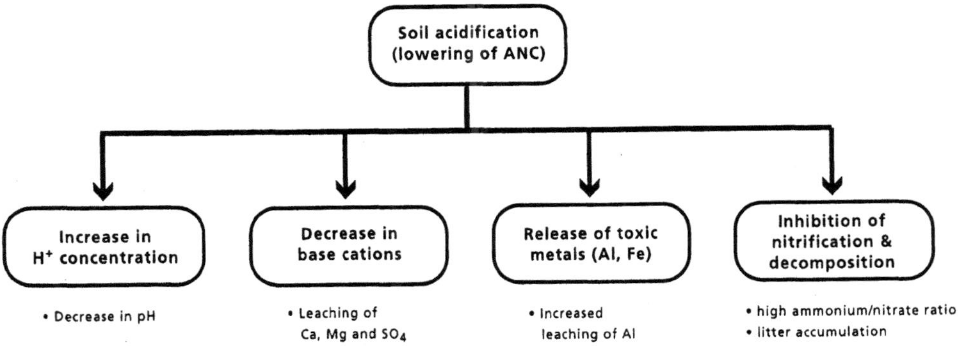

Figure 4.1: Scheme of soil-mediated changes which may occur during soil acidification

leaches from the system because of the inputs of acidifying compounds, but the pH remains the same till almost all calcium carbonate has been depleted. In soils largely dominated by silicate minerals (pH 6.5–4.5) buffering is taken over by cation exchange processes of the soil adsorption complexes (clay minerals and humus particles). In this case protons are exchanged with Ca^{2+} and Mg^{2+} of the adsorption complex, and leached from the soils together with anions, especially nitrate. Because of the limited capacity of this buffering system, the pH of the soil will soon start to decrease, accompanied by losses of several base cations. When ammonium or ammonia is the main deposited N pollutant, soil acidification is accelerated by nitrification and root ammonium uptake. Both these processes lead to local production of protons in the soil. The soil proton concentrations increase in this way, which causes ion competition between H^+ and other cations. At low pH (< 4.5) soil clay minerals are broken down and hydrous oxides of several metals are dissolved. This results in strong increases of the concentrations of toxic Al^{3+} and other metals in the soil solution. As a result of the decrease in soil pH nitrification is strongly inhibited or even completely absent. This leads to accumulation of ammonium, whereas nitrate decreases to almost zero at these or lower pH values (e.g. Roelofs *et al.*, 1985). Furthermore, the decomposition rate of organic matter in the soil is lower in these acidified soils, which leads to an increased accumulation of litter. As a consequence of this complex of changes, growth of plant species and species composition of the vegetation can be seriously affected.

Apart from these effects on soil processes, atmospheric N deposition has a strong impact on species diversity in terrestrial ecosystems. This was first recognised in Dutch wet and dry heathlands. As in most European countries, atmospheric N deposition in the Netherlands has doubled between 1950 and 1980 and has become an important N input to many terrestrial ecosystems (Asman *et al.*, 1988; Skeffington, 1990; Aerts and Heil, 1993). In the Netherlands, N deposition is on average 4.7 g N m^{-2} yr^{-1}, which is 2–8 times as high as in other countries (Heij and Schneider, 1991). In the 1970s, rapid changes in the species composition of Dutch heathland vegetation occurred. The evergreen dwarfshrubs *Erica tetralix* and *Calluna vulgaris* were being replaced by the grasses *Molinia caerulea* and *Deschampsia flexuosa* (as these four species after very often referred to in this chapter, we will mostly use their generic names only). This process has resulted in a dramatic decrease in the species diversity of these heathlands, because many (rare) species which were characteristic of the *Erica*- or *Calluna*-dominated heathlands have disappeared as well (Aerts and Heil, 1993). In wet heathlands this involves e.g. plant species such as *Drosera rotundifolia*, *Gentiana pneumonanthe* and *Narthecium ossifragum*. Also animal species such as butterflies (e.g. *Maculinea alcon*), amphibians (e.g. *Rana arvalis*) and several species of *Carabidae* also disappeared from these heathlands. On dry heathlands higher plant species such as *Cuscuta epithymum*, *Scorzonera humilis*, *Lycopodium clavatum* and *Lycopodium annotinum* disappeared from sites where *Molinia* or *Deschampsia* have become the dominant species. Apart from the plant species, also many insect species, some reptiles (e.g. *Vipera berus*) and rare bird species (such as *Lyrurus tetrix*) have disappeared from many dry heathlands. Since the end of the 1970s, the bryophyte and lichen flora has also declined, both in species number and in cover percentage (De Smidt and Van Ree, 1991). However, Van Dobben and De Bakker (1996) showed that the decline of epiphytic lichens in the Netherlands is most probably caused by high concentrations of SO_2 and not by increased NH_3 concentrations.

The negative effects of increased atmospheric N deposition on plant species diversity became also apparent in other non-forest ecosystems, such as (chalk)grasslands (Bobbink, 1991; Bobbink *et al.*, 1988; Berendse *et al.*, 1994) and fens and bogs (Vermeer, 1986; Verhoeven *et al.*, 1988; Morris, 1991; Aerts *et al.*, 1992a). As a result of these and other studies, our knowledge of the impact of atmospheric N deposition on vegetation processes has steadily increased during the past decade.

The aim of this chapter is to provide a conceptual framework which summarizes the main changes which occur in vegetation processes as a result of increased atmospheric nitrogen deposition. To that end, we elaborate on a scheme in which a sequence of events is presented, together with possible feed-backs, which will occur when N-deposition is increasing in an area with originally low deposition rates. We illustrate the main processes by presenting examples from some important north-west European non-forest ecosystems (heathlands, grasslands and peatlands).

4.2 A Schematic Representation Of The Effects Of Atmospheric N Deposition On Vegetation Processes

The analysis of N deposition effects on vegetation processes is complicated by the fact that many processes interact with each other and operate at different time-scales. Moreover, part of these effects are mediated through changes in soil chemistry. To simplify the analysis, we have developed a conceptual diagram in which a sequence of events is depicted, including possible feed-backs, which will occur when N-deposition is increasing in an area with originally low deposition rates (Figure 4.2). Although we realise that this diagram represents a simplification of reality, nevertheless it provides a structure for the analysis of N-deposition effects on vegetation processes. First, we will discuss this diagram briefly and in the remainder of this chapter we will illustrate our ideas with examples from some important north-west European non-forest ecosystems (heathlands, grasslands and peatlands).

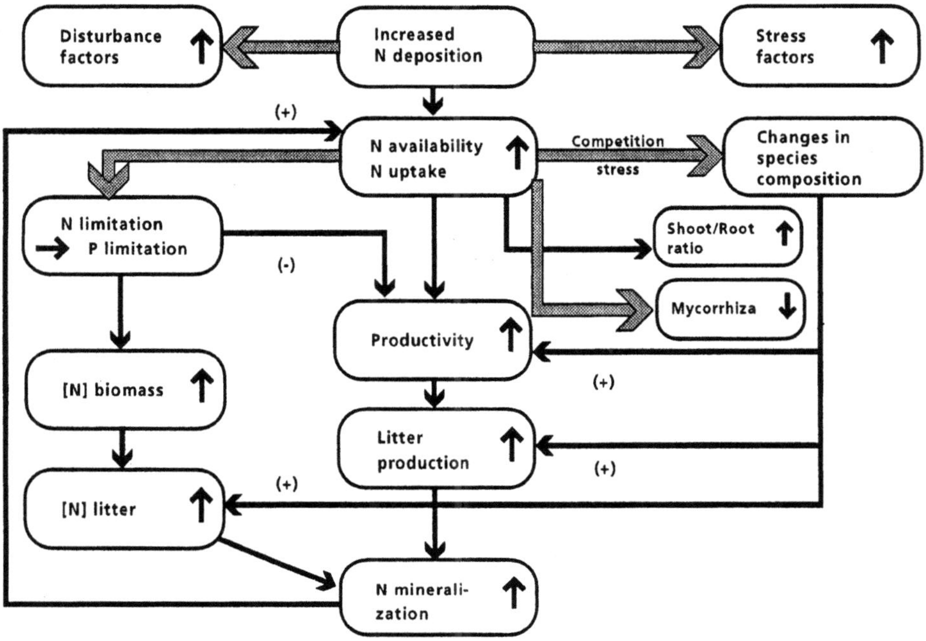

Figure 4.2: A schematic representation of the main effects of increased atmospheric N deposition on vegetation processes in terrestrial ecosystems. ↑ indicates increase; ↓ indicates decrease; → indicates effect which will occur already in the short-term (< 5 years); ➡ indicates effects which already occur in the long-term (> 5 years); (+) indicates positive feed-back; (–) indicates negative feed-back.

In the short-term, increased N deposition leads to higher soil nitrogen availability and increased plant N uptake. The central column of the diagram shows that, with a given species composition of the vegetation, increased N uptake will lead to increased productivity. This in turn leads to higher litter production and to higher N mineralisation in the soil. This is a positive feed-back, because higher N mineralisation leads to higher N uptake, etc. However, most of the deposited nitrogen in NW Europe originates from ammonia/ammonium (Heij and Schneider, 1991). This can cause a change in the dominant nitrogen form in the soil from nitrate to ammonium, especially in acidic habitats. The growth and development of plant species may be significantly influenced by this change. Species of acidic habitats generally prefer ammonium as sole nitrogen source, whereas species of less acidic or calcareous soils favour nitrate or a combination of both nitrogen forms (e.g. Gigon and Rorison, 1972; Kinzel, 1983; Troelstra *et al.*, 1995a,b). One of the effects of enhanced ammonium uptake is the reduced uptake or even exclusion of potassium, magnesium and calcium (Marschner, 1995). Ultimately this can lead to severe nutritional imbalances, which is a main cause of reduced tree growth in ammonium affected areas in NW Europe (Nihlgård, 1985; van Dijk *et al.*, 1990). High ammonium concentrations are also toxic for the roots of sensitive plant species, causing very poor root development, and finally, inhibited shoot growth. Correlative field studies and hydroculture experiments have shown that several endangered heathland and grassland species are very sensitive to high ammonium concentrations and high NH_4^+/NO_3^- ratios (Roelofs *et al.*, 1996; De Graaf *et al.*, 1998).

In the longer term, higher N availability will lead to changes in the species composition of the vegetation (right column of the diagram). This is both due to changes in the competitive balance between plant species and to stress for some species (see above). Changes in species composition may lead to higher ecosystem productivity, to higher litter production and to higher N concentrations in the litter. All these processes will indirectly lead to an acceleration in the rate of N cycling.

Another long-term effect of high levels of N deposition is that plant growth will become P limited instead of N limited (left column of the diagram). As a result, plant productivity will not further increase. However, as N uptake will continue to increase this shift from N limitation to P limitation will lead to increased N concentrations in plant biomass and in plant litter. This in turn leads to higher N mineralisation, which leads to higher soil N availability etc. Moreover, higher N concentrations in plant biomass may affect herbivory. However, in the end the ecosystem will become N-saturated which will lead to increased N leaching.

Atmospheric N deposition will also result in increased shoot/root ratios and probably to reduced mycorrhizal infection. This may have serious implications for the uptake of water and nutrients when these resources are in low supply.

Finally, atmospheric N deposition can lead to an increased occurrence of secondary stress factors, such as frost sensitivity, drought sensitivity and disturbance factors such as increased herbivory (top left and top right of the diagram).

4.3 N Deposition Effects On The Species Composition Of The Vegetation

4.3.1 HEATHLANDS

The large-scale replacement of ericaceous species by grasses in Dutch heathlands was first noticed in the 1970s and is still in progress (Aerts and Heil, 1993). This has resulted in a dramatic decrease in the species diversity of these heathlands, because many plant and animal species which were characteristic of the *Erica tetralix*- or *Calluna vulgaris* dominated heathlands have disappeared as well. This change in species composition took place during a period in which nutrient availability in these originally nutrient-poor ecosystems increased. However, the main causes for this process are not only the high levels of atmospheric N deposition in the Netherlands, but also a lack of management and, for dry heathlands, frequent and severe attacks of the heather beetle which leads to dieback of *Calluna* (Aerts and Heil, 1993). Nevertheless, this massive change in plant species composition is a good illustration of the way in which atmospheric N deposition may affect the species composition of the vegetation. In this section we will discuss some experimental competition studies which were performed in the Netherlands.

The effects of competition are most apparent when the availability of a limiting resource is changed in a particular habitat. In early-successional heathlands the availability of nitrogen is clearly limiting plant productivity (Aerts, 1993). Moreover, the growth response of the dominant heathland species to increased levels of nutrient availability is different (section 4.4). Thus, it might be expected that the competitive balance in heathlands changes when nutrient availability is increased. This has been investigated in the wet heathland area 'Kruishaarse Heide' in the central part of the Netherlands, where a site was selected which was co-dominated by *Erica tetralix* and *Molinia caerulea* and with a small proportion of *Scirpus cespitosus*. At this site, we studied the effect of increased nitrogen and phosphorus supply on the competitive interactions between the dominant species. To this end, nutrients (no nutrients, 20 g N m^{-2} yr^{-1}, 4 g P m^{-2} yr^{-1}) were supplied during three consecutive growing seasons. During the experimental period, cover percentage of each species was determined at the end of each growing season. Because percentage cover of *Scirpus* was very low and was not significantly affected by nutrient addition further results will be discussed only for *Erica* and *Molinia*. At the end of the experiment percentage cover of both *Erica* and *Molinia* was found to be positively correlated with the harvested biomass, so cover percentage can be considered as an estimate of biomass. In the unfertilized control, the cover percentage of *Erica* decreased significantly during the experimental period, whereas the cover of *Molinia* did not change ($P < 0.10$) (Figure 4.3). However, in all fertilized series the cover percentage of *Erica* decreased significantly and the cover percentage of *Molinia* increased. The strongest effects were observed in the series fertilized with phosphorus. At first sight it is rather surprising that P has a stronger effect on vegetation dynamics in this wet heathland than N. This seems to contradict the hypothesis that the increased nitrogen deposition in the Netherlands is responsible for the change in species composition in heathlands. However, it should be noticed that in the wet heathland under study secondary succession had proceeded for several decades in an area with high N deposition. This has resulted in a strong accumulation of nitrogen-rich organic matter with a low phosphorus content. This in turn has led to high rates of nitrogen mineralisation (10–13 g N m^{-2} yr^{-1}), while phosphorus mineralisation was extremely low (Berendse, 1990). Another point which deserves attention is that also in the unfertilized control the cover of *Erica* decreased. In our opinion, this reflects the ongoing decrease of dominance of ericaceous species in heathlands as a result of increased N supply (atmospheric N deposition and internal N mineralisation).

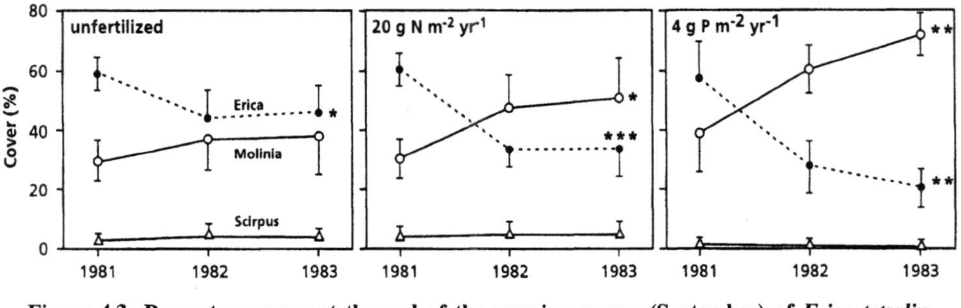

Figure 4.3: Percentage cover at the end of the growing season (September) of *Erica tetralix* (filled circles), *Molinia caerulea* (open circles) and *Scirpus cespitosus* (triangles) in relation to nutrient treatment. Means are given ± S.E (n=5). For each species the percentage change in cover between the start and the end of the experiment was tested using a t-test. * $P < 0.05$; ** $P < 0.01$; *** $P < 0.005$ (Redrawn from Aerts and Berendse, 1988).

A complicating factor in the analysis of the change in species composition in wet heathlands is that also lowering of the water-table may affect the outcome of competition. In a container experiment in which the effect of nutrient availability and water-table depth on competition between *Erica* and *Molinia* was studied it was found that *Erica* was outcompeted by *Molinia* when the availability of phosphorus or nitrogen was increased or when the water-table was lowered (Berendse and Aerts, 1984). Lowering the water-table initially leads to higher mineralisation rates (Berendse *et al.*, 1994; Aerts and Ludwig, 1997). So the change in the competitive balance between *Erica* and *Molinia* after lowering the water-table, might also be due to changes in nutrient availability.

In the wet heathland experiment at the Kruishaarse Heide, the amount of N supplied to the vegetation was very high (20 g N m^{-2} yr^{-1}). Therefore, we performed another field experiment in which N supply was increased more gradually. This competition experiment was conducted at the heathland area Deelense Veld, five years after part of this area had been sod-cut (removal of the vegetation plus organic top-soil). Nitrogen mineralisation rate at this site was 0.5 g N m^{-2} yr^{-1}. The mineralisation of phosphorus was below detection limits. This experiment was carried out according to the replacement principle (De Wit, 1960). During three consecutive growing seasons four levels of nitrogen were supplied (0, 5, 10 and 20 g N m^{-2} yr^{-1}), whereas P and K were supplied in amounts which prevented possible secondary growth limitation by these nutrients. The plots were harvested at the end of the third growing season and the biomass of each species in monoculture and in mixture was determined. A detailed description is given in Aerts *et al.* (1990). The biomass dynamics in this competition experiment are presented in replacement diagrams (Figures 4.4, 4.5). In the low nutrient treatments (control, 5 g N m^{-2} yr^{-1}, 10 g N m^{-2} yr^{-1}) *Erica* was the better competitor, whereas at a N supply of 20 g N m^{-2} yr^{-1} *Molinia* outcompeted *Erica* (Figure 4.4). Thus, at a nitrogen supply of more than 10 g N m^{-2} yr^{-1} *Erica* is outcompeted by *Molinia*. This is in close agreement with earlier experiments (Sheikh, 1969; Berendse and Aerts, 1984) and with the experiment at the Kruishaarse Heide (Figure 4.3).

In dry heathlands, with dominance of *Calluna vulgaris*, the situation is more complicated. This is clearly shown in a competition experiment with *Calluna vulgaris* and *Molinia caerulea* which was designed in a similar way as the previous experiment. In all nutrient treatments, *Calluna* was the superior competitor (Figure 4.5). Thus, *Calluna* was in this experiment even at a nitrogen supply of 20 g N m^{-2} yr^{-1} competitively superior to *Molinia*. The competitive superiority of *Calluna* in this experiment is not in agreement with the results of a competition experiment of Heil and Bruggink (1987) and can not explain the large scale replacement of *Calluna* by *Molinia* in Dutch heathlands. Current knowledge strongly suggests that this replacement of *Calluna* by *Molinia* at high N availability is triggered by stress and disturbance factors which cause opening of the *Calluna* canopy, e.g. senescence, frost, drought, or heather beetle attacks (Aerts and Heil, 1993). Similarly, the replacement of *Calluna* by *Deschampsia* only occurs when the *Calluna* canopy is opened due to stress and disturbance factors (section 4.7).

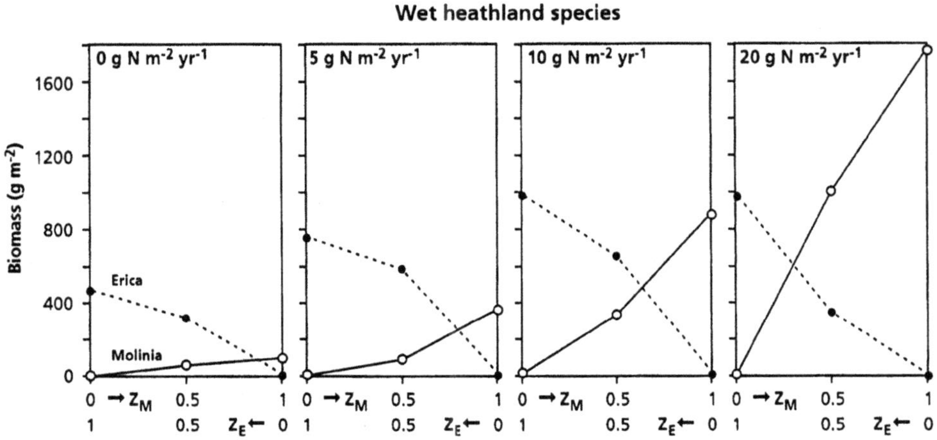

Figures 4.4: Replacement diagrams based on mean aboveground biomass (g m^{-2}) of *Erica tetralix* (triangles) and *Molinia caerulea* (circles), grown at 3 relative plant frequencies (z_x=0: monoculture of the other species; z_x=0.5: mixture with the other species; z_x=1: monoculture of species x) and 4 levels of nitrogen availability at the heathland area Deelense Veld. n=5. Curves can be straight lines (effects of intraspecific competition equal to interspecific competition), convex lines (intraspecific competition is more severe than interspecific competition) or concave lines (intraspecific competition is less severe than interspecific competition). (Redrawn from Aerts and Heil, 1993)

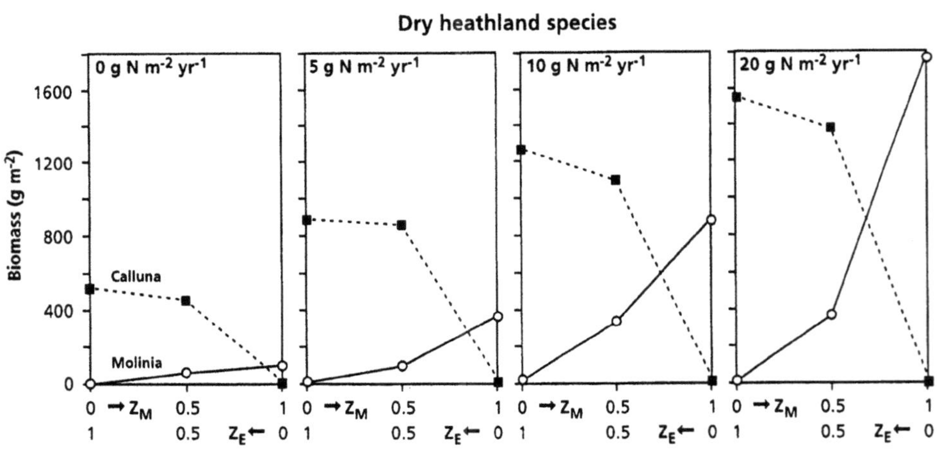

Figure 4.5: Replacement diagrams based on mean aboveground biomass (g m^{-2}) of *Calluna vulgaris* (squares) and *Molinia caerulea* (circles), grown at 3 relative plant frequencies and 4 levels of nitrogen availability at the heathland area Deelense Veld. n=5. See Fig. 4 for further explanation (Redrawn from Aerts and Heil, 1993)

4.3.2. SPECIES-RICH GRASSLANDS

Semi-natural grasslands with traditional agricultural use have been an important part of the landscape of western and central Europe and contain(ed) many rare and endangered plant and animal species (e.g Ellenberg, 1996). A number of these grasslands have been set aside as nature reserves, because of their high species diversity. These semi-natural grasslands of high conservational importance are generally poor in nutrients, because of the long agricultural use with low inputs of manure and with removal of plant parts by grazing or hay making. Many experiments demonstrated that application of fertilizers (N+P+K) alters these grasslands into tall, species-poor stands, dominated by a few highly productive crop grasses (Bakelaar and Odum, 1978; Williams, 1978; Van den Bergh, 1979; Willems, 1980; Van Hecke *et al.*, 1981; Berendse *et al.*, 1992). Especially nitrogen and phosphorus are important in this respect. It is thus likely that the species composition of species-rich grasslands may be affected by increased atmospheric nitrogen input (e.g. Wellburn, 1988; Ellenberg jr., 1988). In this section the effects of nitrogen inputs on species composition in species-rich calcareous and acidic grasslands are summarized.

The effects of increased nitrogen availability were studied in Dutch calcareous (BROMETALIA) grasslands. Subsoils consist of limestone with high contents of calcium carbonate (> 90%), covered by shallow well-buffered rendzina soils low in phosphorus and nitrogen (pH: 7–8). Plant productivity is low and calcareous grasslands are among the most species-rich plant communities in Europe. To maintain the characteristic calcareous vegetation a specific management is needed to prevent their natural succession towards woodland (Wells, 1974, Dierschke, 1985; Willems, 1990). A gradual increase of one grass species (*Brachypodium pinnatum*) has been observed in several Dutch calcareous grasslands in the late 1970s/early 1980s, although the management (hay making in autumn) did not change since the late 1950s. It was hypothesised that the increased atmospheric deposition of nitrogen (from 1.0–1.5 g N m^{-2} yr^{-1} in the 195 0sto 3.0–3.5 g N m^{-2} yr^{-1} in the 1980s) caused this drastic change in vegetation composition (Bobbink and Willems, 1987).

The effects of nitrogen enrichment in Dutch calcareous grasslands on vegetation composition were therefore investigated in field experiments (Bobbink *et al.*, 1988; Bobbink, 1991). Either potassium (10 g K m^{-2} yr^{-1}), phosphorus (3 g P m^{-2} yr^{-1}) or nitrogen (10 g N m^{-2} yr^{-1}) as well as a complete fertilization (N+P+K) were applied during 3 years to study 'long-term' effects on vegetation composition. Total aboveground biomass increased considerably, as expected, after three years of N+P+K fertilization. In the separate application of nutrients, a moderate increase in aboveground dry weight was only seen with nitrogen addition. The relative contribution of species and growth forms to total aboveground dry weight was dramatically affected by nutrient treatments. In the N-treated vegetation the dry weight of the grass species *Brachypodium pinnatum* was ca. 3 times higher than in the control plots (Figure 4.6). With NPK fertilization dry weight of the other grasses increased considerably and became ca. 170 g m^{-2}, compared to about 50 g m^{-2} in the control treatment. Especially productive grasses, such as *Festuca rubra*, *Poa angustifolia*, *Avenula pubescens* and *Dactylis glomerata* increased in these plots. Biomass of legumes (*Fabaceae*, especially *Lotus*, *Ononis* and *Vicia* spp.) significantly increased after P application. This is probably caused by the additional nitrogen capture in their root nodules with nitrogen-fixing micro-organisms. K addition did not affect the distribution of dry weight over the different species in the vegetation (Figure 4.6).

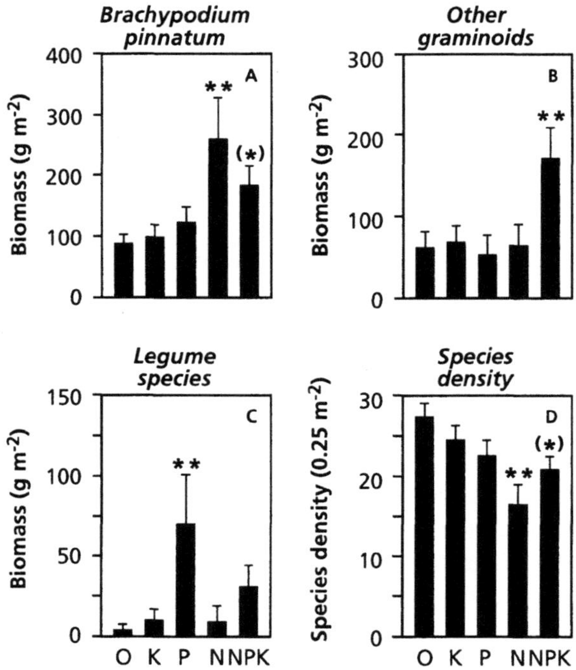

Figure 4.6: Aboveground biomass (g m^{-2}) of *Brachypodium pinnatum*, other graminoids, legume species, and phanerogamic species density (per 0.25 m^2) after 3 years of nutrient additions (O: unfertilised control; K: 10 g K m^{-2} yr^{-1}; P: 3 g P m^{-2} yr^{-1}; N: 10 g N m^{-2} yr^{-1}; NPK: a complete fertilisation) (Redrawn from Bobbink, 1991). $^{(*)}$ $\underline{P} < 0.10$; ** $\underline{P} < 0.01$.

Nitrogen application resulted furthermore in a drastic reduction of the biomass of forb species and of the total number of species (Figure 4.6). Especially the abundance of rosette plants and short-lived species declined in this N-treated vegetation. The observed decrease in species diversity in the nitrogen treated vegetation was certainly not caused by nitrogen toxicity, but by the change in vertical structure of the grassland vegetation. The growth of *B. pinnatum* was strongly stimulated and its overtopping leaves reduced the light quantity and quality in the vegetation. It overshadowed the other characteristic species and growth of these species declined rapidly (Bobbink *et al.*, 1988; Bobbink, 1991). The effects of excess nitrogen supply on the massive expansion of *B. pinnatum* and a drastic reduction in species diversity were also observed in a long-term permanent plot study using a factorial design (Willems *et al.*, 1993). Besides this decrease in phanerogamic plant species, many characteristic lichens and mosses have disappeared in recent years from calcareous grasslands (During and Willems, 1986). This is partly caused by the (indirect) effects of extra nitrogen inputs, as experimentally shown by Van Tooren *et al.* (1990).

In calcareous grassland in England, addition of nitrogen sometimes stimulated dominance of grasses (Smith *et al.*, 1971; Jeffrey and Pigott, 1973). In these studies, with application of 5–10 g N m^{-2} yr^{-1} and additional supply of phosphorus, a strong dominance of the grasses *Festuca rubra*, *F. ovina* or *Agrostis stolonifera* was observed. However, *B.*

pinnatum or *Bromus erectus*, the most frequent species in continental calcareous grassland, were absent from these British sites, so the data are not comparable in this respect. Following a survey of data from a number of conservation sites in southern England, Pitcairn *et al.* (1991) concluded that *B. pinnatum* had expanded in the UK during the last century. They considered that much of the early spread could be attributed to a decline in grazing pressure but that more recent increases in the grass had, in some cases, taken place despite grazing or mowing, and could be related to nitrogen inputs. A study of chalk grassland at Parsonage Downs (UK) showed, however, no substantial change in species composition between 1970 and 1990, a period when nitrogen deposition is thought to have increased to 1.5–2.0 g N m^{-2} yr^{-1} (Wells *et al.*, 1993). *B. pinnatum* was present in the sward but had not expanded as in the Dutch grasslands. It should be noticed, however, that these British calcareous grassland sites are subject to an atmospheric N deposition which does not exceed the critical load for N deposition for this type of vegetation (Bobbink and Roelofs, 1995b).

The effects of nitrogen additions (7–14 g N m^{-2} yr^{-1} as ammonium nitrate or 14 g N m^{-2} yr^{-1} as ammonium sulphate) were studied in a calcareous and in an acidic grassland (see below) in the Peak District in the UK (Morecroft *et al.*, 1994; Lee *et al.*, 1996). Within the first 3-year period Morecroft *et al.* (1994) did not observe a significant change in species composition, caused by nitrogen. However, shoot nitrogen concentrations, nitrate reductase activities and soil nitrogen mineralisation rates clearly increased with enhanced inputs of nitrogen (≥7 g N m^{-2} yr^{-1}). They concluded that plant growth appeared to be co-limited by nitrogen and phosphorus, but various aspects of the nitrogen cycling process were certainly affected by extra N inputs. After six years of treatment a number of statistically significant changes were observed in the both the higher plant and moss species composition. Several plant species (e.g. *Thymus praecox* and *Hieracium pilosella*) declined in abundance after six years of additions with nitrogen, whereas only common species increased (Lee *et al.*, 1996). This study clearly demonstrates the need for long-term field experiments in order to show the long-term effects of increased nitrogen inputs.

A reduction in species diversity was also observed in acidic, weakly buffered grasslands (NARDETALIA) in the Netherlands. Many of the characteristic species of these communities (e.g. *Arnica montana, Antennaria dioica, Dactylorhiza maculata, Gentiana pneumonanthe, Lycopodium inundatum, Narthecium ossifragum, Pedicularis sylvatica, Polygala serpyllifolia* and *Thymus serpyllum*) declined or even became locally extinct. The distribution of these species is related to small-scale, spatial variability of the soils. Dwarf-shrubs as well as grass species are nowadays dominant in former habitats of these endangered species. It has been suggested that atmospheric deposition caused such drastic abiotic changes for these species that they could not survive (Van Dam *et al.*, 1986). Field surveys and experiments in the laboratory or greenhouse demonstrated that these plant species are seriously affected by soil acidification (increased Al concentrations; low base cations) and by the accumulation of ammonium in the soil (Fennema, 1992; Houdijk *et al.*, 1993; Bobbink and Roelofs, 1995; De Graaf *et al.*, 1997a). The impacts of enhanced nitrogen inputs are, however, not studied in the field in the Netherlands at this moment. Fortunately, the effects of long-term nitrogen additions onto a comparable acidic grassland (*Festuca–Agrostis–Galium* community) are quantified in the Peak District in the

UK (Morecroft *et al.*, 1994; Lee *et al.*, 1996) (see above). After the first three years Morecroft *et al.* (1994) did not observe a significant change in species composition, caused by nitrogen inputs. Only the moss *Rhytidiadelphus squarrosus* declined after the addition of ammonium sulphate, compared with the control situation. Shoot nitrogen concentrations, nitrate reductase activities and soil nitrogen mineralisation rates became clearly higher with enhanced inputs of nitrogen (7 g N m^{-2} yr^{-1}), as in the studied calcareous grassland. After six years of nitrogen applications statistically significant changes in the species composition of mosses and higher plants became obvious. Several plant species (e.g. *Agrostis capillaris, Pleurozium scheberi, R. squarrosus*) clearly declined in mass after six years of treatment with nitrogen, whereas two grasses (*Deschampsia flexuosa* and *Nardus stricta*) increased markedly (Lee *et al.*, 1996). Finally, the addition of ammonium sulphate caused greater botanical change in this experiment, than the addition of other nitrogenous compounds, probably through its acidifying effects.

The longest running nutrient addition experiment is the Park Grass experiment at Rothamsted (UK), which was started in 1856. N additions of 4.8 g N m^{-2} yr^{-1} to these neutral grasslands caused a shift to species-poor stands, mostly dominated by tall grass species, such as *Alopecurus pratensis, Arrhenatherum elatius* and *Holcus lanatus* (Williams, 1978; Dodd *et al.*, 1994). In the Park Grass experiment species diversity was related directly with the acidity of the soil and the biomass of the vegetation; low number of species were found in acidic and in high biomass plots. The enrichment with ammonium sulphate resulted through it acidifying effects, in much greater botanical degradation than the other N forms.

The impacts of N additions on species richness were also quantified in flower-rich wet hay meadows in Somerset (UK) during six years (Mountford *et al.*, 1994; Tallowin and Smith, 1994). N additions of 2.5 g N m^{-2} yr^{-1} or higher significantly reduced the number of species, whereas several grasses increased in dominance (*Bromus hordeaceous, Holcus lanatus* and *Lolium perenne*). The number of forbs, characteristic of these old meadows, strongly declined in response to (N) fertilization. Species which disappeared from the N treated plots included *Cirsium dissectum, Lychnis flos-cuculi* and *Lotus pedunculatus.*

4.4 Plant-Mediated Effects Of N Deposition On N Cycling

Plant species play an important role in the cycling of nitrogen in terrestrial ecosystems. Mineralisation of plant-derived organic matter comprises very often more than 70% of the total amount of inorganic N available for plant uptake (Hemond, 1983; Berendse, 1990; Morris, 1991; Koerselman and Verhoeven, 1992; Aerts, 1993). However, the processing of N by plant species shows large interspecific differences (Aerts, 1993). This implies that the effects of atmospheric N deposition on N cycling depend strongly on the species composition of the vegetation, which in turn strongly depends on N availability (see section 4.3). These effects of plant species on N cycling are clearly illustrated by work which has been done in Dutch heathlands. Atmospheric N deposition is the only N input in these ecosystems and amounts to 3–4 g N m^{-2} yr^{-1} (Bobbink and Heil, 1993). This makes these ecosystems very suitable for studying plant-mediated effects of N deposition on N cycling. We will focus on interspecific differences in productivity, litter production, litter decomposition and nutrient mineralisation.

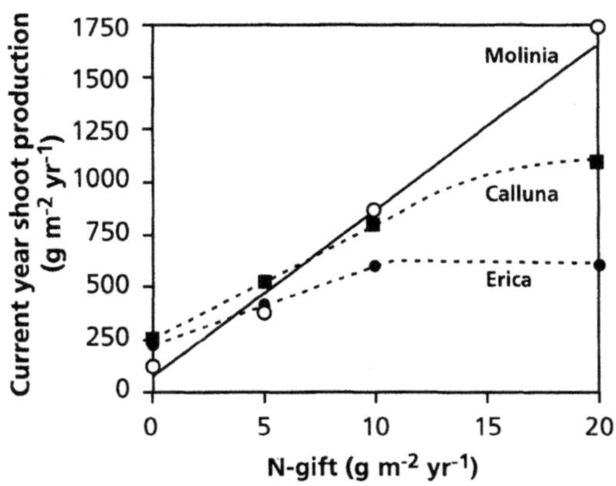

Figure 4.7: Production of current year shoots (g m^{-2} yr^{-1}) of *Erica tetralix, Calluna vulgaris* and *Molinia caerulea* along an experimental gradient of N supply. Data points are means of five replicates. Curves are estimated best fits (Redrawn from Aerts and Heil, 1993).

Current year shoot production of *Erica tetralix* and *Calluna vulgaris* and of the grass *Molinia caerulea* was studied at the same location at the heathland area Deelense Veld where the competition experiments had been performed. Current year shoot production was measured after three years of fertilization (Figure 4.7). In the unfertilized series the biomass of *Erica* and *Calluna* exceeded that of *Molinia.* In all three species current year shoot production showed a significant increase with increasing nutrient availability, but the increase in *Molinia* was much greater than in the evergreens. This resulted in a much higher productivity of *Molinia* in the highest fertilization series. These results clearly show that at a given level of N availability the productivity of the vegetation depends on the species composition.

However, litter production is also strongly species-dependent. This has been demonstrated in two ways : first by a comparative field study in which litter production of monocultures was compared and second by a study in which biomass turnover rates of heathland species were determined. High biomass turnover rates are indicative of a short tissue lifespan and thus of a high litter production per unit plant mass. The field studies were conducted in monocultures of *Erica tetralix*, *Calluna vulgaris*, *Molinia caerulea* and *Deschampsia flexuosa.* The methods to determine litter production are described in full in Aerts (1993). Litter production differed considerably between the species under study. The litter production in the *Molinia* communities exceeded that in the other communities two- to fivefold (Table 4.1).

The aboveground litter production was in all communities more or less equal to the biomass production. Thus, these communities were apparently in a steady state. A clear-cut way to compare the litter production of plant species on a relative scale is by considering the biomass turnover rates (BTR : the ratio between annual litter production and the annual average of the biomass).

Table 4.1: Litter production (g m^{-2} yr^{-1}) of dominant heathland species in the Netherlands (after Aerts, 1993).

	wet heathland		dry heathland		
	Erica	*Molinia*	*Calluna*	*Deschampsia*	*Molinia*
Litter production					
shoots	430	980	570	250	670
roots	370	1080	160	180	1380
Total	800	2060	730	430	2050

The biomass turnover rates were in order of increasing magnitude : *Calluna* < *Erica* < *Deschampsia* < *Molinia* (Table 4.2). So the longevity of a unit of biomass is highest in both ericaceous evergreens. This is mainly due to the synthesis of leaves and woody stems with a high longevity (Aerts, 1989; Aerts and Berendse, 1989). The turnover rate of *Molinia* shoots is extremely high due to the deciduous character of this species. When the biomass turnover rates are compared with the total productivity of each species there seems to be a positive correlation between productivity and biomass turnover. This implies that high-productivity species have also high biomass turnover rates.

Table 4.2: Biomass turnover rates (BTR : yr^{-1}) of shoots, roots and the total plant in field populations of dominant heathland species in the Netherlands (after Aerts, 1993).

	wet heathland		dry heathland		
	Deschampsia	*Molinia*	*Erica*	*Molinia*	*Calluna*
Shoots	0.63	3.32	0.64	1.21	3.32
Roots	1.72	1.27	0.64	0.96	1.68
Total plant	0.88	1.79	0.64	1.09	2.00

Litter production is also stimulated by increased N supply. Power *et al.* (1995) and Uren *et al.* (1997) fertilized a nitrogen-poor dry heathland in southern England with 0.77 and 1.54 g N m^{-2} yr^{-1} (given as ammonium sulphate) during four years. They found that litter production of *Calluna vulgaris* was significantly stimulated by N addition.

The litter which has been produced must be decomposed and the nutrients contained in the litter must be mineralised to make them available for plant uptake again. There are substantial interspecific differences in the mineralisation rates of nutrients which are lost by litter production. French (1988) reported that *Calluna* stem litter decomposed at a much lower rate than did *Molinia* leaf litter. Berendse *et al.* (1989) studied the decomposition of several litter fractions, including roots, of *Erica* and *Molinia* in a wet heathland. For the aboveground litter fractions they found no differences between the decomposition rates of *Erica* and *Molinia* litter. However, the net release of nitrogen and phosphorus from decomposing *Molinia* roots was higher as compared with *Erica* roots. *Molinia* root litter comprises a substantial part of total litter production (Aerts, 1993), so on a whole plant basis the litter decomposition rate of *Molinia* exceeds that of *Erica*. Van Vuuren (1992) conducted a comparative litterbag study on decomposition of shoot litter and root litter and nutrient mineralisation at the same study sites where the measurements on litter production of the heathland species had been carried out. Mass loss data were fitted to the negative exponential decomposition model of Olson (1963) and the decomposition constants (k-values) were calculated. As was to be expected, there were large interspecific differences in decomposition constants. The litter of both *Erica* and *Calluna* decomposed slower (lower k-values) than that of both grass species (Table 4.3). This was mainly due to the higher lignin content of both ericaceous species as compared with the grasses.

Table 4.3: Decomposition constants (per yr, from a negative exponential model) for shoot and root litter of dominant species from Dutch heathlands as determined with the litterbag method (Van Vuuren, 1992).

	wet heathland		dry heathland		
	Erica	*Molinia*	*Calluna*	*Deschampsia*	*Molinia*
Shoot litter	0.10	0.23	0.17	0.34	0.21
Root litter	0.03	0.29	0.12	0.24	0.37

By multiplying the annual litter production by the mean N concentration in the litter it could be calculated how much organic N was transferred annually to the soil N compartment. In these calculations we assumed that N resorption from senescing roots, which we could not determine accurately, was equal to the measured resorption percentages from senescing shoots (Aerts, 1993). It appeared that organic N input into the soil as a result of litter production was highest in the *Molinia* stands and lowest in the *Deschampsia* dominated stand. The stands with ericaceous dominance showed intermediate values (Table 4.4).

Van Vuuren *et al.* (1992) measured the nitrogen mineralisation rates in the communities of this study (Table 4.4). Nitrogen mineralisation was highest in the *Molinia*- or *Deschampsia* dominated stands and lowest in the stands with ericaceous dominance. However, this may partly be due to differences in the amount of organic matter present in the soils of the various sites. When N mineralisation was expressed on a relative basis (mg N g^{-1} soil N yr^{-1}) it still appeared that organic matter produced by the grass species showed consistently higher N mineralisation per unit organic N than the ericaceous species. This implies that 'grass litter' shows a faster N release than 'heather litter'.

Table 4.4: Organic N input into the soil as a result of litter production of dominant heathland species in the Netherlands, and nitrogen mineralisation. (Van Vuuren *et al.* 1992).

	wet heathland		dry heathland		
	Erica	*Molinia*	*Calluna*	*Deschampsia*	*Molinia*
Organic N-input (g N m^{-2} yr^{-1})	7.5	11.3	7.8	3.4	10.6
N-mineralisation[1] (g N m^{-2} yr^{-1})	4.4	7.8	6.2	12.6	10.9
N-mineralisation[1] (mg N g^{-1}soil N yr^{-1})	16	29	25	36	35

Thus, the interspecific differences in the rates of nutrient mineralisation from plant litter probably reinforce the interspecific differences in nutrient input into the soil due to litter production. This emphasises once again the importance of the dominant plant species of an ecosystem in the regulation of ecosystem carbon and nutrient cycling. This effect is probably reinforced at high levels of atmospheric N deposition.

The organic N-losses due to litter production are for most species more or less equal to the sum of nitrogen mineralisation and atmospheric N-input. However, for *Deschampsia* N-mineralisation and atmospheric N-input exceed N-losses due to litter production by far. In contrast with the other stands, nitrification in the *Deschampsia* stand was relatively high (about 40 % of total N-mineralisation) and equalled 5.4 g N m^{-2} yr^{-1} (Van Vuuren *et al.*, 1992). It appears that *Deschampsia* has a relatively high preference for nitrate nutrition, especially at low NO_3^-/NH_4^+ ratios in the soil (Troelstra *et al.*, 1995a,b).

It is clear that atmospheric N deposition influences the species composition of the vegetation and that the species composition of the vegetation has a strong impact on ecosystem N cycling. Using the results presented in this chapter, the work of Chapman and co-workers in British lowland heath (Chapman, 1967, 1979; Chapman *et al.*, 1975a,b) and of Berendse (1990) in Dutch heathlands, we have developed a conceptual scheme of N cycling processes in heathland during secondary succession. The most important basic assumptions are: 1) succession takes place at a constant level of atmospheric N deposition; 2) there is no N leaching to deeper soil layers (cf. Berendse *et al.*, 1987).

During succession the total amount of N in the soil increases linearly with successional age (Figure 4.8a: Berendse, 1990; Chapman *et al.*, 1975a). The slope of this curve is equal to the annual N deposition. Experimental work of Berendse (1990) on N cycling in Dutch heathlands during secondary succession has shown that N mineralisation also increases with successional age, but that during the first 10 years there are no changes, most probably because of high levels of gross N immobilization (Figure 4.8b). When N mineralisation increases, the rate of increase equals atmospheric N deposition. In heathlands, the amount of N available for plant growth equals the sum of atmospheric N deposition and N mineralisation.

4.5 Effects Of N Deposition On Nutrient Limitation For Plant Growth

4.5.1 NITROGEN- VS. PHOSPHORUS-LIMITED PLANT GROWTH

In most terrestrial ecosystems primary productivity is nitrogen-limited (Vitousek and Howarth, 1991). Long-term exposure of ecosystems to high loads of atmospheric nitrogen will lead to a strong increase of N availability (cf. Figure 4.8). Thus, it is to be expected that this will lead to a change of N-limited plant growth to P-limited plant growth (cf. Figure 4.2).

Ombrotrophic raised bogs, which are dominated by *Sphagnum* mosses, receive their mineral nutrients solely by atmospheric deposition and are, therefore, very suitable for monitoring the effects of N deposition on plant growth (Pakarinen and Tolonen, 1977; Ferguson *et al.*, 1984; Woodin and Lee, 1987; Malmer, 1988). High loads of atmospheric nitrogen may, as suggested by Lee and Woodin (1988) and by Malmer (1988, 1990), in the long-term cause a change of the primary element limitation on plant growth in ombrotrophic bogs. This process can very conveniently be studied in Swedish ombrotrophic bogs, because in this country there are profound differences in N deposition rates, both in space and in time. Atmospheric N deposition in southern Sweden has increased during the last decades, while no such change was observed in Swedish Lapland (Malmer and Wallén, 1980; Malmer and Nihlgård, 1980; Asman *et al.*, 1988). In ombrotrophic bogs in southern Sweden the increased nitrogen supply has probably caused an increased productivity of the *Sphagnum* mosses and to a much lesser extent increased nitrogen concentrations in the *Sphagnum* tissues (Malmer, 1988, 1990). Due to the increased productivity other inorganic constituents, such as P, may have been 'diluted' in the biomass of the *Sphagnum* mosses. This phenomenon has been investigated by Malmer (1990) who determined N and P concentrations in *Sphagnum magellanicum* in 1957–1958 in the Åkhult mire in southern Sweden and resampled these locations in 1979 and 1982 (Figure 4.9). From this study it is clear that the N/P mass ratio of these *Sphagnum* mosses has strongly increased in a period of about 25 years. This has resulted in mass ratios of N and P in the capitula of *Sphagnum* mosses as high as 34 in southern Sweden, whereas the mass ratio of N and P is as low as 6 in *Sphagnum* mosses from Swedish Lapland (Malmer, 1988). These values deviate considerably from the N/P ratios in plant biomass for optimal growth (10–14: Van den Driessche, 1974; Ingestad, 1979). In a recent analysis of fertilization experiments in wetlands, Verhoeven *et al.* (1996) have shown that plant growth is N-limited when the N/P ratio is lower than 14 and P-limited when the N/P ratio exceeds 16. For intermediate values plant growth is co-limited by N and P. The data of Malmer (1990) suggest that plant growth in ombrotrophic south Swedish bogs has become P-limited, whereas in northern Sweden it is N-limited. This hypothesis has been investigated by Aerts *et al.* (1992a) who conducted fertilization experiments in *Sphagnum* dominated bogs in Swedish Lapland and in southern Sweden. During one year, they supplied the experimental plots in the bogs with 2 g N m^{-2} yr^{-1} (LN), which is equal to the optimum nitrogen supply for growth of *S. magellanicum* (Rudolph and Voigt, 1986) and with 4 g N m^{-2} yr^{-1} (HN) which corresponds with the average nitrogen deposition in the Netherlands (Heij and Schneider, 1991) and in polluted British regions (Ferguson *et al.*, 1984). Phosphorus was given in a 1:10 mass ratio compared with the N treatments, so the supplied amount were 0.2 g P m^{-2} yr^{-1} (LP) and 0.4 g P m^{-2} yr^{-1} (HP).

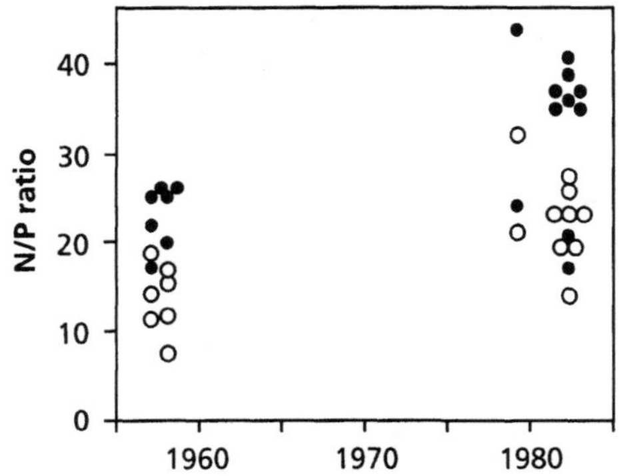

Figure 4.9: N/P mass ratios in *Sphagnum magellanicum* in the Åkhult mire in southern Sweden determined both in 1957–1958 and in 1979 and 1982 (Redrawn from Malmer, 1990).

Nitrogen concentrations in the *Sphagnum* capitula from the unfertilized controls were significantly lower ($\underline{P} < 0.0001$) at the low-N site (Swedish Lapland) than at the high-N site (southern Sweden) (Figure 4.10A). At both sites, N fertilization resulted in higher N concentrations and the relative differences between both sites remained. As was the case for N concentration, the N/P-ratio at the low-N site (4.9±0.4) was significantly lower than at the high-N site (39.5±16.4) ($\underline{P} < 0.0001$). At the low-N site, nitrogen fertilization caused a significant increase of the N/P-ratio in the *Sphagnum* capitula, but there was no effect of P fertilization (Figure 4.10B). At the high-N site, there was a significant increase of the N/P-ratio after N-fertilization and a significant decrease after P-fertilization.

Productivity at the low-N site was strongly affected by nitrogen supply, but not by phosphorus supply (Figure 4.10C). In the HN-treatment the productivity was almost four times as high compared with the control. At the high-N site, productivity increased about three-fold in the P-fertilized treatments, but no productivity response was found after nitrogen addition (Figure 4.10C). These data confirm that high atmospheric nitrogen deposition initially leads to increased productivity and an increased N/P ratio of the plant material. In bogs which are during a long period subject to high N-supply (the South-Swedish bog) the N/P ratio will become so high, that a shift in the primary element limitation on plant growth will occur and phosphorus will become the growth-limiting factor. Such a change has also been observed in Dutch and Danish heathlands (Aerts and Berendse, 1988; Riis-Nielsen, 1997) and in mesotrophic fens (Verhoeven and Schmitz, 1991) in the Netherlands, where the atmospheric nitrogen deposition is about twice as high as in southern Sweden.

Another example of N deposition effects on N/P ratios in plant tissues is provided by studies in Dutch and British chalk grasslands. The N/P ratios in Dutch chalk grassland vegetation were between 13 and 18 in the mid 1980s (Bobbink *et al.*, 1989), which suggests in some cases a relative shortage of P for plant nutrition in this vegetation (cf. Verhoeven *et al.*, 1996). This high ratio could be caused by the high atmospheric nitrogen inputs in the Netherlands (e.g. Van Breemen *et al.*, 1982; Heil *et al.*, 1988).

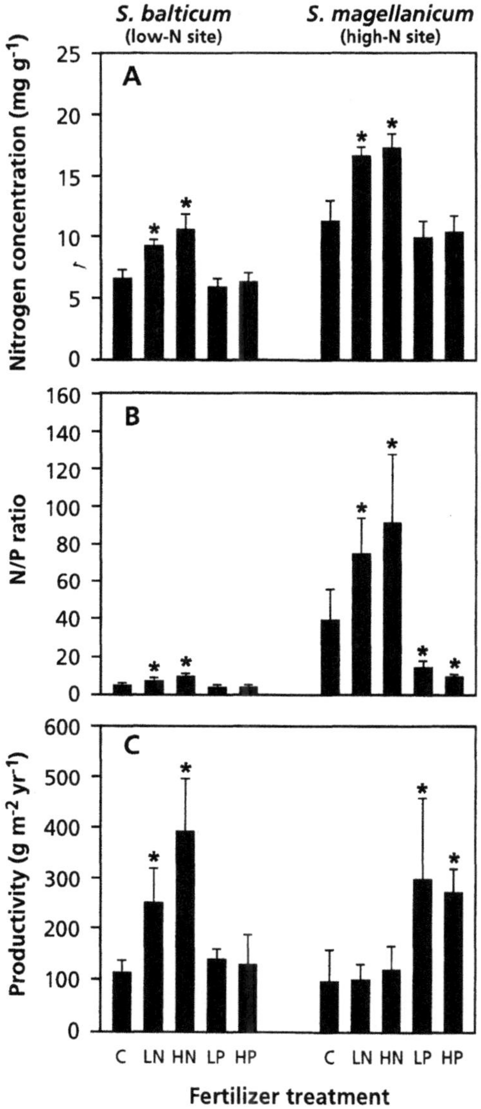

Figure 4.10: Nitrogen concentration in the capitula (A), N/P mass ratio in the capitula (B) and productivity (C) of *Sphagnum balticum* in northern Sweden (low-N site) and of *S. magellanicum* in southern Sweden (high-N site), grown at different nutrient supplies. C: unfertilised control; LN: 2 g N m^{-2} yr^{-1}; HN: 4 g N m^{-2} yr^{-1}; LP: 0.2 g P m^{-2} yr^{-1}; HP: 0.4 g P m^{-2} yr^{-1}. Error bars are S.D. ($\underline{n}$=5). Differences from control tested with Dunnett's test; * $\underline{P} < 0.01$ (Redrawn from Aerts *et al.*, 1992a).

In a three-year fertilization experiment, increased N supply resulted in increased competitive ability of *Brachypodium pinnatum* only, whereas after NPK addition other tall grasses became better competitors (Figure 4.6). This clearly indicates that biomass production of *B. pinnatum* was limited by nitrogen only, whereas the growth of other tall grasses was also limited by phosphorus. Addition of nitrogen resulted in a significant increase in N/P ratio in *B. pinnatum* to more than 30 (Bobbink 1991), whereas in the other graminoids hardly any increase was observed (Figure 4.11). If both N and P were added, the N/P ratio of *B. pinnatum* and the other grasses was in the optimal range for plant growth (10–14; Ingestad, 1979) (Figure 4.10). In that case the other grasses outcompeted *B. pinnatum*. Thus, the potential growth rate of *B. pinnatum* in Dutch chalk grassland is certainly not higher than that of other tall grasses. The key factor in explaining the success of this grass species under long-term N enrichment is its ability to cope with a relative decrease of P availability, caused by N-induced growth stimulation. These observations clearly show the importance of N and P interaction when the supply of both elements is near growth limitation (Latjha and Klein, 1988). Furthermore, it demonstrates that the shift from nitrogen to phosphorus limitation in these grassland is not restricted to the narrow range of N/P ratios (14–16) as Verhoeven *et al.* (1996) suggested, but also depends on the species under consideration.

In the N-addition experiment of Morecroft *et al.* (1994) N/P ratios of nine plants species from the calcareous site were determined in the second year of the experiment. No data of N/P ratios for the vegetation of the acidic grassland are given by them. In the untreated vegetation the N/P ratios of all investigated non-legume species varied between 10–15, which is in the optimal range for plant growth (Ingestad, 1979).

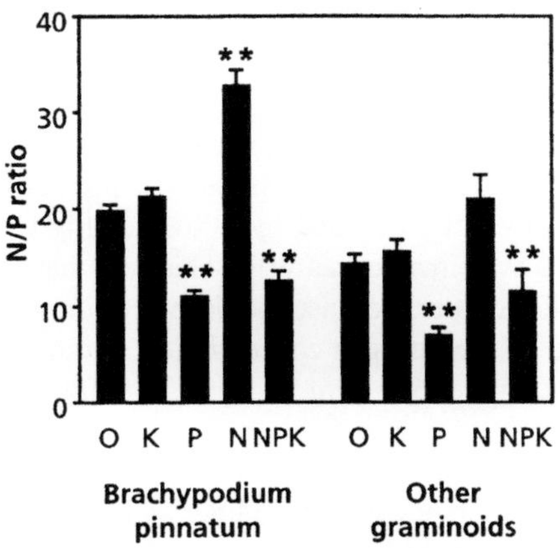

Figure 4.11: N/P mass ratio (± S.E.) of the aboveground biomass of *Brachypodium pinnatum* and the other graminoids after 3 years of nutrient additions (O : unfertilised control; K: 10 g K m^{-2} yr^{-1}; P: 3 g P m^{-2} yr^{-1}; N: 10 g N m^{-2} yr^{-1}; NPK: a complete fertilisation) (Redrawn from Bobbink, 1991). * $\underline{P} < 0.05$; ** $\underline{P} < 0.01$.

In the N-treated plots the N/P ratios of the vegetation increased with enhanced N inputs to values between 13 and 22. This increase was significant for four of the investigated species and points to P-limited plant growth. The enhanced shoot N/P ratios were caused by increased N concentrations in these plants, since foliar P concentrations were not affected. These results suggest that P is more limiting for plant growth than N in the first years of this experiment (see also before). After six years of treatment the species composition of the vegetation was clearly affected by N inputs at both sites, but data on N and P concentrations in plants are unfortunately not available (Lee *et al.*, 1996). Only in the grass *Avenula pratensis* did the P content in the shoots decrease with increasing N inputs in the second year of the experiment. This could be a first sign of N-induced growth stimulation, which may result in a decrease in plant P content at P availabilities near limitation, as shown before in *B. pinnatum* (Bobbink *et al.*, 1988).

Fennema (1990) studied N/P ratios of the vegetation in Danish, German and Dutch NARDETALIA grasslands. The mean N/P ratio of the vegetation in Denmark and Germany was ca. 10–12, compared with 15–17 in the Netherlands. Furthermore, in still species-rich sites with *Arnica montana* Pegtel (1994) observed a N/P ratio of ca. 11 in this species. The differences between the Danish/German and Dutch results may be caused by the significantly higher N loads in the Netherlands and suggested a shift from N to P limitation in these stands. Experimental N and P additions are necessary to verify this conclusion in future, because of inter-species differences in growth responses to N or P. In addition, part of the difference between the Danish/German and the Dutch results may be explained by the fact that these plant species are seriously affected by soil acidification (increased Al concentrations; low base cations) and by the accumulation of ammonium in the soil, which is very manifest in the Netherlands (Fennema, 1992; Houdijk *et al.*, 1993; Bobbink and Roelofs, 1995; De Graaf *et al.*, 1998).

4.5.2 EFFECTS ON PLANT LITTER

When plant growth has become N-limited in areas with high N deposition it is to be expected that the N concentration in plant tissues will increase. As N resorption from senescing tissues is generally not controlled by N availability (Aerts, 1996) this implies that litter N concentrations will increase as well. This may have profound consequences for the decomposability of the litter, because the decomposition of dead organic matter is in many cases determined by the nitrogen concentration or the carbon to nitrogen ratio of dead organic matter (Witkamp, 1966; Berg and Staaf, 1980; Taylor *et al.*, 1989; Tian *et al.*, 1992).

However, there is only very limited evidence for the postulated effect of increased N deposition on N concentrations in newly formed litter and the decay rates of that litter. The most clear-cut results were obtained by Coulson and Butterfield (1978), who found that litter N concentrations in vascular plant species from blanket bogs increased after N fertilization and that this in turn resulted in higher decomposition rates. Pastor *et al.* (1987), however, found that annual decay rates of litter of *Schizachyrium scoparium* were highly correlated with the N content of the litter but not with fertilizer additions. This lack of correlation with fertilizer additions was due to the fact that litter N concentrations were not or only weakly correlated with fertilizer addition. In a recent study, Aerts *et al.* (1995) investigated the effect of increased atmospheric N-deposition on productivity and potential decay rates of *Carex* species growing in peatlands in the Netherlands. Part of the study consisted of fertilization experiments in a P-limited peatland, dominated by *Carex lasiocarpa*. In this peatland there was no significant increase of the net carbon fixation upon enhanced nitrogen supply (Table 4.5), although N-uptake had increased significantly compared with the unfertilized control. Due to the N-fertilization the C/N ratio in the plant biomass decreased significantly, whereas the N/P ratio showed a significant increase. The C/N ratio of leaf litter showed a significant decrease after N-fertilization, but there was no effect on the N/P ratio in the leaf litter. The potential decay rate, measured as CO_2-evolution from the litter under standard laboratory conditions (20°C, 97% relative humidity), was significantly increased upon enhanced N-supply.

Although this study shows that under P limited conditions N deposition can increase the decomposition rate of plant litter which has been produced by plants growing at higher N supply, it does not provide evidence that N mineralisation is also increased (cf. Figure 4.2). We investigated this aspect in a long-term (3 years) decomposition experiment in the field at the same P-limited site where the previous experiment was conducted (Aerts and De Caluwe, 1997). At this site, we incubated three leaf litter types of *Carex lasiocarpa*: litter collected in the field (FLD); and litter from experimental populations grown at low (LN : 3.3 g N m^{-2} yr^{-1}) and high N supply (HN : 20.0 g N m^{-2} yr^{-1}), respectively.

Table 4.5: Results of N fertilisation (10 g N m^{-2} yr^{-1}) on growth and decomposition of *Carex lasiocarpa* from a P-limited fen in the Netherlands. Data are means±S.D. (n=9).

		Unfertilized	N-fertilized
Living plants			
	Carbon fixation (g C $plant^{-1}$)	0.32±0.30	0.53±0.38
	N-uptake (mg $plant^{-1}$)	2.78±2.89	7.08±4.55*
	C/N ratio biomass	54.1±5.1	40.9±6.3***
	N/P ratio biomass	38.0±4.2	49.8±9.8**
Leaf litter			
	C/N-ratio	58.0±4.8	43.4±7.0***
	N/P ratio	122±16	101±12
	Decomposition rate (mg CO_2 g^{-1}leaf hr^{-1})	0.056±0.012	0.070±0.012*

* $P < 0.05$; ** $P < 0.01$; *** $P < 0.001$

Decomposition rates of litter produced at low and high N supply were significantly higher than that of 'natural' litter. Moreover, litter from both N fertilized treatments showed N mineralisation from the start of the experiment onwards, whereas natural litter went through a net N immobilization phase of two years before any net release of N occurred. Thus, increased N supply led to faster N cycling. In addition, it should be noticed that the rate of nutrient cycling is not only determined by the nutrient release rate from the litter (expressed per unit litter mass) but also by the total amount of litter (both above- and belowground) which is produced per unit ground area (Chapin, 1991). In *Carex lasiocarpa*, litter production strongly increases upon enhanced nutrient supply (Aerts and De Caluwe, 1994). Thus, the positive feed-back between high N deposition and the rate of N cycling is reinforced by the increase in litter production in response to increased N supply.

4.5.3 EFFECTS ON N MINERALISATION AND NITRIFICATION

The effects of increased N inputs on soil mineralisation and nitrification rates were also studied with field incubation methods in the Derbyshire grassland experiment (Morecroft *et al.*, 1994). Mineralisation rates varied considerably through the year, but N additions significantly increased the net mineralisation rates in both the acidic and the calcareous site. In the vegetation which received 14 g N m^{-2} yr^{-1}, mineralisation increased to 4.0 in the calcareous site and to 4.9 g N m^{-2} yr^{-1} in the acidic site, compared with 0.7 and 1.1 g N m^{-2} yr^{-1} in the untreated plots, respectively. Yearly N mineralisation and N inputs showed a positive, linear relationship. Treatment with glucose did not affect N mineralisation. Furthermore, no differential effects upon mineralisation of ammonium nitrate or ammonium sulphate were found. After six years of treatment mineralisation rates were still higher in the plots with N additions (Lee *et al.*, 1996). Nitrification was also studied in this experiment (Morecroft *et al.*, 1994). Nitrification rates were negligible on the acidic site, with the exception of the plots which received 14 g m^{-2} yr^{-1} as ammonium nitrate. In the calcareous grassland nitrification rates were higher from March to August and clearly increased with enhanced N inputs. Finally, treatment with ammonium sulphate resulted in higher nitrification, than addition of ammonium nitrate. It is thus obvious from this long-term field study that N mineralisation strongly responds to realistically increased N inputs and becomes a major component of N cycling in these oligotrophic, species-rich grasslands.

4.6 N Deposition Effects On Allocation And Mycorrhizal Infection

4.6.1 ALLOCATION

Atmospheric nitrogen deposition does not only affect the species composition of vegetation, N cycling in ecosystems, and growth-limitation of plant species, but there are also effects on the partitioning of biomass between shoots and roots and on the level of mycorrhizal infection. These effects might have important implications for the capacity of plants species to take up water and nutrients.

Allocation patterns are very plastic and are especially determined by light intensity and nutrient levels (Brouwer, 1962a,b; Robinson, 1986; Robinson and Rorison, 1988; Aerts and De Caluwe, 1989; Boot and Den Dubbelden, 1990; Olff *et al.*, 1990). Here we describe the results of nitrogen supply experiments on the allocation of biomass in species from heathlands, fens and chalk grassland and we try to assess the implications of these experiments for N deposition effects on nutrient and water uptake.

We studied biomass allocation in the heathland species *Erica tetralix, Calluna vulgaris* and *Molinia caerulea* in a pot-experiment in an experimental garden during two years (Aerts *et al.*, 1991). At the end of the experiment, the Root Weight Ratio (RWR : root mass as a fraction of total biomass) showed a significant decrease with increasing nitrogen supply, except in *Erica* (Table 4.6). The Leaf Weight Ratio (LWR : leaf mass as a fraction of total biomass) increased significantly with increasing nitrogen supply, except in *Molinia* (Table 4.6). However, it should be noted that due to the increased total biomass after nitrogen addition the absolute amount of leaf mass increased in all species.

Table 4.6: Root Weight Ratio (RWR : g root g^{-1} plant) and Leaf Weight Ratio (LWR : g leaf g^{-1} plant) of three heathland species in response to nitrogen supply (10 g N m^{-2} yr^{-1}) in a pot experiment in an experimental garden. Data are means±S.D. (n=6).

	Erica tetralix		*Calluna vulgaris*		*Molinia caerulea*	
	0	N	0	N	0	N
RWR						
	0.24±0.04	0.23±0.02	0.27±0.05	0.17±0.05*	0.67±0.05	0.49±0.06*
LWR						
	0.13±0.05	0.20±0.04*	0.22±0.03	0.27±0.04*	0.08±0.02	0.08±0.02

* $P < 0.05$

In a study with species from fens we were able to study the effects of nitrogen supply on allocation in more detail. We investigated the relation between allocation parameters and productivity in closely-related perennial *Carex* species, which are characteristic for habitats with different levels of nitrogen availability. We established experimental populations of four *Carex* species which are the dominant vascular species in mesotrophic or eutrophic fens in the Netherlands. These species were, in order of potential productivity of these species : *Carex diandra* ≤ *C. rostrata* < *C. lasiocarpa* ≤ *C. acutiformis* (Aerts *et al.*, 1992b). Each species was grown at two levels of nitrogen supply. These levels were 3.3 g N m^{-2} yr^{-1} (LN) and 20 g N m^{-2} yr^{-1} (HN), respectively. Other nutrients were given in non-limiting amounts. Further details of the experimental design are given in Aerts *et al.*

(1992b). For all species the shoot–root ratio increased with increasing nitrogen supply, with the high-productivity species having the highest S/R ratio both at low and at high N supply (Table 4.7). The ratio between the total leaf area and the total root length increased with increasing N-supply for the high-productivity species only.

Table 4.7: Shoot/root ratio (S/R) and ratio between total leaf area and total root length (LA/RL: m^2 leaf m^{-1} root) in July in experimental populations of two mesotrophic and two eutrophic *Carex* species grown for two years at two nitrogen supplies (LN : 3.3 g N m^{-2} yr^{-1}; HN : 20 g N m^{-2} yr^{-1}) in an experimental garden. Data are means ± 1 S.D. in parentheses ($\underline{n}$=5) (after Aerts *et al.*, 1992b).

	C. diandra		*C. rostrata*		*C. lasiocarpa*		*C. acutiformis*	
	LN	HN	LN	HN	LN	HN	LN	HN
S/R	1.07	2.20^{***}	1.16	1.75^{**}	1.93	4.75^{***}	1.68	3.74^{***}
	(0.16)	(0.40)	(0.20)	(0.40)	(0.28)	(1.26)	(0.15)	(0.54)
LA/RL	0.25	0.31	1.09	1.68	0.83	1.56^{*}	0.91	2.49^{**}
	(0.14)	(0.04)	(0.26)	(0.79)	(0.57)	(0.62)	(0.34)	(1.26)

* $\underline{P} < 0.05$; ** $\underline{P} < 0.01$; *** $\underline{P} < 0.001$

The effects of nutrients upon the allocation of biomass of *Brachypodium pinnatum* and *Dactylis glomerata* were studied in a two way factorial design (N and P; each with three nutrient levels) in pots in an experimental garden during two years (Bobbink *et al.*, unpublished). An almost ten-fold difference in dry weights of these grasses was observed in this experiment, with significant effects of both N or P treatments. The allocation of biomass was also influenced by the nutrient application. In the low-N treatments (1.2 g N m^{-2} yr^{-1}) Root Weight Ratios of the calcareous grassland plant *B. pinnatum* were 0.50–0.55, which decreased to ca. 0.35 in the high-N-treated (29.7 g N m^{-2} yr^{-1}) pots (Figure 4.12). The RWR of the common crop grass *D. glomerata* responded in a similar way as *B. pinnatum* on N treatment, but the ratios were, in general, a little lower than those of *B. pinnatum*. Remarkably, P additions hardly influenced the RWR's of these two grass species (Figure 4.12).

Because of the restrictions of pot experiments, the allocation of biomass between roots and shoots was also studied at the end of the 3-year nutrient addition experiment in Dutch calcareous grasslands (Bobbink, 1991; see before). Shoot:root ratios are normally low in nutrient-poor ecosystems (e.g. Chapin, 1980). This certainly holds for the investigated Dutch calcareous grasslands: in the unfertilized vegetation the shoot:root ratio was c. 0.3 (Bobbink, 1991). Effects of single nutrient additions on root mass and shoot/root ratios in nutrient-poor ecosystems grasslands are less predictable. Plant productivity generally became higher after NPK addition, and therefore it is likely that below-ground biomass increases, too.

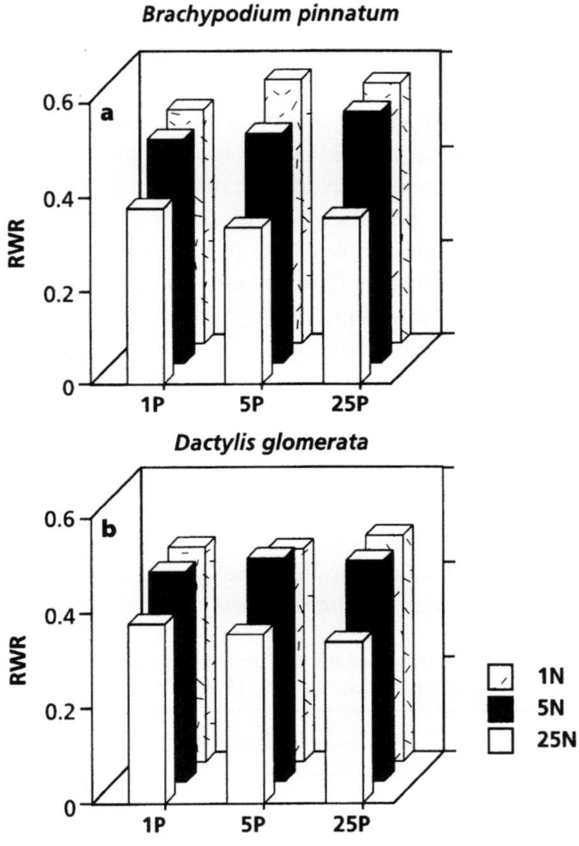

Figure 4.12: Root Weight Ratios (RWR : g root g^{-1} plant) of *Brachypodium pinnatum* and *Dactylis glomerata* after two years of factorial treatment with N or P (amounts in g m^{-2} yr^{-1}).

Only NPK addition had caused a higher total root mass, and shoot:root ratio increased slightly to ca. 0.5, compared with 0.3 in the untreated plots. N additions resulted in an increase to 0.75 of the shoot/root ratio of the vegetation. This was, however, caused by a decrease in root weight of the other species, whereas shoot:root ratio of *B. pinnatum*, the only species that benefited from N enrichment (Figure 4.6), hardly increased after three years of N additions. Thus, an increase in the shoot/root ratios with enhanced N inputs, was less clear in the field, than in most pot studies.

These experiments show that increased nitrogen availability generally leads to an increased shoot-root ratio, especially in high-productivity species which are more plastic in their allocation patterns (Aerts *et al.*, 1991). This implies that the ratio between transpiring surfaces (the leaves) and water uptake surfaces (the roots) will increase. As long as water is in sufficient supply, this will not cause any problem to the plant.

However, in cases of severe droughts this may lead to substantial growth reduction. This growth reduction will be reinforced by the fact that during drought periods mass transport of nitrogen is blocked.

4.6.2 MYCORRHIZAL INFECTION

In the symbiotic relation of mycorrhizal fungi and host plants both derive at least some benefit from the other. In the most basic picture the host plant supplies the fungi with carbohydrates and the fungi supplies the plant with nutrients. Mycorrhizal associations are especially known to increase the absorption of immobile nutrients, such as PO_4^{3-}. However, there are more benefits for the plants than increased P uptake alone, such as increased resistance to root pathogens (Newsman *et al.*, 1995), increased water uptake (Allen *et al.*, 1981, Jackson and Mason, 1984), increased resistance to herbivores as a result of the alteration of the secondary chemistry of plants (Newsman *et al.*, 1995), reduced uptake of toxic heavy metals and increased tolerance to extreme temperatures and very acid soils (Harley and Smith, 1983; Newsman *et al.*, 1995). These observations make clear that mycorrhizal infection can be very beneficial to plants, especially in stressed environments. This raises the question of how increased N supply through atmospheric N deposition might change mycorrhizal infection. We have tried to answer this question by searching the literature on experiments in which the effects of N supply on mycorrhizal infection were studied (Table 4.8). To ensure the robustness of the analysis, we have also included the results of experiments with tree species.

Table 4.8: Summary of nitrogen addition experiments on mycorrhizal infection of different plant species (adapted from Ludwig, 1996).

Type	n	Effect of N supply
VA mycorrhiza	15	6 expts : negative effect
		5 expts : positive effect
		4 expts : no effect
Ecto-mycorrhiza	7	6 expts : negative effect[1]
		1 expt : no clear effect
Ericoid mycorrhiza	4	2 expt : negative effect
		1 expt : positive effect
		1 expt : no effect

1 More negative effects of NH_4^+ compared with NO_3^-

Our analysis of literature data clearly shows that the effects of N deposition on mycorrhizal infection are dependent on the type of mycorrhizal infection. In ectomycorrhizal species the number of fungal infections decreases with increasing N supply, whereas in VA mycorrhizal species and in ericoid mycorrhizal species the results are very variable. Another important point is that the nitrogen effects depend on the nitrogen form: in almost all cases where both the effects of ammonium and nitrate were studied separately the most pronounced effects were found when N was supplied as ammonium. As ammonium is the dominant N form in atmospheric N deposition (Bobbink and Heil, 1993), this implies that atmospheric N deposition generally has negative effects on ectomycorrhizal species. This may have negative consequences for P uptake and, under conditions of low water supply, water uptake.

From these observations it can be concluded that atmospheric nitrogen deposition may affect both biomass allocation to the roots and, in ectomycorrhizal species, the degree of mycorrhizal infection. The combination of both factors may lead to P deficiency and, in dry periods, to an increased susceptibility to drought stress.

4.7 N Deposition Effects On Herbivory And Frost Sensitivity

So far, we have focused on the primary effects of nitrogen deposition on vegetation processes. However, it has become clear that atmospheric nitrogen deposition also affects secondary stress and disturbance factors. Stress consists of the phenomena which restrict photosynthetic production, such as shortages of light, water, and mineral nutrients, or suboptimal temperatures. Disturbance is associated with the partial or total destruction of the plant biomass and arises from the activities of herbivores, pathogens, man (trampling, mowing, and ploughing), and from phenomena such as wind-damage, frosting, droughting, soil erosion, and fire (Grime, 1979). In this section we confine ourselves to herbivory as an example of a disturbance factor and frost sensitivity as a stress factor. We will illustrate the effects of N deposition on these factors by examples from studies on the ecology of *Calluna vulgaris* which were performed in the Netherlands, the United Kingdom and Denmark.

4.7.1 EFFECTS ON HERBIVORY

In *Calluna*-dominated heathland outbreaks of the chrysomelid heather beetle (*Lochmaea suturalis*) occur frequently (Berdowski, 1993). The heather beetle feeds monophagously on the green parts of *Calluna* (Cameron *et al.*, 1944) and when it occurs in high densities, the intensive insect grazing in July and August causes brown-colouring of *Calluna* over large areas. Heavily infested plants may develop some new shoots in autumn, but almost all of the damaged plants do not survive the subsequent winter and die next spring. In this way the closed *Calluna* canopy is opened over large areas and the interception of light by *Calluna* is strongly decreased. Consequently, the growth of the understorey grasses (*Deschampsia flexuosa* and *Molinia caerulea*) can enhance significantly and this may lead to dramatic changes in the species composition of the vegetation (section 4.3).

In general, herbivory is affected by the palatability of the plant material, which is strongly determined by the nitrogen content (e.g. Crawley, 1983). Experimental applications of nitrogen to Dutch heathland vegetation have demonstrated that the N concentrations in the (green) parts of *Calluna* increased (e.g. Brunsting and Heil, 1985). It is thus very likely that the frequency and the intensity of the heather beetle grazing are stimulated by the increased atmospheric nitrogen inputs in Dutch heathland. This hypothesis is corroborated by work of Van der Eerden *et al.* (1990). They were (accidentally) able to investigate the effects of ammonium sulphate and ammonia upon the growth of the heather beetle, due to an outbreak of the beetle in a *Calluna* vegetation which was subject to artificial N deposition. They found no significant effect of the ammonium sulphate treatments on the total number and on the biomass of the larvae. This was probably caused by the large variation in the number of larvae and the sampling technique. However, the development of subsequent larval stages was accelerated by the application of ammonium sulphate in the artificial rain: the percentage of 3rd stage instars increased by 20 %, compared with the control treatment. In another experiment, heather beetle larvae were fed with *Calluna* shoots taken from plants which had been fumigated during a year with ammonia in open top chambers (application of 4 to 105 $\mu g\ m^{-3}$; Van der Eerden *et al.*, 1991). After 7 days the larvae were counted and weighed. Both the biomass and the development rate of the larvae clearly increased with increasing concentrations of ammonia.

Also in Danish heathlands, where atmospheric N deposition is about 50% of that in the Netherlands, the larval development of heather beetles was stimulated by increased N deposition (Riis-Nielsen, 1997). Despite the fact that at very high rates of N supply (7 $g\ N\ m^{-2}\ yr^{-1}$) larval development was lower than at intermediate N supply rates (1.5–3.5 $g\ N\ m^{-2}\ yr^{-1}$), the damage to the vegetation increased linearly with increased N supply. Similarly, experimental addition of ammonium sulphate and ammonium to *Calluna vulgaris* in British lowland heath stimulated the growth of the heather beetle (Uren, 1992; Power *et al.*, 1998).

Historical data also point to the crucial role of atmospheric N deposition in the dynamics of the heather beetle. Atmospheric N deposition in the Netherlands has steadily increased during this century and in the period 1950–1980 it doubled from about 2 $g\ N\ m^{-2}\ yr^{-1}$ to 4 $g\ N\ m^{-2}\ yr^{-1}$ (Asman *et al.*, 1988). Blankwaardt (1977) reported that from 1915 onwards heather beetle outbreaks have occurred in the Netherlands around 1927, 1945 and 1967 (with an interval of c. 20 years), whereas in the last 30 years the outbreaks occur at intervals of 5 to 10 years. Furthermore, it was also observed that during a heather beetle outbreak *Calluna* plants were more severely damaged in vegetation that was experimentally fertilized with nitrogen (Heil and Diemont, 1983).

From these observations we conclude that nitrogen inputs influence the frequency and intensity of the outbreaks of the heather beetle. However, the exact relationship between both processes is no yet quantified and needs further research.

4.7.2 EFFECTS ON FROST SENSITIVITY

Some studies have shown that frost sensitivity in tree species, especially conifers, increased with increasing concentrations of air pollutants (e.g. Aronsson, 1980; Dueck *et al.*, 1991). Also for *Calluna*-dominated heathlands there are indications that atmospheric N deposition and other pollutants lead to increased frost sensitivity.

Caporn *et al.* (1994) studied frost sensitivity in a stand of *Calluna vulgaris* which had received ammonium nitrate applications (0, 4, 8 or 13 g N m^{-2} yr^{-1}) in the field for 30 months. They found that frost sensitivity in October, assessed both by scores of visible injury and electrolyte leakage, was marginally reduced in response to the experimental nitrogen additions. Their observations suggest that pollutant inputs of nitrate and ammonium to heather may promote the hardening process in autumn. However, after 6 to 7 years more late frost damage was observed in N-fertilized upland *Calluna* plants due to drought damage (Caporn, personal communication).

The effects of sulphur dioxide, ammonium sulphate and ammonia upon frost sensitivity in *Calluna vulgaris* were studied by Van der Eerden *et al.* (1990). They used electrolyte leaching from *Calluna* leaves (attached to the stems) as an indicator for frost injury. After fumigation with 90 μg m^{-3} sulphur dioxide during 3 months increased frost injury in *Calluna* was only observed at temperatures lower than –20 °C. However, such low temperatures hardly occur in the Netherlands. Fumigation with 100 μg m^{-3} ammonia of *Calluna* plants in open top chambers during 4 to 7 month periods revealed that frost sensitivity was not affected in autumn (September or November), whereas in February, just before growth started, frost injury increased significantly at –12 °C (Van der Eerden *et al.*, 1991). They also studied the frost sensitivity in *Calluna* vegetation which was artificially treated with "rainfall" with 6 levels of ammonium sulphate under an experimental 'roof' in the field after 5 months (in September) and 7 months (in November), respectively. The frost sensitivity of *Calluna* increased slightly, although significantly, after 5 months in vegetation treated with the highest level of ammonium sulphate (450 μmol l^{-1}, which was equivalent to 9.1 g N m^{-2} yr^{-1}), compared with the control treatments. However, frost sensitivity of *Calluna* decreased again two months later and no significant effects of the ammonium sulphate application upon frost hardiness were found at that time. These data show that the effects of ambient nitrogen and sulphur loads on frost sensitivity of *Calluna* plants are still uncertain. However, high levels of ammonia or ammonium sulphate may have a significant impact on frost sensitivity.

The data presented here suggest that *Calluna*-dominated heathland vegetation is seriously affected by secondary stress and disturbance factors. High loads of atmospheric nitrogen (and sulphur) seem to increase the frequency and intensity of these factors. Heather beetle attacks and frost can lead to severe damage to the *Calluna* canopy. In the end, this may lead to the complete death of *Calluna* over large areas. Outbreaks of the heather beetle, or similar canopy damaging processes, such as frost and summer drought damage, catalyse the transition from heathland dominated by *Calluna* into grassland. Grasses cannot increase under a closed canopy of *Calluna*, but they are able to survive in its shade. After opening up of a *Calluna* vegetation, grasses can become established and those already present increase and eventually become dominant.

4.8 Conclusions And Topics For Further Research

This chapter has made clear that atmospheric nitrogen deposition has a wide variety of effects on vegetation processes. The most obvious effect is that atmospheric nitrogen deposition causes changes in the species composition of (semi-)natural plant communities. In general, slow growing species with low rates of tissue turnover and low nutrient loss rates are being replaced by relatively fast growing species with high rates of tissue turnover and high nutrient loss rates. As a result, internal nitrogen cycling in the ecosystem is accelerated, which affects in turn the rate of change in species composition. In our opinion, quantification of these long-term plant-mediated feed-backs on soil nitrogen availability is essential for our understanding of the effects of atmospheric N deposition on vegetation processes. These long-term effects can hardly be analysed by means of experimental studies, so long-term field experimentation or modelling studies are clearly required here. This is inevitably necessary when critical loads are to be determined for particular ecosystems. A good example of an approach in which the results of experimental field studies are integrated in a simulation model is provided by the Dutch HEATHSOL model (Bakema *et al.*, 1994).

In the long-term, increased atmospheric N deposition will lead to a shift from N- to P-limited plant growth. This shift has consequences for both ecosystem productivity and for ecosystem nutrient cycling. The severity of P-limitation may be increased by the fact that mycorrhizal infection, which is important for P uptake, is in many cases decreased as a result of increased atmospheric N deposition. However, there is clearly a need for further study on this topic.

Both biomass allocation and mycorrhizal infection are affected by atmospheric N deposition. These effects (higher shoot/root ratios, lower degree of mycorrhizal infection) probably makes natural vegetation more susceptible to drought stress and phosphorus deficiency. There is, however, little evidence from field experiments to support this hypothesis. Due to the potentially large impacts on ecosystem productivity more research effort should be dedicated to these topics.

There are strong indications that atmospheric N deposition makes vegetation more susceptible to secondary stress and disturbance factors, such as frost sensitivity, drought sensitivity and susceptibility to herbivore attacks. Currently, many questions remain on the generality of these observations and on their ecological significance. This presents a further challenge for those interested in the effects of atmospheric N deposition on vegetation processes in terrestrial ecosystems.

References

Aerts, R. (1993) Biomass and nutrient dynamics of dominant plant species in heathlands, in *Heathlands, patterns and processes in a changing environment*, (eds R. Aerts and G.W. Heil), Kluwer Academic Publishers, Dordrecht, pp. 51–84.

Aerts, R. (1996) Nutrient resorption from senescing leaves of perennials: are there general patterns? *Journal of Ecology*, **84**, 597–608.

Aerts, R. and Berendse, F. (1988) The effect of increased nutrient availability on vegetation dynamics in wet heathlands. *Vegetatio*, **76**, 63–69.

Aerts, R. and Berendse, F. (1989) Above-ground nutrient turnover and net primary production of an evergreen and a deciduous species in a heathland ecosystem. *Journal of Ecology*, **77**, 343–356.

Aerts, R. and De Caluwe, H. (1989) Aboveground productivity and nutrient turnover of *Molinia caerulea* along an experimental gradient of nutrient availability. *Oikos*, **54**, 320–324.

Aerts, R. and De Caluwe, H. (1994) Nitrogen use efficiency of *Carex* species in relation to nitrogen supply. *Ecology*, **75**, 2362–2372.

Aerts, R. and De Caluwe, H. (1997) Nutritional and plant-mediated controls on leaf litter decomposition of *Carex* species. *Ecology*, **78**, 244–260.

Aerts, R., Heil, G.W. (eds) (1993) *Heathlands, patterns and processes in a changing environment*. Kluwer Academic Publishers, Dordrecht.

Aerts, R., Ludwig, F. (1997) Water table changes and nutritional status affect trace gas emissions from laboratory columns of peatland soils. *Soil Biology and Biochemistry*, **29** 1691–1698.

Aerts, R., Boot, R.G.A. and Van der Aart, P.J.M. (1991) The relation between above- and below ground biomass allocation patterns and competitive ability. *Oecologia*, **87**, 551–559.

Aerts, R., Wallén, B., Malmer, N. (1992a) Growth-limiting nutrients in *Sphagnum*-dominated bogs subject to low and high atmospheric nitrogen supply. *Journal of Ecology*, **80**, 131–140.

Aerts, R., De Caluwe, H., Konings, H. (1992b) Seasonal allocation of biomass and nitrogen in four *Carex* species from mesotrophic and eutrophic fens as affected by nitrogen supply. *Journal of Ecology*, **80**, 653–664.

Aerts, R., Berendse, F., De Caluwe, H. and Schmitz, M. (1990) Competition in heathland along an experimental gradient of nutrient availability. *Oikos*, **57**, 310–318.

Aerts, R., Van Logtestijn, R., Van Staalduinen, M., Toet, S. (1995) Nitrogen supply effects on productivity and potential leaf litter decay of *Carex* species from fens differing in nutrient limitation. *Oecologia*, **104**, 447–453.

Allen, M.F., Smith, W.K., Moore, Jr. T.S. and Chritensen, M. (1981) Comparative water relations and photosynthesis of mycorrhizal and non-mycorrhizal *Bouteloua gracilis* H.B.K. Lag ex steud. *New Phytologist*, **88**, 683–693.

Aronsson, A. (1980) Frost hardiness in Scots pine. II. Hardiness during winter and spring in young trees of different mineral status. *Studies in Forestry Sueca*, **155**, 1–27.

Asman, W.A.H. and Maas, J.J.M. (1986) Estimation of the deposition of ammonia and ammonium in the Netherlands (in Dutch). Report R–86–8. Institute for Meteorology and Oceanography, Utrecht University.

Asman, W.A.H., Drukker, B. and Janssen, A.J. (1988) Modelled historical concentrations and depositions of ammonia and ammonium in Europe. *Atmospheric Environment*, **22**, 725–735.

Bakelaar, R.G. and Odum, E.P. (1978) Community and population level responses to fertilization in an old-field ecosystem. *Ecology*, **59**, 660–665.

Bakema, A.H., Meijers, R., Aerts, R., Berendse, F. and Heil, G.W. (1994) HEATHSOL: a heathland competition model. National Institute of Public Health and Environmental Protection, Bilthoven. Report 259102009.

Berdowski, J.J.M. (1993) The effect of external stress and disturbance factors on *Calluna*-dominated heathland vegetation, in *Heathlands, patterns and processes in a changing environment*, (eds R. Aerts and G.W. Heil), Kluwer Academic Publishers, Dordrecht, pp. 85–124.

Berendse, F. (1990) Organic matter accumulation and nitrogen mineralization during secondary succession in heathland ecosystems. *Journal of Ecology*, **78**, 413–427.

Berendse, F. and Aerts, R. (1984) Competition between *Erica tetralix* L. and *Molinia caerulea* (L.) Moench as affected by the availability of nutrients. *Acta Oecologica/Oecologia Plantarum*, **5**(19), 3–14.

Berendse, F., Beltman, B., Bobbink, R., Kwant, R. and Schmitz, M. (1987) Primary production and nutrient availability in wet heathland ecosystems. *Acta Oecologica/Oecologia Plantarum*, **8**(22), 265–279.

Berendse, F., Bobbink, R. and Rouwenhorst, G. (1989) A comparative study on nutrient cycling in wet heathland ecosystems. II. Litter decomposition and nutrient mineralization. *Oecologia*, **78**, 338–348.

Berendse, F., Oomes, M.J.M., Altena, H.J. and de Visser, W. (1994) A comparative study on nitrogen flows in two similar meadows affected by different groundwater levels. *Journal of Applied Ecology*, **31**, 40–48.

Berg, B. and Staaf, H. 1980. Decomposition rate and chemical changes in decomposing needle litter of Scots pine. II. Influence of chemical composition. *Ecological Bulletins*, **32**, 373–390.

Blankwaardt, H.F.H. 1977. The occurrence of the heather beetle (*Lochmaea suturalis* Thomson) in the Netherlands since 1915. *Entomologische Berichten*, **37**, 34–40 (in Dutch).

Bobbink, R. (1991) Effects of nutrient enrichment in Dutch chalk grassland. *Journal of Applied Ecology*, **28**, 28–41.
Bobbink, R. and Willems, J.H. (1987) Increasing dominance of *Brachypodium pinnatum* (L.) Beauv. in chalk grasslands: a threat to a species-rich ecosystem. *Biological Conservation*, **40**, 301–314.
Bobbink, R. and Heil, G.W. (1993) Atmospheric deposition of sulphur and nitrogen in heathland ecosystems, in *Heathlands, patterns and processes in a changing environment*, (eds R. Aerts and G.W. Heil), Kluwer Academic Publishers, Dordrecht, pp. 25–50.
Bobbink, R. and Roelofs, J.G.M. (1995a) Ecological effects of atmospheric deposition on non-forest ecosystems in western Europe, in: *Acid rain research: Do we have enough answers?* (eds G.J. Heij and J.W. Erisman), Elsevier, Amsterdam, pp. 279–292.
Bobbink, R. and Roelofs, J.G.M. (1995b) Nitrogen critical loads for natural and semi-natural ecosystems: the empirical approach. *Water, Air and Soil Pollution*, **85**, 2413–2418.
Bobbink, R., Bik, L. and Willems, J.H. (1988) Effects of nitrogen fertilization on vegetation structure and dominance of *Brachypodium pinnatum* (L.) Beauv. in chalk grassland. *Acta Botanica Neerlandica*, **37**, 231–242.
Bobbink, R., Den Dubbelden, K. and Willems, J.H. (1989) Seasonal dynamics of phytomass and nutrients in chalk grassland. *Oikos*, **55**, 216–224.
Bobbink, R., Boxman, D., Fremstad, E., Heil, G., Houdijk, A. and Roelofs, J. (1992a) Critical loads for nitrogen eutrophication of terrestrial and wetland ecosystems based upon changes in vegetation and fauna. *Nord (Miljörapport)*, **41**, 111–159.
Bobbink, R., Heil, G.W. and Raessen, M.B.A.G. (1992b) Atmospheric deposition and canopy exchange processes in heathland ecosystems. *Environmental Pollution*, **75**, 29–37.
Boot, R.G.A. and Den Dubbelden, K.C. (1990) Effects of nitrogen supply on growth, allocation and gas exchange characteristics of two perennial grasses from inland dunes. *Oecologia*, **85**, 115–121.
Brouwer, R. (1962a) Distribution of dry matter in the plant. *Netherlands Journal of Agricultural Science*, **10**, 361–376.
Brouwer, R. (1962b) Nutritive influences on the distribution of dry matter in the plant. *Netherlands Journal of Agricultural Science*, **10**, 399–408.
Brunsting, A.M.H. and Heil, G.W. (1985) The role of nutrients in the interactions between a herbivorous beetle and some competing plant species in heathlands. *Oikos*, **44**, 23–26.
Buijsman, E. Maas, J.M. and Asman, W.A.H. 1987. Anthropogenic NH_3 emissions in Europe. *Atmospheric Environment*, **21**, 1009–1022.
Cameron, A.E., McHardy, J.W. and Bennet, A.H. (1944) The heather beetle (*Lochmaea suturalis*). An inquiry into its biology and control. *British Fields Sports Society*, **53**, 69.
Chapin, F.S. (1980) The mineral nutrition of wild plants. *Annual Review of Ecology and Systematics*, **11**, 233–260.
Chapin, F.S. (1991) Effects of multiple stresses on nutrient availability and use, in *Response of plants to multiple stresses* (eds H.A. Mooney, W.E. Winner and E.J.), Academic Press, San Diego, pp. 67–88.
Chapman, S.B. (1967) Nutrient budgets for a dry heath ecosystem in the south of England. *Journal of Ecology*, **55**, 677–689.
Chapman, S.B. (1979) Some interrelationships between soil and root respiration in lowland *Calluna* heathland in southern England. *Journal of Ecology*, **67**, 1–20.
Chapman, S.B., Hibble, J. and Rafarel, C.R. (1975a) Net aerial production by *Calluna vulgaris* on lowland heath in Britain. *Journal of Ecology*, **63**, 233–258.
Chapman, S.B., Hibble, J. and Rafarel, C.R. (1975b) Litter accumulation under *Calluna vulgaris* on a lowland heath in Britain. *Journal of Ecology*, **63**, 259–271.
Coulson, J.C. and Butterfield, J. (1978) An investigation of the biotic factors determining the rates of plant decomposition on blanket bog. *Journal of Ecology*, **66**, 631–650.
Crawley, M.J. 1983. *Herbivory – the dynamics of animal/plant interactions*. Blackwell, Oxford.
De Graaf, M.C.C., Bobbink, R., Verbeek, P.J.M. and Roelofs, J.G.M. (1997a) Aluminium toxicity and tolerance in three heathland species. *Water Air and Soil Pollution*, **98**, 229–239.
De Graaf, M.C.C., Bobbink, R., Roelofs, J.G.M. and Verbeek, P.J.M. (1997b) Differential effects of ammonium and nitrate on three heathland species. *Plant Ecology* **135**, 186–196
De Smidt, J.T. and Van Ree, P. (1991) The decrease of bryophytes and lichens in Dutch heathland since 1975. *Acta Botanica Neerlandica*, **40**, 379.
De Wit, C.T. (1960) On competition. *Agricultural Research Report*, **66**, 1–82.

Dierschke, H. (1985). Experimentelle Untersuchungen zur Bestandesdynamik von Kalkmagerrasen (Mesobromion) in Südniedersachsen. I. Vegetationsentwicklung auf Dauerflachen 1972–1984. *Münsterische Geographischen Arbeiten*, **20**, 9–24.

Dodd, M.E., Silvertown, J., McConway, K., Potts, J. and Crawley, M. (1994) Application of the British national vegetation classification to the communities of the Park Grass experiment through time. *Folia Geobotanica and Phytotaxonomica*, **29**, 321–334.

During, H.J. and Willems, J.H. (1986) The impoverishment of the bryophyte and lichen flora of the Dutch chalk grasslands in the thirty years 1953–1983. *Biological Conservation*, **36**, 143–158.

Dueck, Th.A. and Elderson, J. (1992) Influence of NH_3 and SO_2 on the growth and competitive ability of *Arnica montana* L. and *Viola canina* L.. *New Phytologist*, **122**, 507–514.

Dueck, Th. A., Van der Eerden, L.J., Beemsterboer, B. and Elderson, J. (1991) Nitrogen uptake and allocation by *Calluna vulgaris* (L.) Hull and *Deschampsia flexuosa* (L.) Trin. exposed to $^{15}NH_3$. *Acta Botanica Neerlandica*, **40**, 257–267.

Ellenberg, H. jr. (1988) Floristic changes due nitrogen deposition in central Europe. *Nord (Miljörapport)*, **15**, 375–383.

Ellenberg, H. (1996) *Vegetation Mitteleuropas mit den Alpen.* Ulmer, Stuttgart.

Fennema, F. (1990) Effects of exposure to atmospheric SO_2, NH_3 and $(NH_4)_2SO_4$ on survival and extinction of *Arnica montana* L. and *Viola canina* L.. RIN, Arnhem, The Netherlands. Report no. 90/14, pp. 1–61.

Fennema, F. (1992) SO_2 and NH_3 deposition as possible causes for the extinction of *Arnica montana* L. *Water, Air and Soil Pollution*, **62**, 325–336.

Ferguson, P., Robinson, R.N., Press, M.C. and Lee, J. (1984) Element concentrations in five *Sphagnum* species in relation to atmospheric pollution. *Journal of Bryology*, **13**, 107–114.

French, D.D. (1988) Some effects of changing soil chemistry on decomposition of plant litters and cellulose on a Scottish moor. *Oecologia*, **75**, 608–618.

Gigon, A. and Rorison, I.H. (1972) The response of some ecologically distinct plant species to nitrate- and to ammonium-nitrogen. *Journal of Ecology*, **60**, 93–102.

Grime, J.P. (1979) *Plant strategies and vegetation processes*. Wiley, Chichester.

Harley, J.L. and Smith, S.E. (1983) *Mycorrhizal Symbiosis*. Academic Press, London.

Heij G.J. and Schneider T. (eds) (1991) *Acidification research in the Netherlands: Final report of the Dutch Priority Programme on Acidification*. Elsevier, Amsterdam.

Heil, G.W. and Bruggink, M. (1987) Competition for nutrients between *Calluna vulgaris* (L.) Hull and *Molinia caerulea* (L.) Moench. *Oecologia*, **73**, 105–108.

Heil, G.W. and Diemont, W.H. (1983) Raised nutrient levels change heathland into grassland. *Vegetatio*, **53**, 113–120.

Heil, G.W., Werger, M.J.A., De Mol, W., Van Dam, D. and Heijne, B. (1988) Capture of atmospheric ammonium by grassland canopies. *Science*, **239**, 764–765.

Hemond, H.F. (1983) The nitrogen budget of Thoreau's bog. *Ecology*, **64**, 99–109.

Houdijk, A.L.F.M., Verbeek, P.J.M., van Dijk, H.F.G. and Roelofs, J.G.M. (1993) Distribution and decline of endangered herbaceous heathland species in relation to the chemical composition of the soil. *Plant and Soil*, **148,** 137–143.

Ingestad, T. (1979) Nitrogen stress in birch seedlings. II. N, K, P, Ca, and Mg nutrition. *Physiologia Plantarum*, **45**, 149–157.

Jackson, R.M. and Mason, P.A. (1984) *Mycorrhiza*. The Institute of Biology's Studies in Biology. Edward Arnold Publishers, London.

Jeffrey, D.W. and Pigott, C.D. (1973) The response of grasslands on sugar-limestone in Teesdale to application of phosphorus and nitrogen. *Journal of Ecology*, **61**, 85–92.

Kinzel, H. (1983) Influence of limestone, silicates and soil pH on vegetation, in: *Physiological plant ecology IV C. Responses to the chemical and biological environment*, (eds O.L. Lange, P.S. Nobel, C.B. Osmond and H. Ziegler), Springer Verlag, Berlin, pp. 201–244.

Koerselman, W. and Verhoeven, J.T.A. (1992) Nutrient dynamics in mires of various trophic status: nutrient inputs and outputs and the internal nutrient cycle, in *Fens and Bogs in the Netherlands: Vegetation, History, Nutrient Dynamics and Conservation* (ed J.T.A. Verhoeven), Kluwer Academic Publishers, Dordrecht, pp. 397–432.

Latjha, K. and Klein, M. (1988) The effect of varying nitrogen and phosphorus availability on nutrient use by *Larrea tridentata*, a desert shrub. *Oecologia*, **75**, 348–353.

Lee, J.A. and Woodin, S.J. (1988) Vegetation structure and the interception of acidic deposition by ombrotrophic mires, in *Vegetation structure in relation to carbon and nutrient economy*, (eds J.T.A. Verhoeven, G.W. Heil and M.J.A. Werger), SPB Academic Publishing, The Hague, pp. 137–147.

Lee, J.A., Carroll, J.A. and Caporn, S.J.M. (1996) Effects of long-term nitrogen addition to acidic and calcareous grassland in Derbyshire. Internal report, University of Sheffield.

Ludwig, F. (1996) The effect of an increased nitrogen deposition on mycorrhiza. Doctoral thesis, Utrecht University.

Malmer, N. and Nihlgård, B. (1980) Supply and transport of mineral nutrients in a subarctic mire. *Ecological Bulletins*, **30**, 63–95.

Malmer, N. and Wallén, B. (1980) Figures for the wet deposition of mineral nutrients in Southern Sweden. *Meddelningen Växtekologiska Institutionen*, **43**, Lunds Universitet.

Malmer, N. (1988) Patterns in the growth and the accumulation of inorganic constituents in the *Sphagnum* cover on ombrotrophic bogs in Scandinavia. *Oikos*, **53**, 105–120.

Malmer, N. (1990) Constant or increasing nitrogen concentrations in *Sphagnum* mosses in mires in Southern Sweden during the last few decades. *Aquilo Seria Botanica*, **28**, 57–65.

Marschner, H. (1995) *Mineral nutrition of higher plants.* Academic Press, London.

Morecroft, M.D., Sellers, E.K. and Lee, J.A. (1994) An experimental investigation into the effects of atmospheric nitrogen deposition on two semi-natural grasslands. *Journal of Ecology*, **82**, 475–483.

Morris, J.T. (1991) Effects of nitrogen loading on wetland ecosystems with particular reference to atmospheric deposition. *Annual Review of Ecology and Systematics*, **22**, 257–279.

Mountford, M.O., Lakhani, K.H. and Holland, R. (1994) The effects of nitrogen on species diversity and agricultural production on the Somerset Moors, Phase II: (a) after seven years of fertilizer application; (b) after cessation of fertilizer input for three years. Report to the Institue for Grassland and Environmental Research. Abbots Ripton.

Newsman, K.K., Fitter, A.H. and Watkinson, A.R. (1995) Multi-functionality and biodiversity in arbuscular mycorrhizas. *Trends in Ecology and Evolution*, **10**, 407–411.

Nihlgård, B. (1985) The ammonium hyphothesis – an additional explanation to the forest dieback in Europe. *Ambio*, **14**, 2–8.

Olff, H., Van Andel, J. and Bakker, J.P. (1990) Biomass and shoot/root allocation of five species from a grassland succession series at different combinations of light and nutrient supply. *Functional Ecology*, **4**, 193–200.

Olson, J.S. (1963) Energy storage and the balance of producers and decomposers in ecological systems. *Ecology*, **44**, 322–331.

Pakarinen, P. and Tolonen, K. (1977) Distribution of lead in *Sphagnum fuscum* profiles in Finland. *Oikos*, **28**, 69–73.

Pastor, J., Stillwell, M.A. and Tilman, D. (1987) Little bluestem litter dynamics in Minnesota old fields. *Oecologia*, **72**, 327–330.

Pegtel, D.M. (1994) Habitat characteristics and the effect of various nutrient solutions on growth and mineral nutrition of *Arnica montana* L. grown on natural soil. *Vegetatio*, **114**, 109–121.

Power, S.A., Ashmore, M.R., Cousins, D.A. and Ainsworth, N. (1995) Long term effects of enhanced nitrogen deposition on a lowland dry heath in southern Britain. *Water, Air and Soil Pollution*, **85**, 1701–1706.

Power, S.A., Ashmore, M.R., Cousins, D.A. and Sheppard, L.J. (1998) Effects of nitrogen addition on the stress sensitivity of *Calluna vulgaris. New Phytologist*, in press.

Prins, A.H., Berdowski, J.J.M. and Latuhihin, M.J. (1991) Effect of NH_4-fertilization on the maintenance of a *Calluna vulgaris* vegetation. *Acta Botanica Neerlandica*, **40**, 269–279.

Riis-Nielsen, T. (1997) Effects of nitrogen on the stability and dynamics of Danish heathland vegetation. Thesis, University of Copenhagen

Robinson, D. and Rorison, I.H. (1988) Plasticity in grass species in relation to nitrogen supply. *Functional Ecology*, **2**, 249–257.

Robinson, D. (1986) Compensatory changes in the partitioning of dry matter in relation to nitrogen uptake and optimal variations in growth. *Annals of Botany*, **86**, 841–848.

Roelofs, J.G.M., Kempers, A.J., Houdijk, A.L.F.M. and Jansen, J. (1985) The effect of air-borne ammonium on *Pinus nigra* var. *maritima* in the Netherlands. *Plant and Soil*, **84**, 45–56.

Roelofs, J.G.M. (1986) The effect of airborne sulphur and nitrogen deposition on aquatic and terrestrial heathland vegetation. *Experientia*, **42**, 372–377.

Roelofs, J.G.M., Bobbink, R., Brouwer, E. and De Graaf, M.C.C. (1996) Restoration ecology of aquatic and terrestrial vegetation on non-calcareous sandy soils in The Netherlands. *Acta Botanica Neerlandica*, **45**, 517–542.

Sheikh, K.H. (1969) The effects of competition and nutrition on the inter-relations of some wet-heath plants. *Journal of Ecology*, **57**, 87–99.

Skeffington, R.A. (1990) Accelerated nitrogen inputs – A new problem or a new perspective ? *Plant and Soil*, **128**, 1–11.

Smith, C.T. Elston, J. and Bunting, A.H. (1971) The effects of cutting and fertilizer treatment on the yield and botanical composition of chalk turfs. *Journal of the British Grassland Society*, **26**, 213–219.

Tallowin, J.R.B. and Smith, R.E.N. (1994) The effects of inorganic fertilizers in flower-rich hay meadows on the Somerset Levels. Report English Nature, Petersborough.

Taylor, B.R., Parkinson, D. and Parsons, W.F.J. (1989) Nitrogen and lignin content as predictors of litter decay rates: a microcosm test. *Ecology*, **70**, 97–104.

Tian, G., Kang, B.T. and Brussaard, L. (1992) Effects of chemical composition on N, Ca and Mg release during incubation of leaves from selected agroforestry and fallow plant species. *Biogeochemistry*, **16**, 103–119.

Troelstra, S.R., Wagenaar, R. and Smant, W. (1995a) Nitrogen utilization by plant species from acid heathland soils I. Comparison between nitrate and ammonium nutrition at constant low pH. *Journal of Experimental Botany*, **46**, 1103–1112.

Troelstra, S.R., Wagenaar, R. and Smant, W. (1995b) Nitrogen utilization by plant species from acid heathland soils II. Growth and shoot/root partitioning of NO_3^- assimilation at constant low pH and varying NO_3^-/NH_4^+ ratio. *Journal of Experimental Botany*, **46**, 1113–1121.

Ulrich, B. (1983) Interaction of forest canopies with atmospheric constituents: SO_2, alkali and earth alkali cations and chloride. In: Ulrich, B. and Pankrath, J. (eds.) *Effects of accumulation of air pollutants in forest ecosystems.* D. Reidel Publ., Dordrecht, pp.33–45.

Ulrich, B. (1991) An ecosystem approach to soil acidification. *Soil acidity*, (eds B. Ulrich and M.E. Summer), Springer Verlag, Berlin.

Uren, S.C. (1992) The effects of wet and dry deposited ammonia on *Calluna vulgaris*. Thesis, University of London.

Uren, S.C., Ainsworth, N., Power, S.A., Cousins, D.A., Huxedurp, L.M. and Ashmore, M.R. (1997) Long-term effects of ammonium sulphate on *Calluna vulgaris*. *Journal of Applied Ecology*, **34**, 208–216.

Van Breemen, N., Burrough, P.A., Velthorst, E.J., Dobben, H.F. van, Wit, T. de, Ridder, T.B. and Reijnders H.F.R. (1982) Soil acidification from atmospheric ammonium sulphate in forest canopy throughfall. *Nature*, **299**, 548–550.

Van Dam, D. (1990) Atmospheric deposition and nutrient cycling in chalk grassland. PhD thesis, University of Utrecht.

Van Dam, D., Van Dobben, H.F., Ter Braak, C.F.J. and De Wit, T. (1986) Air pollution as a possible cause for the decline of some phanerogamic species in The Netherlands. *Vegetatio*, **65**, 47–52.

Van den Bergh, J.P. (1979) Changes in the composition of mixed populations of grassland species, in: *The study of vegetation*, (ed M.J.A.Werger), Junk, The Hague, pp. 59–80.

Van den Driessche, R. (1974) Prediction of mineral nutrient status of trees by foliar analysis. *The Botanical Review*, **40**, 347–394.

Van der Eerden, L.J., Dueck, Th.A., Elderson, J., Van Dobben, H.F., Berdowski, J.J.M. and Latuhihin, M. (1990) Effects of NH_3 and $(NH_4)_2SO_4$ on terrestrial semi-natural vegetation on nutrient-poor soils. Report of project 124/125 Dutch Priority Programme on Acidification, 169 pp.

Van der Eerden, L.J., Dueck, Th. A., Berdowski, J.J.M., Greven, H. and Van Dobben, H.F. (1991) Influence of NH_3 and $(NH_4)_2SO_4$ on heathland vegetation. *Acta Botanica Neerlandica*, **40**, 281–296.

Van Dobben, H. and De Bakker, A.J. (1996) Re-mapping epiphytic lichen biodiversity in the Netherlands: effects of decreasing SO_2 and increasing NH_3. *Acta Botanica Neerlandica*, **45**, 55–72.

Van Dijk, H.F.G. De Louw, M.H.J. Roelofs, J.G.M and Verburgh, J.J. (1990) Impact of artificial ammonium-enriched rainwater on soils and young coniferous trees in a greenhouse. Part 2–Effects on the trees. *Environmental Pollution*, **63**, 41–60.

Van Hecke, P., Impens, I. and Behaeghe, T.J. (1981) Temporal variation of species composition and species diversity in permanent grassland plots with different fertilizer treatments. *Vegetatio*, **47**, 221–232.

Van Tooren, B.F., Odé, B., During, H.J. and Bobbink, R. (1990) Regeneration of species richness of the bryophyte layer of Dutch chalk grasslands. *Lindbergia*, **16**, 153–160.

Van Vuuren, M.M.I. (1992) Effects of plant species on nutrient cycling in heathlands. PhD Thesis, Utrecht University.

Van Vuuren, M.M.I., Aerts, R., Berendse, F. and De Visser, W. (1992) Nitrogen mineralization in heathland ecosystems dominated by different plant species. *Biogeochemistry*, **16**, 151–166.

Verhoeven, J.T.A. and Schmitz, M.B. (1991) Control of plant growth by nitrogen and phosphorus in mesotrophic fens. *Biogeochemistry*, **12**, 135–148.

Verhoeven, J.T.A., Koerselman, W. and Beltman, B. (1988) The vegetation of fens in relation to their hydrology and nutrient dynamics: a case study, in *Vegetation of inland waters*, (ed J.J. Symoens), Kluwer Academic Publishers, Dordrecht, pp. 249–282.

Verhoeven, J.T.A., Koerselman, W. and Meuleman, A.F.M. (1996) Nitrogen or phosphorus limited growth in herbaceous mire vegetation: relations with atmospheric inputs and management regimes. *Trends in Ecology and Evolution*, **11**, 494–497.

Vermeer, J.G. (1986) The effect of nutrients on shoot biomass and species composition of wetland and hayfield communities. *Acta Oecologica/Oecologia Plantarum*, **7**, 31–41.

Vitousek, P.M. and Howarth, R.W. (1991) Nitrogen limitation on land and in the sea: How can it occur? *Biogeochemistry*, **13**, 87–115.

Wellburn, A. (1988) *Air pollution and acid rain.* Longman, Harlow.

Wells, T.C.E. (1974) Some concepts of grassland management. in: *Grassland ecology and wildlife management*, (ed E. Duffey), Chapman and Hall, London, pp. 163–74.

Wells, T.C.E., Sparks, T.H., Cox, R. and Frost, A. (1993) Critical loads for nitrogen assessment and effects on southern heathlands and grasslands. Report to National Power, Institute of Terrestrial Ecology, Monks Wood, UK.

Willems, J.H. (1980) An experimental approach to the study of species diversity and above ground biomass in chalk grassland. *Proceedings Koninklijke Nederlandse Akademie van Wetenschappen Series* **C–83**, 279–306.

Willems, J.H. (1990). Calcareous grasslands in Continental Europe, in: *Calcareous grasslands–Ecology and Management* (eds S.H. Hillier, D.W.H. Walton and D.A. Wells), Bluntisham Books, Bluntisham, pp. 3–10.

Willems, J.H., Peet, R.K. and Bik, L. (1993) Changes in chalk grassland structure and species richness resulting from selective nutrient additions. *Journal of Vegetation Science*, **19**, 203–212.

Williams, E.D. (1978) Botanical composition of the Park Grass plots at Rothamsted 1856–1976. Rothamsted Experimental Station Internal Report, Harpenden.

Wilson, E.J., Wells, T.C.E. and Sparks, T.H. (1995) Are calcareous grasslands in the UK under threat from nitrogen deposition? – an experimental determination of a critical load. *Journal of Ecology*, **83**, 823–832.

Witkamp, M. (1966) Decomposition of leaf litter in relation to environment, microflora and microbial respiration. *Ecology*, **47**:194–201.

Woodin, S.J. and Lee, J.A. (1987) The fate of some components of acidic deposition in ombrotrophic mires. *Environmental Pollutution*, **45**, 61–72.

FACTORS INFLUENCING NITROGEN SATURATION IN FOREST ECOSYSTEMS: ADVANCES IN OUR UNDERSTANDING SINCE THE MID 1980s

E. J. WILSON[1] AND B. A. EMMETT[2]

[1]*National Power PLC, Windmill Hill Business Park, Swindon, Wiltshire, SN7 7PE, UK*

[2]*Institute of Terrestrial Ecology, Deniol Road, Bangor, Wales, LL57 2UP, UK*

5.1. Introduction

It was first proposed that atmospheric inputs of nitrogen could be damaging forests and semi-natural ecosystems over 10 years ago (Nihlgard, 1985). The strongest empirical evidence came from the Netherlands where typical symptoms of forest dieback such as needle yellowing and crown thinning were correlated with ammonium deposition originating from intensive livestock farming (Van Breemen and Van Dijk, 1988). Many damaged trees had high foliar N concentrations and N:K, P and Mg ratios indicative of unbalanced nutrition (Dijk and Roelofs, 1988). However, in other areas of Europe, away from the influence of point agricultural sources there was little to substantiate a role for N deposition in forest decline.

In 1987 an International Workshop on Excess Nitrogen Deposition was held at the Central Electricity Research Laboratories, Leatherhead (UK). The aim was to review evidence from all over Europe for the role of N deposition in ecosystem damage and attempt to reach a consensus on the underlying mechanisms. The workshop was structured around a number of key issues, fundamental to understanding whether atmospheric inputs of N were a damaging pollutant or a beneficial fertilizer (Skeffington and Wilson, 1988). This question was particularly relevant for forest ecosystems which were traditionally viewed as N-limited. Evidence was presented which revealed that forests in southern Germany had been growing better since 1960, an effect partly attributed to increases in atmospheric N deposition (Kenk and Fisher, 1988). Many fertilizer experiments had shown that trees responded positively to N addition (Albrekston *et al.*, 1977; Tamm, 1991) and it was proposed that increases in atmospheric N deposition could be acting in a similar way. We were aware of the limitations of these experiments as an adequate simulation of atmospheric N inputs (Skeffington and Wilson, 1988), but since more realistic data were scarce, they provided a useful insight into the potential effects of increases in anthropogenic N emissions. Over the last ten years we have seen a proliferation of manipulation experiments both in Europe and the US which have attempted to simulate atmospheric N inputs to mature forests in terms of level, frequency and chemical composition. Together with deposition gradient and nutrient cycling studies, these investigations have significantly advanced our understanding of the fate and impacts of N deposition on forest ecosystems.

Perhaps the most fundamental shift over the last ten years has been our perception of what is likely to be the primary impact of elevated N inputs to forests. The focus of the 1987 Workshop was clearly effects on the trees themselves, exploring mechanisms to explain the luxury consumption of N which resulted in high foliar N concentrations, imbalanced nutrition, increased susceptibility to fungal pathogens and decline symptoms in Dutch forests. It is now recognised that the conditions experienced by these forests are somewhat atypical of most of Europe. N deposition is extremely high – in many cases in excess of 60 kg N ha^{-1} yr^{-1} in throughfall – predominantly in the form of ammonium and with a high dry deposition component (Van Breemen and Van Dijk, 1988). The damaged forests were usually growing on very base-poor, sandy soils with a history of litter removal (Van Breemen and Van Dijk, 1988). Although similar damage symptoms have subsequently been observed close to large emission sources elsewhere in Europe (eg. Bonneau and Nys, 1993; Heinsdorf, 1993), these effects on forest health have not been reproduced at the more moderate deposition rates typical of most of Europe (5–25 kg N ha^{-1} yr^{-1}) (Tuovinen *et al.*, 1994).

In Table 5.1 we have summarised the results of deposition gradient studies and recent N addition experiments which attempt to simulate atmospheric inputs. The original aim was to collate evidence for effects at moderate N inputs, but since even the first treatment level in most manipulation experiments is > 30 kg N ha^{-1} yr^{-1} we have had to include data for N inputs up to 58 kg N ha^{-1} yr^{-1}. It is clear that elevated N generally has very little effect on the tree component of the forest ecosystem. Increases in foliar N concentration have been reported in some experiments (e.g. Magill *et al.*, 1996; Boxman *et al.*, 1995), often after a delay of several years due to competition for N between trees and soil microbes. An increase over a gradient of N deposition has also been shown (e.g. McNulty *et al.*, 1991; Tietema and Beier, 1995), although effects can be confounded with other factors (e.g. climate, soil type, elevation). Foliar N concentration does not always respond to N addition however, particularly if needle N status is already high (Emmett *et al.*, 1995c) or an increased growth rate has 'diluted' the %N in foliage (Boxman *et al.*, 1995). In some cases, changes in %N are accompanied by a decrease in nutrient/N ratios in the foliage, indicating an altered nutritional balance (Table 5.1, Balsberg Påhlsson, 1992). Moderate rates of N deposition (12–30 kg N ha^{-1} yr^{-1}) have been implicated in nutritional imbalances and decline in Norway spruce in the Bohemian Forest, Austria, although topography and elevation are clearly contributory factors (Katzensteiner *et al.*, 1992).

Some studies have demonstrated effects more readily perceived as damage; changes in wood accumulation rates, mortality and recruitment patterns have been reported for certain species (Boxman *et al.*, 1998b; Magill *et al.*, 1997; McNulty *et al.*, 1996). There is also evidence that mycorrhizal fruiting bodies and species composition of the ground flora may be affected by changes in N deposition (Brandrud and Timmerman, 1998; Boxman *et al.*, 1998b; Falkengren–Grerup 1986). Many of these of these studies can be difficult to interpret, however, due to synchronous changes in S deposition and soil acidification.

Current evidence suggests that effects on the tree component of the forest ecosystem are uncommon and vary with species, soil type and geographical location. In contrast,

Table 1: Summary of statistically significant responses to changing nitrogen deposition in recent manipulation experiments and surveys. Inclusion of N addition experiments was limited to studies with several N applications per year with final loading <58 kg N ha^{-1} yr^{-1}. Only monitoring surveys with data on both water and tree responses were included. '+' indicates significant increase, '–' significant decrease and N/A data not available. For roots/mycorrhizae, 'yes' indicates some change (qualitative or quantitative) and 'no' indicates no change.

Site/study	Species sampled /dominant species	Understorey	Change in N inputs due to treatment (kg N ha^{-1} yr^{-1})	Change in leaching losses due to treatment (kg N ha^{-1} yr^{-1})	Change in foliar %N (+/-)	Change in foliar nutrient/N ratios?	Change in wood accumulation rates (+/-)	Roots/ mycorrhizae changes
Manipulation experiments								
Gardsjön[1] (SW)	Norway spruce	dwarf shrub & mosses	11 to 51	0.1 to 2	0	0	0	yes
Klosterhede[2] (DK)	Norway spruce	mosses and grasses	23 to 58	0.3 to 2.3	0	–	0	N/A
Aber[3] (UK)	Sitka spruce	none	14 to 52	8 to 24	0	0	0	no
Solling[4] (Ger)	Norway spruce	none	36 to <5	22 to 5	0	0	0	no
Speuld[5] (NL)	Douglas fir	none	50 to <5	27 to 3	0	0	0	no
Ysselsteyn[6](NL)	Scots pine	ferns and grasses	45 to <5	35 to 16	–	+	+	yes
Bearbrook[7] (Maine, USA)	Beech	mixed spruce/ hardwood	5 to 33	1 to 3	+	N/A	0	N/A
Harvard Forest (MA, USA)[8]	Oak	mixed hardwood	8 to 58	<1 to >1	0	N/A	0	yes
Harvard Forest (MA, USA)[8]	Red pine	none	8 to 58	<1 to >1	+	N/A	0	yes
Mt Ascutney (SE Vermont, USE)[9]	Spruce–fir mixture	fir/maple/birch	5 to 20 5 to 36	N/A N/A	+ +	N/A N/A	+ –	N/A
Hubburd Brook (NH, USA)[10]	Beech.	mixed hardwood	7 to 47	no increase	0	N/A	0	N/A
Surveys								
23 sites in southern Sweden [11]	Norway spruce	N/A	2.4 to 30.5	no nitrate leaching <10 (kg N ha^{-1} yr^{-1})	+ve relationship to deposition	N/A	N/A	N/A
27 European Monitoring sites (EMEP)[12]	Mixed conifer	variable	0.6 to 22.7	0.1 to 6.3 (+ve relationship to deposition)	+ve relationship to output (data available for 10 sites only)	N/A	N/A	N/A
12 NITREX and EXMAN sites (Europe)[13]	Mixed conifer	variable	9 to 61	0.1 to 44 (+ve relationship to deposition)	+ve relationship to deposition	–ve relationship to ammonium deposition	N/A	N/A

1 Moldan and Wright (1998), Stuanes et al.1998a, b); Boxman et al. (1998a), 2 Gundersen and Rasmussen (1995), Gundersen (1998); 3 Emmett et al. (1998), Boxman et al. (1998,a), Gundersen et al. (1998); 4 Bredemeier et al. (1998), Boxman et al. (a) (1998,a); 5 & 6 Boxman et al. (1998,a&b) Gundersen et al. (1998); 7 Magill et al. (1996); 8 Magill et al. (1997); 9 McNulty et al. (1996); 10 Christ et al. (1995); 11 Näsholm et al (1997); 12 Kleemola and Forsius (1996); 13 Tietema and Beier (1995).

soil waters show a clearer and more consistent response to elevated N. What we generally see is an increase in nitrate leaching to streams and groundwater as forests lose the ability to retain incoming N (Table 5.1). Evidence that forests will leach N when inputs are in excess of requirements has existed for quite some time (Abrahamsen, 1980), and led to the development of the 'nitrogen saturation' concept (Ågren, 1983). What recent studies have confirmed however, is that nitrate leaching is often the first measurable response to elevated N inputs, is geographically widespread, and can occur in the absence of any visible damage to the trees themselves. Some nitrate leaching is of course a normal feature of the N cycle in forest ecosystems, even at low 'background' levels of N. At elevated N inputs, we tend to see an increase in leaching losses compared with annual fluxes in pristine areas (1–4 kg N ha^{-1} yr^{-1}), and in some cases a loss of the seasonal pattern of nitrate leaching (Stoddard, 1994; Bringmark and Kvarnas, 1995; Kahl *et al.,* 1993). In the short-term, nitrate leaching in excess of background fluxes demonstrates that the normally tight N cycle of the forest has been disrupted; in the longer term it will almost certainly reduce forest productivity in soils where the weathering rate is insufficient to replenish the soil store of base cations. There will also be adverse effects on downstream water quality, in terms of both eutrophication and acidification (Christie and Smol, 1993; Nodvin *et al.,* 1995). Nitrate leaching is thus a useful indicator of potential long-term damage and critical load exceedence, and as such it is important that we understand what controls its onset and magnitude.

In this chapter we examine the relationship between atmospheric N deposition and nitrate leaching in forests and evaluate the role of moderate N inputs as the causal factor of N saturation (as defined by Aber *et al.,* 1989). We then go on to explore some of the many other factors which can influence nitrate leaching. Drawing on the limited evidence currently available, we hope to demonstrate the need to take account of these factors if we are to understand and predict leaching behaviour in forests as diverse as a dense, single-aged spruce stand in the UK and a mixed-age, mixed-species deciduous forest in the USA. In the final section, we suggest some remediation alternatives to reduce N leaching and protect forest health and water quality.

5.2. Is There A Relationship Between N Deposition And Nitrate Leaching?

A correlation between wet N deposition and N loss in run-off was demonstrated as early as 1980 when Abrahamsen showed an approximately linear relationship between N inputs and outputs for a number of forested catchments in Europe and North America (Abrahamsen, 1980). Adding later data reinforced the relationship, although it appeared that there was a threshold input of about 10 kg N ha^{-1} yr^{-1} below which forests did not leach N (Grennfelt and Hultberg, 1986). Skeffington and Wilson (1988) argued that a relationship would not be expected since output will depend on the condition of the major N stores and the range of site-specific factors which determine the size of soil and vegetation sinks for N. As a result of extensive catchment and plot studies, far larger and more representative data sets exist which have been used to analyze input–output budgets. A particularly thorough analysis was carried out for 65 forested sites from the ENSF database (Evaluation of Nitrogen and Sulphur Fluxes) by Dise and Wright (1995). The relationship between N inputs and outputs is shown in Figure 5.1. The first thing that stands out is that leaching at these sites is consistent with the 10 kg N ha^{-1} yr^{-1} 'threshold' suggested by Grennfelt and Hultberg (1986), although other data from low deposition sites in the US suggest a threshold of about 5 kg N ha^{-1} yr^{-1} (Peterjohn *et al.,* 1996). Considering the entire ENSF data set, N deposition (as throughfall) was the most important single predictor of N output and the correlation between the two factors was highly significant (R^2=0.69, $p<0.001$). The good correlation relies partly on the inclusion of sites with very high inputs which we would expect to leach large amounts of N. The region of real interest however, is between inputs of 10 and 25 kg N ha^{-1} yr^{-1}, being both more typical of most of Europe and the sort of range currently being proposed as the critical load of N for forests (Werner and Spranger, 1996). Dise and Wright looked specifically at this section of the data and found the two factors with the greatest influence on N leaching were B-horizon soil pH and % (A horizon) soil N*. N output was still significantly correlated with N deposition although the relationship was weak (R^2=0.28, $p<0.002$). From Figure 5.1, this is not surprising; at inputs between 10 and 25 kg N ha^{-1} yr^{-1} we can see that some sites at the upper end of the deposition range are leaching virtually no N at all while sites at the low end of the range are leaching almost as much N as is coming in.

Other data sets also show that the relationship between N inputs and outputs at these rates of N deposition is weak or non-existent. N inputs and outputs were measured as part of the Integrated Forest Study which looked at nutrient cycling in 17 forests, predominantly in the US (1 site in Norway and one in Canada) (Johnson, 1992). The

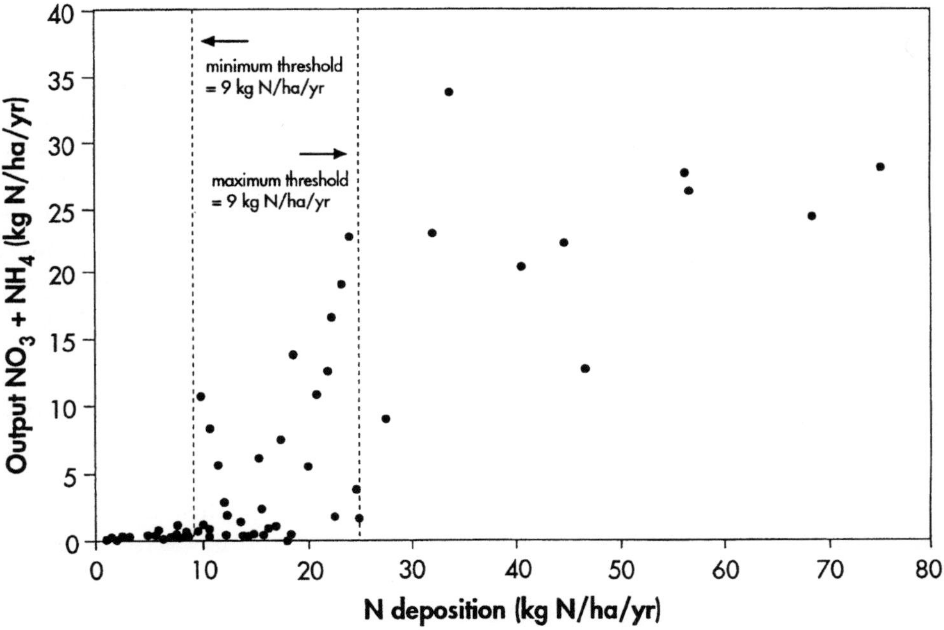

Figure 5.1: The relationship between N deposition (as throughfall) and N outputs in European forested catchments and plots in the ENSF database (From Dise and Wright, 1995).

*(increased N out at higher pH and% N)

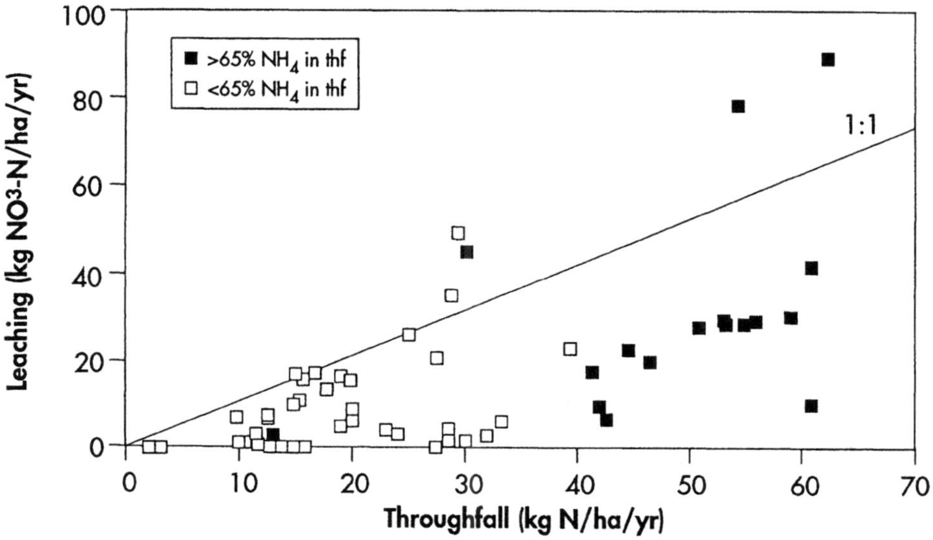

Figure 5.2: The relationship between N inputs (as throughfall) and leaching in European forest plots (ECOFEE database) (Modified from Gundersen, 1995)

stands covered a range of climatic conditions, tree age and species, including both deciduous and coniferous. Analysis of N budgets for these sites showed that atmospheric N inputs were insufficient to explain the variation in nitrate leaching across the sites and the two factors were only weakly positively correlated (R^2=0.095) (Van Miegroet *et al.*, 1992). Deposition at the sites ranged from approximately 5–28 kg N ha^{-1} yr^{-1}, but was less than 14 kg N ha^{-1} yr^{-1} in many sites, most of which showed no measurable leaching. When only those sites showing considerable nitrate leaching were included in the analysis, the relationship between N inputs and outputs improved but was still weak (R^2=0.26). The single most important factor explaining differences in N leaching was N mineralisation potential (44% of the variation).

Input–output relationships have also been examined for the ECOFEE database, a compilation of 64 forest plots sites in Europe, incorporating both deciduous and coniferous forests (Gundersen, 1995) (Figure 5.2). Apart from two Dutch sites (on calcareous soils with high nitrification rates) which are leaching over 75 kg N ha^{-1} yr^{-1} at high inputs, there appears to be very little relationship between deposition (as throughfall) and outputs. At the other end of the deposition scale, 9 forested watersheds in the Chesapeake Bay region of the mid-Appalachians showed a wide range of nitrate export (0.04–5.12 kg N ha^{-1} yr^{-1}), but no correlation with N deposition which was similar across all catchments (8–11 kg N ha^{-1} yr^{-1}) (Williard, 1996). Certain soil properties were able to explain the variation in nitrate leaching (see below).

It is clear that differences in N inputs cannot adequately explain the current variation in N leaching that we see at more moderate rates of deposition. This scatter across sites reflects differences in the size of the forest's two major sinks for N – soil microbes and vegetation. Any factors which influence the size of these sinks or the rate of N transformation processes within the soil, will affect the ability of the ecosystem to retain N – and hence the degree to which incoming N is leached. The size of the soil sink will be determined by the soil's physical, chemical and biological properties. While largely a function of geology and development history, these properties will also depend on local climate, previous land use and practices such as litter raking, ploughing and burning. N uptake by vegetation will be a function of stand age, species, nutrient availability and forest management. In turn, climatic factors and the form of deposition (nitrate or ammonium) will influence N retention by both soil and vegetation sinks. The complex interaction of all these factors is the key to understanding what makes some sites vulnerable to N leaching at quite low inputs and others resistant even at high inputs. Below we consider the evidence for their influence on N saturation and leaching, focusing particularly on more recent studies.

5.3. Factors Influencing Nitrate Leaching

5.3.1 SOIL PROPERTIES

The balance between mineralisation and immobilisation depends on the *quantity* of the soil's organic C and N pool and its *quality*, in terms of C/N ratio and %N. C and N pools will be smaller in more recently developed soils, thin soils and soils with a history of disturbance or organic matter removal. In general, we can expect these soils to have a high capacity for N retention since they are still building up the C–N store and are essentially N limited. They will not necessarily be a perfect sink for incoming N as microbial uptake can become limited by C availability. At the other extreme, soils with a *high* percentage of organic matter will also tend to immobilise N due to the large amount of C available. N retention may be limited by hydrological factors such as residence time (e.g. in very thin soils) or waterlogging which will reduce mineralisation and nitrification irrespective of the size of the C and N store. In terms of organic matter *quality*, soils with high %N and low C:N ratio will tend to have lower immobilisation rates and so be more likely to have a source of ammonium in excess of biological demand which can be nitrified and leached.

A number of studies show that these soil properties can exert a significant control on N leaching. In the Integrated Forest Study (IFS), nitrate leaching ranged from undetectable to over 20 kg N ha^{-1} yr^{-1}, but all the sites which leached nitrate had large amounts of N accumulated in the mineral soil and forest floor (Van Miegroet *et al.*, 1992). The two soil properties viewed as the best index of nitrate leaching at these sites were a high total N content and a C/N ratio <20. They were also associated with high mineralisation rates – one of the two factors which best explained the variation in nitrate-N leaching across the sites. The % total N in the upper 10 cm of mineral soil was found to explain most of the variation (46%) in nitrate export form the forested watersheds on the Chesapeake Bay Region, with the ratio of net nitrification: net mineralisation rate coming a close second (explaining 42% of the variation) (Williard, 1996). High rates of N mineralisation and nitrification were associated with low C/N ratios (<16). Similar results have emerged from European studies within coniferous stands. A negative correlation between nitrate leaching and the C/N ratio of the forest floor has been observed for a range of coniferous forest sites in Europe (Matzner and Grosholz 1997; Gundersen *et al.*, 1998; Emmett *et al.*, 1995c) (Figure 5.3). Considering the 17 Norway spruce sites alone suggested a threshold C/N ratio of about 30 for the onset of nitrate leaching (Figure 5.3) (Matzner and Grosholz 1997), but this threshold becomes less clear when different coniferous species are also included (Figure 5.3).

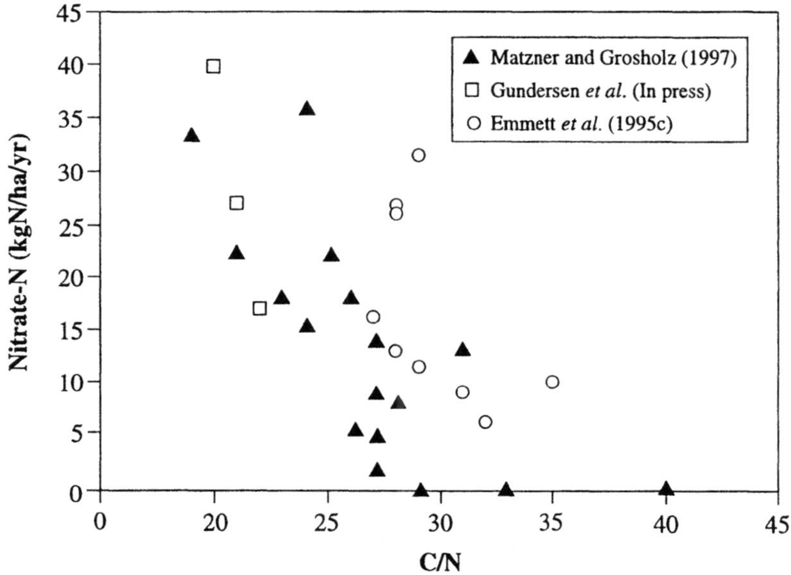

Figure 5.3: The relationship between C/N in Oh horizons and nitrate leaching in 20 European Norway spruce sites (redrawn from Matzner and Grosholz, 1997). Three sites from NITREX have been added: Sitka spruce (Aber, UK), Douglas fir (Speuld, NL) and Scots pine (Ysselsteyn, NL) (Gundersen *et al.*, in press), and nine mature Sitka spruce sites in Wales (Emmett *et al.*, 1995c).

Evidence for a threshold relationship between nitrification rates and soil %N is more convincing. Figure 5.4 is a plot of N concentration in the forest floor horizon and nitrification rates in a combination of N addition experiments and deposition gradient studies in the US and Europe (McNulty *et al.*, 1991 and 1996; Warmerdam 1993; Emmett *et al.*, 1995c). Nitrification increases sharply once the %N falls to around 1.2–1.5. Similar conclusions have been drawn from other studies (e.g. Kriebitzsch, 1978), although as we see from Figure 5.4, more data at low %N are needed to improve confidence in this threshold. Although mineralisation and nitrification appear to control nitrate leaching, recent evidence suggests that nitrate assimilation by saprophytic organisms could also be an important N retention mechanism in forests, reducing leaching even at high nitrification rates (Stark and Hart, 1997). These results have not been confirmed by measurements at the NITREX sites, a series of N addition and removal experiments in Europe (Wright and Van Breemen, 1995): nitrate immobilisation was negligible irrespective of the presence or absence of nitrate leaching (Tietema, 1998).

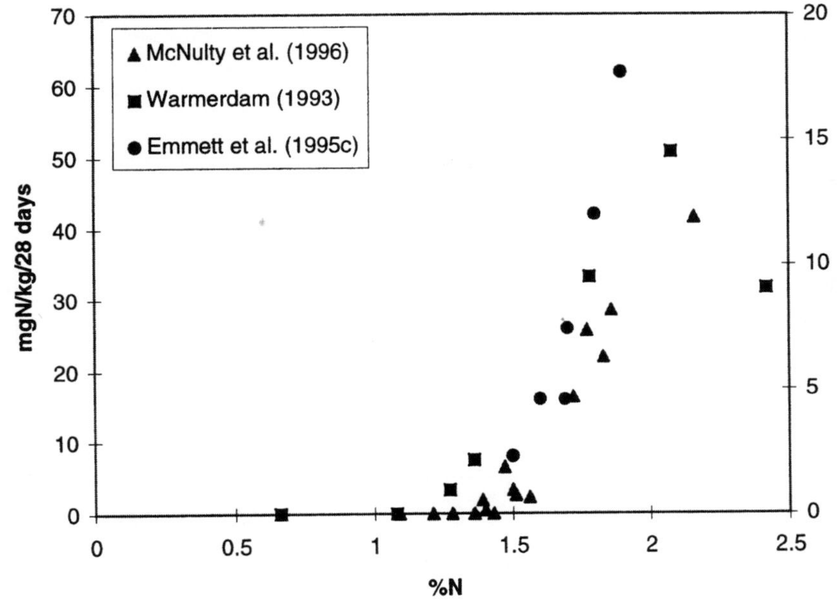

Figure 5.4: The relationship between %N in forest floor and soil nitrification rates. Data is from a nitrogen deposition gradient study in New England and ten experimental spruce–fir plots on Mt Ascutney (redrawn from McNulty *et al.,* 1996), NITREX sites across a nitrogen deposition gradient in Europe (nitrification rate expressed on a per kg basis) (Warmerdam, 1993) and six spruce stands in Wales (expressed on a per m^2 basis) (Emmett *et al.,* 1995c). See original papers for differences in methodology.

Although more data are needed to provide a robust case, there is reasonable evidence that net mineralisation and nitrification rates (and hence N leaching) are strongly influenced by the %N of the soil and/or its C:N ratio. The key question then, is whether N deposition at around the proposed critical load in recent years could have had a significant impact on the %N and C:N ratio of forest soils, acknowledging that the N store in the soil is large compared with annual atmospheric inputs. Data from spruce–fir sites in New England, USA (McNulty *et al.,* 1991) and an analysis of data from the NITREX and EXMAN projects (Tietema and Beier, 1995) suggest that current N deposition may indeed have altered the nitrogen characteristics of the soil (Figure 5.5). In the US sites a good correlation was found between N deposition and both C:N ratio and %N in the forest floor (McNulty *et al.,* 1991) while total N in precipitation was correlated with N % in the ecto-organic layer in the NITREX and EXMAN sites (Tietema and Beier, 1995).

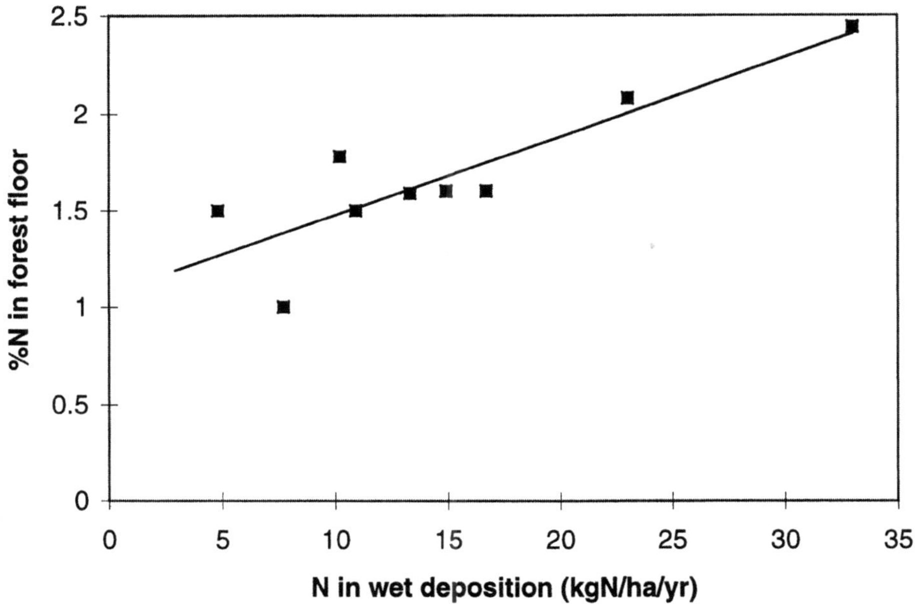

Figure 5.5: The relationship between nitrogen in wet deposition and %N in the forest floor across a European deposition gradient (data from Tietema and Beier, 1995).

5.3.2 CLIMATE

Both long-term climatic changes (e.g. those associated with global warming) and localised extreme events can alter N dynamics in forests. Gradients in climatic factors such as temperature and rainfall may potentially explain much of the geographic variability in nitrate leaching through the effect they have on biological activity. In practice, it is often difficult to evaluate the role of these factors in N leaching as they are frequently correlated with pollution gradients (Dise and Wright, 1995).

An increase in nitrate loss has been observed in response to freezing events in forested watersheds (mixed deciduous and coniferous) in north-eastern US (Mitchell *et al.,* 1996). These catchments, with low N inputs in wet deposition (4–8 kg N ha^{-1} yr^{-1}), had shown no significant increase in nitrate leaching over the last 10 years. However, in 1989–90, high nitrate concentrations in drainage water were recorded during the dormant period following an extreme cold event when soils probably froze. The freezing–thawing cycle is thought to have enhanced microbial activity, resulting in a nitrate pulse. Nitrate events and acidification pulses are also associated with periods of rainfall following periods of drought (Ulrich, 1983). Stream and soilwater nitrate peaks have been related to drought in both moorland and forests in Wales (Reynolds *et al.,* 1992 and Reynolds *Pers. Comm*).

This effect could not be reproduced experimentally however, when droughts were induced in mature forests at EXMAN sites (Lamersdorf *et al.*, 1998). Nitrate pulses have also been associated with snowmelt waters, although it is not always clear whether the nitrate originates from the snowpack itself (e.g. Brown, 1988) or is flushed from the soil beneath the snowpack as it melts (Williams *et al.*, 1996).

Even small differences in air temperature have been shown to have marked effects on nitrate leaching from forested catchments in Finland and Sweden (Lepistö *et al.*, 1995). A positive correlation was obtained between mean annual air temperature and nitrate-N export, and it was the single factor explaining most of the variation in N leaching. As the authors pointed out, climatic factors are often intercorrelated with other factors, complicating the interpretation of results. In this study, higher temperatures were also correlated with more productive stands, higher N deposition (larger canopy) and higher turnover of N. Rainfall may also be an important control on nitrate leaching through its influence on denitrification (by creating a water-logged, anaerobic environment), especially in forests at northern latitudes with highly organic/peat soils (Lepistö *et al.*, 1995; Black *et al.*, 1993).

One of the greatest areas of uncertainty is how the climatic changes expected as a result of global warming will interact with N saturation in forests. Increases in greenhouse gases are predicted to have long term effects on both soil temperature and moisture. The largest temperature changes are expected at high latitudes (Houghton *et al.*, 1990), where quite small increases in temperature are likely to stimulate soil microbial activity, increasing the turnover rate of both C and N (Nadelhoffer *et al.*, 1992). Depending on the degree of saturation of vegetation and microbial sinks, nitrate leaching may result. Theoretically, the extra N available could be retained as a result of higher tree growth rates since increases in CO_2 concentration and temperature are predicted to stimulate productivity. Even if this was the case, nitrate leaching could still result as a consequence of higher rates of N uptake, since this will lower C/N ratios in foliage and litter and faster and enhance N cycling. Changes in soil moisture will also affect the size of the soil and vegetation sinks for N (Billings *et al.*, 1983; Johnson *et al.*, 1996). Clearly, any long-term changes in climate will have complex implications for N cycling and it is impossible to predict the net effect on N leaching from the information we currently have available. This is a new area of research and experiments have only got under way in recent years. There is some evidence however, to suggest that changes in climatic factors will affect N availability and nitrate leaching. As part of the CLIMEX experiments, the soil in a forested catchment in southern Norway has been warmed to 3–5 °C above ambient. The result was a significant increase in nitrate concentration in runoff (Lükewille and Wright, 1997). Figure 5.6 shows the net retention of nitrate in the untreated reference catchment (ROLF) and the catchment with soil warming (EGIL); by 1995 we see that retention is significantly less in the warmed catchment. The catchment was not N-limited and presumably unable to utilise all the additional N produced by the increase in microbial activity. This response is consistent with results from other soil warming studies which have generally shown an increase in N availability (van Cleve *et al.*, 1990; Mitchell *et al.*, 1994, Rustad *et al.*, 1995), with one exception (Peterjohn *et al.*, 1993). These experiments obviously produce a rapid increase in soil temperature compared with global warming where changes will occur gradually over some years, allowing some selection and adaptation in the soil flora and fauna.

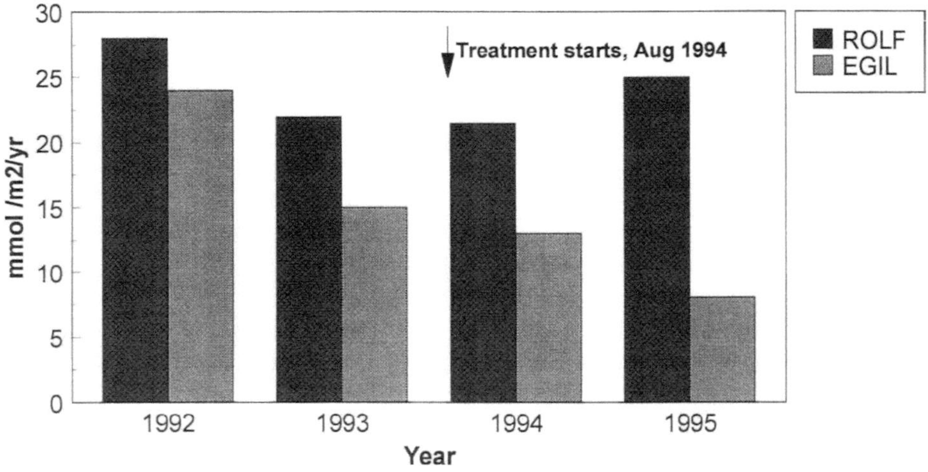

Figure 5.6: Net nitrate retention in the ROLF catchment (untreated reference) and EGIL catchment (soil warming) of the CLIMEX experiment during frost free periods, 1992–1995. (Redrawn from Lükewille and Wright, 1997)

5.3.3 MANAGEMENT PRACTICES AND LAND USE

Historical management practices such as litter raking, burning and agricultural cropping are likely to influence N leaching through their effect on the C and N content and C/N ratio of the upper soil horizons. Litter raking and burning tend to reduce the organic C and N pool, resulting in a greater sink for active accumulation of incoming N. Where forests have been ploughed prior to planting or previously supported agricultural crops, the C/N ratio is likely to have been reduced and the rate of N cycling increased. Thus the management history of a site, going back decades and even centuries, may affect the N retention capacity of an existing forest. Current-day management practices such as species choice, planting density, fertilization and harvesting methods can also influence the way in which a forest responds to N.

Ploughing is thought to affect N dynamics in forest soils, even many years after the land was in agricultural use. Land use history has been invoked to explain high rates of nitrate leaching in forests with low N deposition in north-eastern and south-eastern US. In nutrient cycling studies in Harvard Forest, MA, ploughing and pasturing up to 1926 was viewed as a key factor explaining the higher mineralisation rates and nitrate leaching in red pine plots compared with mixed hardwood plots (Magill *et al.*, 1997). In the UK NITREX site, ploughing of a relatively nitrogen-rich clay horizon prior to afforestation is thought to have increased N cycling and nitrate leaching above that predicted from N deposition (Emmett *et al.*, 1998a).

There have been few controlled studies to investigate the effect of burning on N leaching, but work in the Coweeta Watershed in the southern Appalachians suggests that results may not always be what we would predict. Two burning prescriptions (cut and burn and stand replacement burning) are employed in the area to maintain mixed pine–hardwood forests. The treatments had minimal effect on nitrate leaching, despite as much of 60% of the above-ground pool of N being lost on some sites (Swank and Vose, In press).

The type of trees previously grown on a site may affect N cycling in existing forests. Differences in N leaching from two Norway spruce stands in the Black Forest have largely been explained by previous land use (Feger, 1992). The two experimental watersheds, Villingen and Schluchsee, have similar inputs of atmospheric N but show very different leaching responses. The Schluchsee stand has a much higher mineralisation rate and is essentially N-saturated, leaching N although inputs in bulk precipitation are only 10.7 kg N ha^{-1} yr^{-1}. The higher rate of N cycling at this site is thought to be a relic of its former use as a beech stand. In some cases, a history of deciduous planting has been shown to effect N leaching in coniferous forests even in the 3rd rotation (Kreutzer, 1989). Other management practices are also considered to be important in explaining the differences in N leaching at the Black Forest sites. Litter raking and forest grazing at Villingen has reduced the fertility of the site, resulting in a much higher C:N ratio than at Schluchsee (48 and 28, respectively) so N is still being incorporated into the organic pool. Input–output budgets for the EXMAN and NITREX manipulation experiments also suggest that nitrate leaching from coniferous forests may be much greater where a deciduous forest has been grown previously. A reasonable correlation between N inputs in bulk precipitation and N in drainage waters was obtained for these sites with the exception of a Norway spruce stand at Högewald, Germany (Tietema and Beier, 1995). This forest showed uncharacteristically high N leaching at fairly moderate inputs. The site had previously supported a beech forest and the rate of N cycling in the accumulated deciduous litter was assumed to be much higher compared with coniferous litter. However, the role played by an earlier deciduous forest in present-day leaching patterns is not always clear; several of the other stands in NITREX-EXMAN had also been dominated by deciduous trees in the past but did not show unexpectedly high leaching.

Utilisation of N by trees plays an important role in preventing or reducing N leaching. Uptake of N by vegetation should increase with higher rates of N deposition until growth becomes limited by some other requirement such as water, light or mineral nutrients. By providing more of the limiting factor, we should theoretically be able to increase the tree sink for N. Nutrient availability is known to limit tree growth even under low rates of N deposition, and PK fertilizer is often applied in the early growth stages of commercial forests. It has been demonstrated experimentally that fertilization can significantly affect N status and nitrate leaching by improving the tree's nutritional status. When P and K was applied to a mature Sitka spruce stand in Wales at 120 and 200 kg ha^{-1}, respectively, leaching losses of nitrate were significantly reduced compared with the control (Figure 5.7) (Stevens *et al.,* 1993). Foliar analysis indicated that fertilization had significantly increased N uptake by the trees. In general, we can expect forests growing on nutrient-poor soils to be more likely to leach N at a given deposition than forests growing on soils with an optimum nutrient supply. However, where high N deposition (particularly as ammonium-N) leads to soil acidification, forests may enter a positive feedback loop, with acidification and leaching losses of base cations inducing nutrient deficiency which then reduces the capacity of the vegetation for N uptake and increases nitrate leaching and acidification (e.g. Mohren *et al.*, 1986).

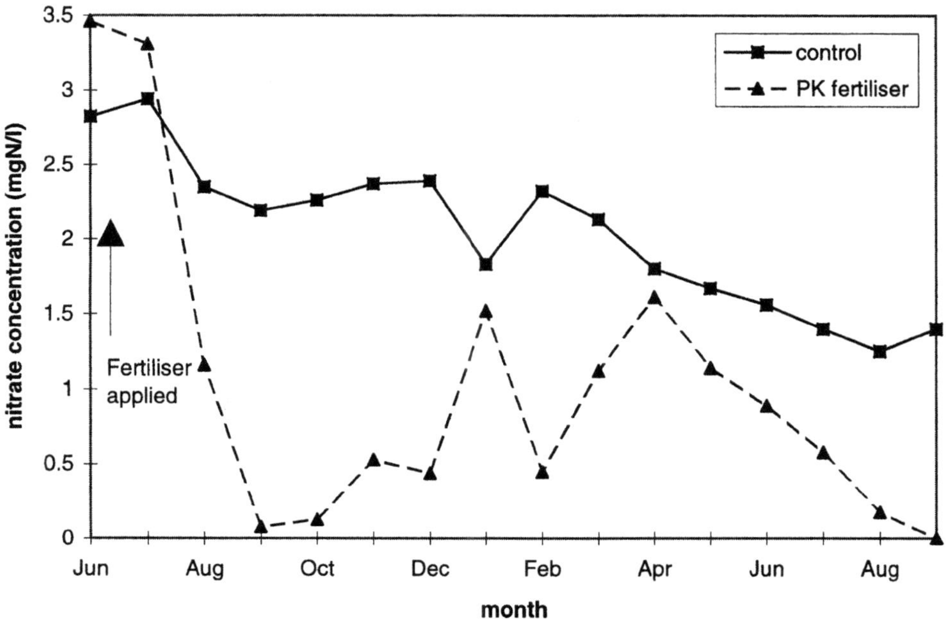

Figure 5.7: The effect of PK fertilizer on soil water nitrate concentrations at 50–70cm depth in a mature Sitka spruce stand, N.Wales. P was applied at 120 kg ha^{-1} and K at 200 kg ha^{-1} in June 1990. (Redrawn from Stevens *et al.*, 1993).

Forest management practices which affect light penetration such as planting density, choice of species, thinning and selective harvesting, have a direct influence on the growth of understorey vegetation. Forest ground vegetation may represent an important additional sink for N, affecting both the magnitude and onset of nitrate leaching. Relative to trees, ground vegetation may be more efficient in exploiting different forms of N (nitrate-N, ammonium-N and organic N) and different temporal and spatial niches. Ground flora may be able to use available N during the tree's dormant periods (e.g. early spring), reducing seasonal peaks in nitrate leaching. Ground vegetation may be able to utilise N at lower nutrient levels than trees, acting as a sink for N when tree uptake has become limited. While ground flora cannot be treated as a permanent sink (since it will die, decay and return N to the soil), the rate of recycling may be slow enough for it to significantly reduce N leaching during a forest rotation. Data quantifying the ground flora sink for N and its impact on N leaching in established stands are scarce. However, there is evidence from a long-term study in a Ponderosa pine forest in Oregon that the understorey (which in this case included N-fixers) can increase the soil store of C and N (Busse *et al.*, 1996). In plots where ground vegetation was manually removed, nitrogen and carbon pools were significantly lower to soil depths of 4 cm and 12 cm, respectively, compared with plots where the ground flora was allowed to establish (Figure 5.8). This may have important implications for the timing of N saturation due to the effect on the mineralisation–immobilisation balance in the soil profile.

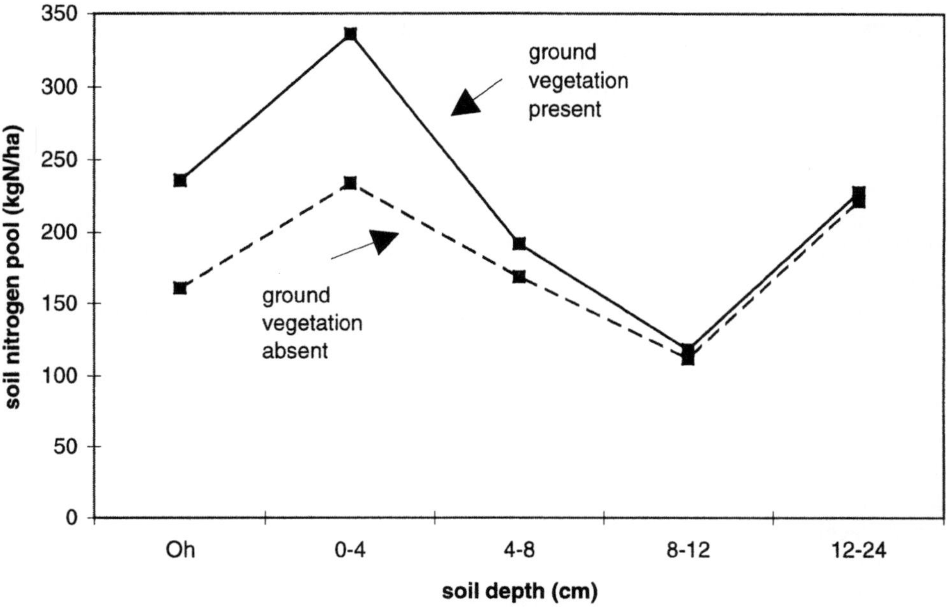

Figure 5.8: Influence of the presence of understorey which included N-fixers under old-growth Ponderosa pine on soil nitrogen pools, following 35 years of manipulation. Understorey was manually controlled in plots with five different tree spacings between 1959 and 1994. Data taken from Busse *et al.* (1996)

The choice of harvesting method can influence forest N dynamics in a number of ways. The amount of biomass removed from the ecosystem and the extent of soil disturbance will determine the size of the C and N store, C/N ratio and rate of N cycling. Whether brash and debris is removed or left on the forest floor is a key factor affecting re-growth of ground vegetation and may be particularly important in controlling nitrate pulses following felling. A study which compared whole tree harvesting with bole-only removal showed that where brash was left on the forest floor the ground flora did not re-establish. Nitrate-N leaching was significantly higher in these plots compared with whole tree harvested plots where ground vegetation was re-establishing (Stevens and Hornung, 1995). In a lysimeter study at the same site, the re-establishing ground vegetation was estimated to have reduced nitrate leaching losses following felling by 27 kg N ha^{-1} over a 15 month period (relative to controls) while the presence of surface brash increased nitrate leaching losses by 30 kg N ha^{-1} relative to controls (Emmett *et al.*, 1991).

5.3.4 STAND AGE

As discussed above, factors which affect tree growth such as nutrient availability (Section 5.3.3) or CO_2 concentration (Section 5.3.2) will influence vegetation demand for N and hence the amount of N available for leaching at a given deposition. It has been hypothesised for some time that the tree uptake sink for N will also vary according to the age of the forest (Vitousek and Reiners, 1975), although this factor has received little attention in recent studies of N saturation. N accumulation increases rapidly during the early stages of rotation, slows down as the plantation approaches canopy closure then declines once the trees have reached maturity (Miller and Miller, 1988). For a given rate of N deposition (and assuming the size of other sinks for N are comparable) we would expect older forests to leach more N.

This has been tested in a UK study where N inputs (in throughfall) and outputs were surveyed for a number of moorland and forested catchments on similar soils and geology in Wales with plantations ranging in age from 10–55 years (Emmett *et al.,* 1993, Stevens *et al.,* 1994) (Figure 5.9). Below 30 years of age, inorganic-N outputs were less than 5 kg N ha^{-1} yr^{-1} in stands. Above this age threshold, N leaching could be as high as 30 kg N ha^{-1} yr^{-1}. The magnitude of leaching losses in these mature stands were found to be related to the %N of the organic horizon and soil drainage characteristics at the individual sites (Emmett *et al.,* 1995c). Higher nitrate losses from mature stands were confirmed in a more extensive survey of 136 upland catchments in Wales, on 3 different soil types. Reynolds *et al.* (1994) found a significant relationship between nitrate leaching and area cover of trees greater than 30 years of age, although all the plantations were Sitka spruce and managed in the same way. However, no evidence for either a threshold or positive relationship between stand age and nitrate leaching was identified by Dise and Wright (1995) or Gundersen (1995) in their analysis of the ENSF and ECOFEE databases, respectively. These studies covered a range of forest 'types' (in terms of species, age and soil), but did not include stands within the age range in which N accumulation rates change so dramatically (0–30 years of age) (Miller *et al.,* 1993; Binkley and Johnson, 1992). It is difficult to assess how important stand age is in controlling N leaching across a range of forest types, but it clearly needs to be considered at the 'country level', where species and management practices are often common.

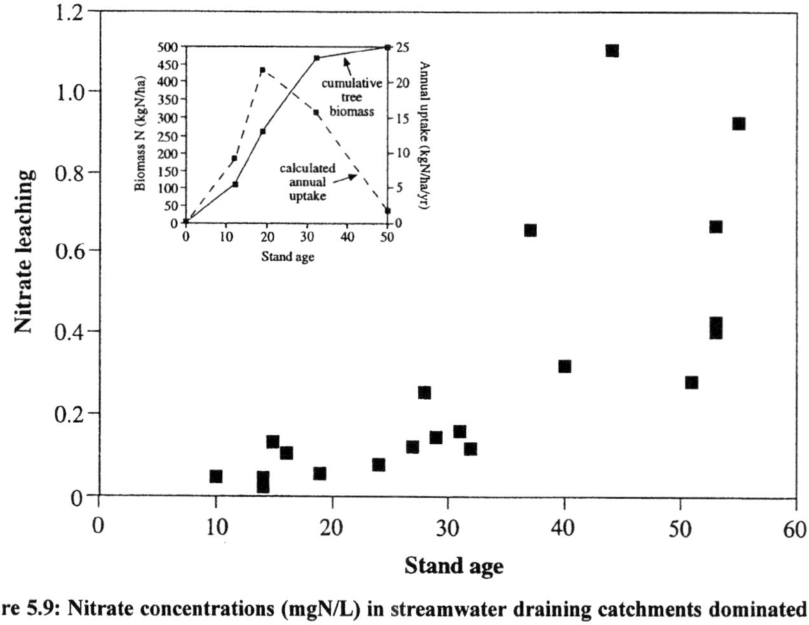

Figure 5.9: Nitrate concentrations (mgN/L) in streamwater draining catchments dominated by single-aged Sitka spruce stands in Wales (redrawn from Emmett *et al.*, 1993). Insert is redrawn from Miller *et al.* (1993) and illustrates changes in tree biomass N recorded across an age sequence of Sitka spruce stands in Scotland. Calculated annual N uptake rates are also shown to indicate the significant change in net uptake rates after 20–25 years of age.

The age of a forest can affect parts of the N cycle other than tree uptake. In older forests, the gaps created by fallen trees result in localised soil warming and wetting as light and water penetrate the disturbed canopy, enhancing mineralisation and nitrification. Fir mortality and gap creation is thought to be one of the key factors responsible for higher rates of nitrate leaching at high-elevation sites in the northern Appalachians compared with the southern-Appalachians, where both forests have similar N inputs and total N content (Van Miegroet *et al.*, 1992). New gaps could potentially have the opposite effect: N retention may be enhanced due to the additional supply of C from woody debris and stimulated growth of ground vegetation (see Section 5.3.3).

5.3.5 TREE SPECIES

The size of the tree uptake sink for N can potentially be very different depending on the species grown as well as the stand age. Uptake rates are generally higher in deciduous than in coniferous forests because of the N required to replace the entire foliage each year. In hardwoods, uptake rates are typically 40–50 kg N ha^{-1} yr^{-1} compared with 20–30 kg N ha^{-1} yr^{-1} for conifers (Van Miegroet *et al.*, 1992). Higher uptake rates of nitrate-N by deciduous species have been shown experimentally at Bear Brook, Maine (USA). N was added to the mixed hardwood forest at two treatment levels (28 and 56 kg N ha^{-1} yr^{-1}) as a spray of ^{15}N labelled nitric acid (Nadelhoffer *et al.*, 1995). This enabled the fate of the N to be traced within the various compartments of the ecosystem. Microbial and plant sinks for the nitrate-N were shown to be approximately equal, but uptake differed significantly between the tree species (Figure 5.10).

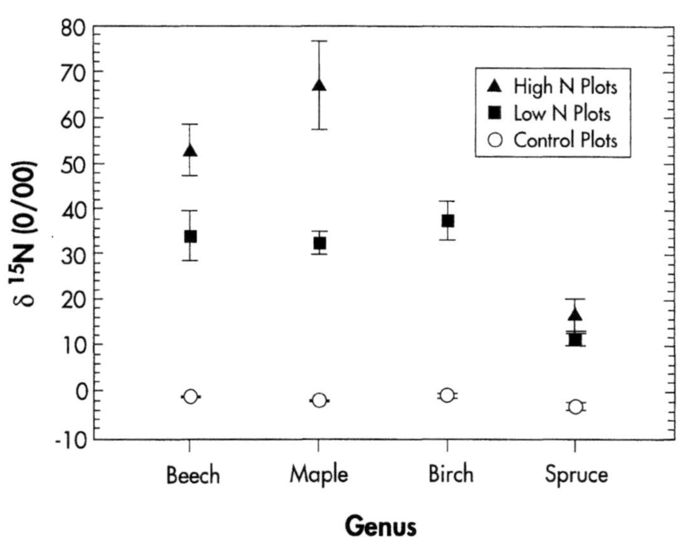

Figure 5.10: ^{15}N contents of bolewood of different species in Bear Brook Watershed (Maine), following N addition. Symbols are means and vertical lines are SEs (n=3 plots). (From Nadelhoffer *et al.*, 1995).

N assimilation into bolewood was far lower in red spruce than in the hardwood species throughout the treatment year, and the same pattern was observed in the foliage by the final year of treatment. Similar results were obtained from an N addition experiment with red pine and mixed hardwood stands in Harvard Forest, MA. The deciduous forests responded to N inputs with slightly higher rates of wood accumulation and N retention, and lower rates of leaching compared with the pine stands (Magill *et al.,* 1997). The plots were not replicated in this study, so these trends cannot be verified statistically.

While higher rates of N uptake in deciduous forests may reduce nitrate leaching compared with coniferous forests, the long-term story is more complex. Deciduous systems have been studied less intensively than coniferous ones, but it appears that they may cycle N more rapidly as a result of higher rates of N uptake. This means that %N in foliage and litter quality (C/N ratio) in deciduous forests should respond more quickly to increasing N deposition. There is some experimental evidence that this is the case. When N was added to a high elevation forest on Mount Ascutney, Vermont (USA), a correlation between N applied and N concentration in litter was obtained for deciduous species (*Betula* and *Acer*) but not for coniferous ones (McNulty and Aber, 1993). This is consistent with an analysis of European coniferous and deciduous forests, which showed that deciduous forests are more likely to show increased return of N in litterfall with increasing N deposition than coniferous forests (Figure 5.11) (Gundersen, 1995). Whether deciduous species are more vulnerable to N leaching in the long term than coniferous ones will depend on the delicate balance between lower scavenging rates of pollution combined with higher uptake rates, and faster decomposition, mineralisation and nitrification rates.

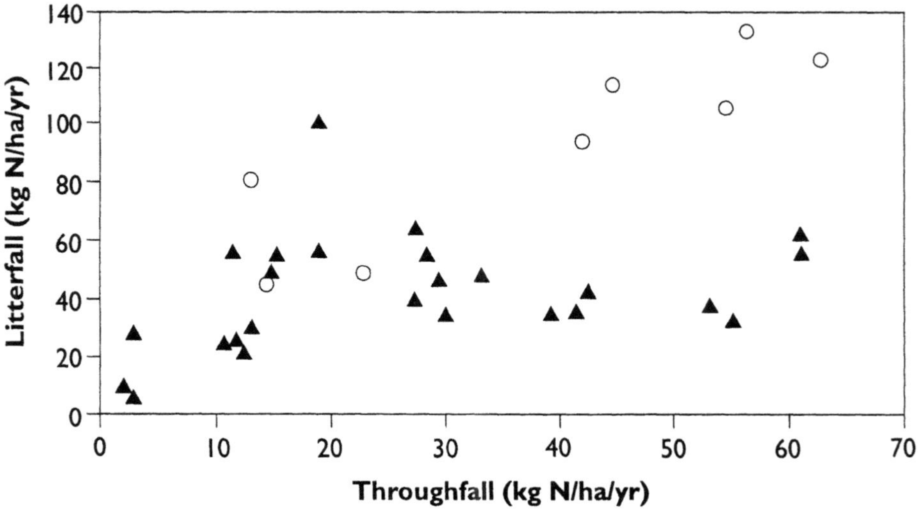

Figure 5.11: The effect of N inputs in throughfall on N flux in litterfall in deciduous (○) and coniferous (▲) forest plots in Europe (ECOFEE database). (From Gundersen, 1995).

5.3.6 NITROGEN FORM

The 1987 conference on excess nitrogen addressed the need to consider the impacts of ammonium and nitrate separately, particularly from the point of view of soil and water acidification (Skeffington and Wilson, 1988). The ability of the two forms to induce N-saturation is also likely to be very different, depending on the retention capacity of both soil and vegetation. In general, the soil sink will be greater for ammonium than nitrate. The more mobile nitrate ion tends to be leached, while positively charged ammonium inputs can be retained on negative exchange sites in the soil. Soil microbes will favour ammonium-N as a source of N since assimilation is energetically more efficient compared with nitrate-N. Nitrate-N assimilation rates are rarely measured but a recent study suggests that it may in fact represent a substantial sink in mature forests (Stark and Hart, 1997). N uptake by vegetation has also been shown to depend on the form of N, but whether ammonium or nitrate is preferentially utilised appears to be species-specific (Kronzucker *et al.*, 1997; Bigg and Daniel, 1978) and in some cases dependent on water availability (Gijsman, 1990).

In theory, forests with a similar N status could show different leaching responses at a given input depending on the ratio of nitrate:ammonium in deposition. In general, it would appear that coniferous forests are more likely to leach N when the nitrate fraction is higher, since forest floor microbial populations are a more effective sink for ammonium-N than nitrate-N. There have been very few studies designed to look specifically at forest response to different forms of N. In a pot experiment with Norway spruce seedlings where N was applied as $NH_4(SO_4)_2$ or $NaNO_3/HNO_3$, the plant–soil system was able to utilise almost all of the applied ammonium, up to inputs of over 120 kg N ha^{-1} yr^{-1}, while about 40 % of added nitrate was leached even in lower treatments (30 and 60 kg N ha/yr).

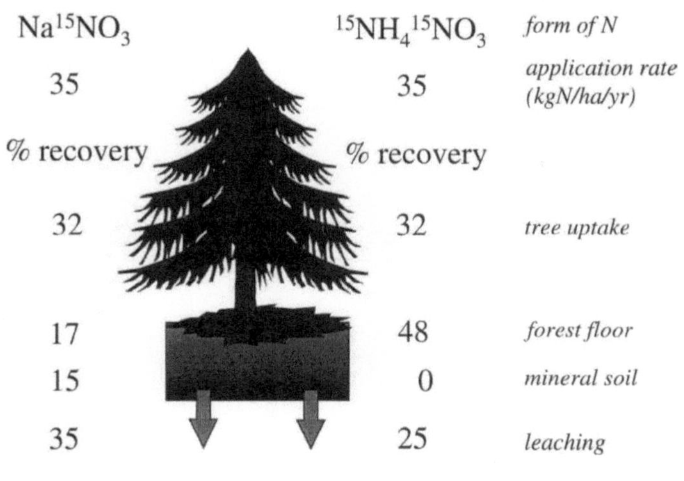

Figure 5.12: The fate of ^{15}N-labelled additions in a 32 yr old Sitka spruce stand. Nitrogen was applied weekly to the forest floor of replicated plots for one year and the recovery of the labelled-N into the different soil, water and tree pools recorded. Total recovery was 99% (s.e. ± 5%) in the sodium nitrate treatments and 105% (s.e. ± 12%) in the ammonium nitrate treatment.

The soil was seen as the key sink for ammonium since N uptake by the trees was not significantly affected by either the form of N or dose applied (Wilson and Skeffington, 1994). Similar results have subsequently been obtained with mature trees. When ammonium and nitrate were applied (as 35 kg N ha^{-1} yr^{-1}) as either NH_4NO_3 or $NaNO_3$ to a Sitka spruce forest in Wales, nitrate inputs were generally leached while most of the applied ammonium was retained (Emmett *et al.*, 1995a). The trees showed no preference for uptake of ammonium or nitrate; the difference in N retention between the two treatments lay in the forest floor sink. Soil microbes retained 17 kg N ha^{-1} yr^{-1} or 48% of inputs in the ammonium treatment compared with 6 kg N ha^{-1} yr^{-1} or 17% of inputs in the nitrate treatment, leaching the excess (Figure 5.12) (Tietema *et al.*, 1998). This demonstrates that forests such as Aber may be specifically 'nitrate saturated' rather than simply N saturated.

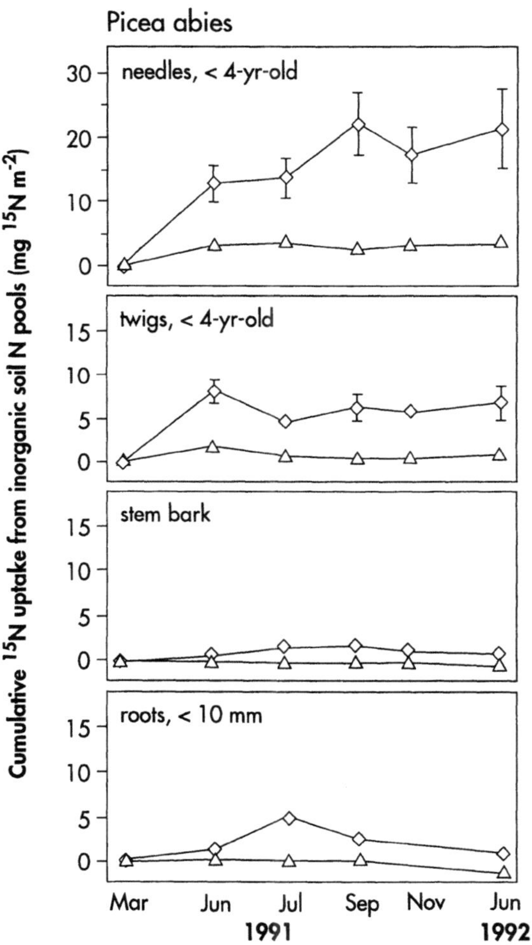

Figure 5.13: Cumulative ^{15}N uptake from inorganic soil N pools in different tissues following field-scale applications of $K^{15}NO_3$ (△) and $^{15}NH_4Cl$ (◇) to Norway spruce stands in the Fichtelgebirge, Germany. (From Buchmann *et al.*, 1995).

Although the studies above show that the capacity of vegetation to retain incoming N is independent of its chemical form, other studies have demonstrated preferential utilisation of either ammonium or nitrate. When ^{15}N labelled ammonium and nitrate (as NH_4Cl and KNO_3 respectively) was added to a 15 yr old Norway spruce plantation in the Fichtelgebirge, Bavaria, the trees took up about 2–4 times as much ammonium as nitrate (Figure 5.13) (Buchmann *et. al.*, 1995). In experiments using an ^{13}N radiotracer, Kronzucker *et al.* (1997) showed that uptake of ammonium-N by white spruce seedlings was 20 times greater than uptake of nitrate-N. Other studies also support the hypothesis that coniferous species are more likely to use ammonium-N than nitrate-N (Lavoie *et al.*, 1992; Marschner *et al.*, 1991; Van den Driessche and Dangerfield, 1975), although results of solution culture experiments may not be applicable to the field.

Nitrogen form can also indirectly affect the magnitude of the tree and soil sinks through effects on soil acidification and thus the availability of other nutrients. Both tree uptake and nitrification of ammonium result in the production of H^+. In contrast, nitrate assimilation produces OH^- and thus net alkalinisation of the soil (Van Breemen *et al.*, 1983; Skeffington and Wilson 1988). As the soil acidifies, base cations may be depleted by leaching and phosphorus becomes less available due to precipitation with aluminium or iron (e.g. Mohren *et al.*, 1986). Even in non-nitrifying soils, ammonium may still have adverse effects on nutrient uptake. High NH_4^+/BC ratios in the soil solution have been shown to depress assimilation of other nutrients due to competition at root surfaces (Van Breemen and Van Dijk, 1988). While there is more evidence to suggest that ammonium-N could indirectly affect tree nutrition, some data show that nitrate can have a more detrimental impact on the formation and persistence of mycorrhiza relative to ammonium, although the mechanism is unclear (Holopainen and Heinonen–Tanski, 1993). A reduction in mycorrhizal infection may have important implications for uptake of N by trees due to the many other benefits that mychorrizae have on tree growth and health. These possible mechanisms are little more than hypotheses since there is no experimental evidence that one form of N has more adverse effects on tree health than the other at realistic inputs. When N was applied to a mixed spruce–fir forest in Mt Ascutney, Vermont as either $NaNO_3$ or NH_4Cl (at rates of 15.7 to 31.4 kg N ha^{-1} yr^{-1}), no effects of N form on N saturation or forest health were observed over the 6 year period (McNulty *et al.*, 1996). Similarly, no differential effects of N form on tree growth or health were reported when N was applied to Sitka spruce stand in Wales as $NaNO_3$ or NH_4NO_3 (see above) (Emmett *et al.*, 1995b and 1998b). Further work on mature stands at the field-scale is clearly required in order to assess the relative impacts of the different N forms.

5.4. Discussion

Research during the last decade has certainly confirmed the highly complex nature of forest N dynamics. In this chapter we have shown how a whole range of factors including soil type, species and age structure, current and past management and climate can influence the way a forest responds to N deposition. While recent studies have answered many of the questions posed 10 years ago, they have inevitably raised new ones concerning the exact nature of these interactions. The aim of this section is to outline some of these questions and highlight the areas where we need more data.

There is little evidence from deposition gradient studies and field experiments that more moderate rates of N deposition have an adverse effect on tree health and growth. Of course such studies always leave us with the uncertainty of what would have happened if N had been applied for longer or if a site had been monitored for a further decade. Current rates of N deposition may be producing changes which are not easily detected but will impact on tree health some years down the line (e.g. an increase in soil N store or rate of N cycling). What is clear, however, is that N deposition is affecting N retention and loss in many forests. Although it may not be an ideal measure of 'damage', nitrate leaching in excess of 'background' levels serves as a pragmatic indicator of ecosystem imbalance.

There is a good correlation between N deposition and N leaching when sites with higher rates of N deposition are included in a dataset, but only a relatively weak one at inputs around the critical load. The need to understand what controls nitrate leaching in forests at this deposition is two fold: firstly to provide the knowledge and data for models so that we can predict the magnitude and onset of leaching beyond the timescales of research projects, and secondly, so that we can implement some simple management measures to reduce leaching and protect both down-stream waters and forest health.

Current evidence points towards soil properties as the common link controlling N dynamics in a range of forest types across the world. The key parameters appear to be the C and N content and C/N ratio of the soil due to their close relationship with microbial transformations. Evidence suggests that soils will leach nitrate when the N% is above about 1.2–1.5 and/or the C/N ratio is less than ca. 30, but this raises a number of questions. In view of the other influences on microbial N transformations, can we expect a simple predictor for the onset of saturation? Is there really a threshold C/N ratio and N% at which soils begin to nitrify and leach, or is this just an artifact of current data points? Will further studies 'fill in the gaps' and reveal a linear response? If there are thresholds, can we predict how long it will take for a soil to reach them at a given N deposition? Should we be focusing on soil properties in the organic or mineral horizons? Data from European experiments identified the C/N ratio or N% of the organic horizons as the key determinant of nitrate leaching (Tietema and Beier, 1995; Gundersen *et al.*, 1998; Matzner and Grosholz, 1997), while in some US forests N% in the mineral horizon was more important (Johnson *et al.*, 1992; Williard, 1996). How relevant is the *availability* of the carbon source to microorganisms relative to the absolute amount?

The rates of microbial transformations such as immobilisation, mineralisation and nitrification largely determine the size of the soil sink for N and play a pivotal role in net leaching. Hence climatic factors which influence microbial activity, could potentially explain a large amount of the temporal and spatial variability in leaching. While a lot is known about the effects of temperature and moisture on microbial transformations, it needs to be applied to the complexity of forest ecosystems to improve our understanding of both year-on-year and geographic variation in N leaching. Predicting the long-term impacts of global warming on N saturation is even more challenging. The complex interaction between the effects of changing CO_2 concentration, temperature and rainfall on both soil microbes and tree productivity need to be addressed. This is clearly an area where more data from ecosystem-scale experiments are required to improve the reliability of existing models.

Superimposed on the common factor of soil properties are the more site-specific influences on N leaching. Current forest management and past land use are often invoked to explain uncharacteristically high or low rates of N leaching, although rarely substantiated experimentally. If more was known in this area, forest management could be optimised to reduce N leaching or prevent its onset and minimise other potential impacts (eg. changes in ground flora, nutritional imbalances etc). Based on the limited data available, we suggest a number of management options in Table 5.2. There are clearly question marks over some of these. Will burning result in a flush of nitrate due to disturbance and increased temperature, or will the reduction in the C–N pool stimulate long-term N retention? The long-term effects of planting deciduous species are also uncertain. Higher uptake rates may reduce N leaching compared with coniferous stands initially, but what will be the impact of higher rates of N cycling and better quality litter on future rotations? Even less is known about how individual species could be selected to minimise leaching or reduce the risk of nutritional imbalance, although recent manipulation studies show that their response to N inputs can be very different.

It is evident from recent studies that forests are likely to be very heterogeneous in their response to N; to treat this complexity as a generalised 'forest ecosystem' is clearly an oversimplification. This has important implications for setting critical loads and raises the question of whether it is realistic to aim for anything other than a broad range.

Table 5.2: Possible management options to reduce/prevent nitrate leaching losses and other potentially deleterious effects. See Sections 5.3.3–5.3.5 for more details on effects of individual management practices.

Effect	Management option
Nitrate leaching	Plant deciduous species Shorten rotation time Reduce stand density Remove brash after felling Encourage ground flora within mature forests, after felling and in riparian zones. Remove N-rich organic layer Fertilize with limiting nutrients (e.g. P, K) Burning
Tree health and nutrition	Grow N tolerant species Apply limiting nutrient fertilizer
Change in species composition of ground flora	Removal of N rich organic layer
Eutrophication of soils	Removal of N rich organic layer
Acidification of soils and waters	Liming, addition of other base cations

Alternatively, have we enough information to calculate critical loads separately for certain 'forest types' depending on species, soil and climate/geography? It is acknowledged that critical loads should be set independently for ammonium-N and nitrate-N, but we still have insufficient data to do this (Hornung *et al.,* 1995). The evidence reviewed in this chapter indicates that the impact of N deposition can be very different depending on its form, but there is still a dearth of studies where ammonium and nitrate have been applied separately. Forests will rarely receive 100% ammonium-N or nitrate-N; what is required are experiments with the range of ratios and levels experienced across Europe and North America.

The discussion above shows that there are still gaps in our knowledge of how forest ecosystems currently respond to N deposition. Not unexpectedly, trying to predict the way they will behave in the future raises additional questions. Where forests are already leaching N above background levels, will we definitely see effects on tree health and growth in the future? What will be the response time? We still know very little about long-term recovery although there are roof experiments which have shown the effect of reducing inputs from very high levels to very low levels (Boxman *et al.*, 1995). What would be the response to a reduction from more moderate levels (e.g. 30 kg N ha^{-1} yr^{-1}) to just under the proposed critical load? How would this compare with changes to forest management? SO_2 emissions have decreased dramatically in the last decade and will continue to do so as countries meet targets agreed under the 2nd S Protocol (United Nations, 1994). What, if any, effect is this likely to have on N saturation in forests? Some studies indicate an interaction between nitrate and sulphate leaching (Moldan, 1995, Nodvin *et al.,* 1988), which may merit further research.

The focus of most studies over the last decade has been the forest itself, with the relationship between N saturation and effects on downstream waters receiving less attention. A key question is: how long will it take to see an effect on river nitrate levels after a forest starts to leach and what will be its magnitude? How will the nature and timing of the response be modified by the hydrological and physical properties of the catchment and denitrification characteristics of the river? Will the primary impact be acidification or eutrophication of downstream waters?

Many of these questions concerning long-term responses can only be answered by process-driven models. Where possible, such models should address the large number of factors which can influence N leaching if they are to predict forest response to changing N deposition with any accuracy. A number of N models are currently being developed both in the US and Europe (eg. Thornley and Cannell, 1992; Postek *et al.*, 1995; Whitehead *et al.*, 1998; Cosby *et al.*, 1998). It is important that we have the capability to address how strategies covering aspects from land use and management to pollution and climate change, are likely to influence N dynamics in our forests. These models, which integrate the complex combination of sources and sinks for N in a changing environment, have become a key priority for future research.

References

Abrahamsen, G. (1980) Acid Precipitation, plant nutrients and forest growth. In: D. Drablos and A. Tollen (Editors), *Ecological Impact of Acid Precipitation. SNSF Project Norway.* pp 58–63.

Aber, J.D., Nadelhoffer, K.J., Steudler, P. and Melillo, J.M. (1989) Nitrogen saturation in northern forest ecosystems. *Bioscience,* **29** (6), 378–387.

Ågren, G.I. (1983) Model Analysis of some consequences of acid precipitation on forest growth. In: *Ecological Effects of Acid Deposition,* Swedish National Environment Protection Board, Stockholm, Report PM 1636, pp. 233–44.

Albrekston, A., Aronsson, A. and Tamm, C.O. (1977) The effect of fertilization on primary production and nutrient cycling in the forest ecosystem. *Silva Fenn,* **11**, 233–238.

Balsberg Påhlsson, A.M. (1992) Influence of nitrogen fertilization on minerals, carbohydrates, amino acids and phenolic compounds in beech (*Fagus sylvatica* L.) leaves. *Tree Physiology*, **10**, 93–100.

Bigg, W.L and Daniel, T.W. (1978) Effects of nitrate and ammonium and pH on the growth of conifer seedlings and their production of nitrate reductase. *Plant and Soil*, **50,** 371–385.

Billings, W.D., Luken, J.O., Mortensen, D.A. and Petersen, K.M. (1983) Increasing atmospheric carbon dioxide: Possible effects on arctic tundra. *Oecologia,* **58**, 286–289.

Binkley, D. and Johnson, D.W. (1992) Southern pines. In: D.L. Johnson and S.E. Lindberg (Editors). *Atmospheric Deposition and Forest Nutrient Cycling. A Synthesis of the Integrated Forest Study.* Springer, New York, pp 534–543.

Black, K.B., Lowe, J.A.H., Billett, M.F. and Cresser, M.S. (1993) Observations on the changes in nitrate concentrations along streams in seven upland moorland catchments in northeast Scotland. *Water Resources*, **27**, 1195–1199.

Bonneau M. and Nys, C. (1993) A nitrogen cycle model for calculating the reduction of N-input necessary to reduce soil acidification and nitrate leaching and the consequences of this for wood production. *Water Air and Soil Pollution*, **69**,1–20.

Boxman, A.W., Blanck, K., Brandrud, T.E., Emmett, B.A., Gundersen, P., Hogervorst, R., Kjønaas, O.J., Persson, H.A. and Timmermann, V. (1998a) Vegetation and soil biota response to experimentally changed nitrogen inputs in coniferous forest ecosystems of the NITREX project. *Forest Ecology and Management.* **101,** 65–80.

Boxman, A.W., Van der Ven, P.J.M. and Roelofs, J.G.M. (1998b) Ecosystem recovery after a decrease in nitrogen input at Ysselsteyn, the Netherlands. *Forest Ecology and Management.* **101,** 155–164.

Brandrud, T.A. and Timmerman, V. (1998) Ectomycorrhizal fungi in the NITREX site at Gardsjon, Sweden; below and above ground responses to experimentally changed nitrogen inputs 1990–1995. *Forest Ecology and Management.* **101**, 297–314

Bredemeier, M., Blanck, K., Xu, Y.-J., Tietema, A., Boxman, A., Emmett, B.A., Moldan, F., Gundersen, P., Schleppi, P. and Wright, R.F. (1998) Input/Output budgets at the NITREX sites. *Forest Ecology and Management.* **101**, 57–64.

Bringmark, L. and Kvarnas, H. (1995) Leaching of nitrogen from small forest catchments having different deposition and different stores of nitrogen. *Water, Air and Soil Pollution,* **85**, 1167–1172.

Brown, D.J.A (1988) Effect of atmospheric N deposition on surface water chemistry and the implications for fisheries. *Environmental Pollution*, **54**,, 275–285.

Buchman, N., Schulze, E.D. and Gebauer, G. (1995) ^{15}N-ammonium and ^{15}N-nitrate uptake of a 15-year-old *Picea abies* plantation. *Oecologia*, **102**, 361–370.

Busse M.D., Cochran P.H., and Barrett, J.W. (1996) Changes in Ponderosa Pine site productivity following removal of understorey vegetation. *Soil Science Society of America Journal*, **60**, 1614–1621.

Christ, M., Hang, Y., Likens, G.E. and Driscoll, C.T. (1995) Nitrogen retention capacity of a northern hardwood forest soil under ammonium sulphate additions. *Ecological Applications*, **5**, 802–812.

Christie, C.E. and Smol, J.P. (1993) Diatom assemblages as indicators of lake trophic status in southeastern Ontario lakes. *Journal of Phycology*, **29**, 575–586.

Cosby, B.J., Ferrier, R.C., Jenkins, A., Emmett, B.A., Wright, R.F. and Tietema, A. (1998) Modelling the ecosystem effects of nitrogen deposition: Model of Ecosystem Retention and Loss of Inorganic Nitrogen (MERLIN). *Hydrology and Earth System Sciences.* **1**, 137–158.

Dise, N.B. and Wright, R.F. (1995) Nitrogen leaching from European forests in relation to nitrogen deposition. *Forest Ecology and Management,* 71, 153–162.

Emmett, B.A., Anderson, J.M. and Hornung M. (1991) The controls on dissolved nitrogen losses following two intensities of harvesting in a Sitka spruce forest (N. Wales). *Forest Ecology and Management,* **41**, 65–80.

Emmett, B.A., Reynolds, B., Stevens, P.A., Norris, D.A., Hughes, S., Gorres, J. and Lubrecht, I. (1993) Nitrate leaching from afforested Welsh catchments – interactions between stand age and nitrogen deposition. *Ambio,* **22**, 366–394.

Emmett B A., Brittain A., Hughes S., Gorres J., Kennedy V., Norris D., Rafarel R., Reynolds B. and Stevens P. (1995a) Nitrogen additions ($NaNO_3$ and NH_4NO_3) at Aber forest, Wales: I. Response of throughfall and soil water chemistry. *Forest Ecology and Management,* **71**, 45–59.

Emmett B.A., Brittain A., Hughes S. and Kennedy V. (1995b) . Nitrogen additions ($NaNO_3$ and NH_4NO_3) at Aber forest, Wales: II. Response of trees and soil nitrogen transformations. *Forest Ecology and Management,* **71**, 61–73.

Emmett B.A., Stevens P.A. and Reynolds B. (1995c) Factors influencing nitrogen saturation in Sitka spruce stands in Wales. *Water Air and Soil Pollution,* **85**, 1629–1634.

Emmett, B.A., Kjønaas, O.J., Gundersen, P., Koopmans, C.J., Tietema, A. and Sleep, D. (1998a) Natural abundance of ^{15}N in forest across a nitrogen deposition gradient. *Forest Ecology and Management.* **101**, 9–18.

Emmett, B.A., Reynolds, B., Silgram, M., Sparks, T. and Woods, C. (1998b) The consequences of chronic nitrogen additions on N cycling and soilwater chemistry in a Sitka spruce stand, N.Wales. *Forest Ecology and Management.* **101**, 165–176.

Falkengren-Grerup, U. (1986) Soil acidification and vegetation changes in deciduous forest in southern Sweden. *Oecologia,* **70**, 339–347.

Feger, K.H. (1992) Nitrogen cycling in two Norway spruce (*Picea abies*) ecosystems and effects of a $(NH_4)_2$ SO_4 addition. *Water, Air and Soil Pollution,* **61**, 295–307.

Gijsman, A.J. (1991) Soil water content as a key factor determining the source of nitrogen (NH_4^+ or NO_3^-) absorbed by Douglas-fir (*Pseudotsuga menziesi*) and the pattern of rhizosphere pH along its root. *Canadian Journal of Forest Research,* **21**, 616–625.

Grennfelt, P. and Hultberg, H. (1986) Effects of nitrogen deposition on the acidification of terrestrial and aquatic ecosystems. *Water, Air and Soil Pollution,* **30**, 945–63.

Gundersen, P. (1995) Nitrogen deposition and leaching in European forests – preliminary results from a data compilation. *Water, Air and Soil Pollution,* **85**, 1179–1184.

Gundersen, P. and Rasmussen, L. (1995) Nitrogen mobility in a nitrogen limited forest at Klosterhede, Denmark, examined by NH_4NO_3 addition. *Forest Ecology and Management,* **71**, 75–88.

Gundersen, P. (1998) Effects of enhanced N deposition in a spruce forest at Klosterhede, Denmark, examined by moderate NH_4NO_3 addition. *Forest Ecology and Management.* **101**, 251–268.

Gundersen, P., Emmett, B.A., Kjønaas, O.J., Koopmans, C.J. and Tietema, A. (1998) Impact of nitrogen deposition on N cycling in forests; a synthesis of NITREX data. *Forest Ecology and Management.* **101**, 37–56.

Heinsdorf, D. (1993) The role of nitrogen in declining Scots pine forests *(Pinus sylvestris)* in the lowland of East Germany. *Water, Air and Soil Pollution,* **69**, 21–35.

Holopainen, T. and Heinonen-Tanski, H. (1993) Effects of different nitrogen sources on the growth of Scots pine seedlings and the ultrastructure and development of their mycorrhizae. *Canadian Journal of Forest Research,* **23**, 362–372.

Hornung, M., Sutton, M.A. and Wilson, R.B (1995) Mapping and modelling of critical loads for nitrogen – a workshop report. *Report of a UN–ECE Convention on Long Range Transboundary Air Pollution workshop, Grange-over-Sands, Cumbria, UK, 24–26 October, 1994.* Publ. Institute of Terrestrial Ecology, Penicuik, Scotland.

Houghton, J.T., Jenkins, G.T., Ephraums, J.J. (1990*) Climate change: the IPCC Scientific Assessment.* Cambridge University Press.

Johnson, D.W. (1992) Nitrogen retention in forest soils. *Journal of Environmental Quality*, **21** (1), 1–12.

Johnson, L.C., Shaver, G.R., Giblin, A.E., Nadelhoffer, K.J., Rastetter, E.R., Laundre, J.A. and Murray, G.L. (1996) Effects of drainage and temperature on carbon balance of tussock tundra microcosms. *Oecologia,* **108**, 737–748.

Jonasson, S., Havstrom, M., Jensen, M. and Callaghan, T.V. (1993) In situ mineralization of nitrogen and phosphorus or arctic soils after perturbations simulating climate change. *Oecologia,* **95**, 179–186.

Kahl, J.S. Fernanadez, I.J., Nadelhoffer, K.J., Driscoll, C.T.,and Aber, J.D. (1993) Experimental inducement of nitrogen saturation at watershed scale. *Environmental Science and Technology,* **27**, 565–568.

Katzensteiner, K., Glatzel, G. and Kazda, M. (1992) Nitrogen induced nutritional imbalances – a contributing factor to Norway spruce decline in the Bohemian Forest (Austria). *Forest Ecology and Management*, **51**, 29–42.

Kenk, G. and Fisher, H. (1988) Evidence from nitrogen fertilization in the forests of Germany. *Environmental Pollution*, **54**, 199–218.

Kleemola, S. and Forsius, M. (Eds) (1996) International co-operative programme on integrated monitoring of air pollution effects on ecosystems. UN ECE Convention on Long-Range Transboundary Air Pollution. 5th Annual Report 1996. Publ. The Finnish Environment Institute, Helsinki.

Kriebitzsch, W.U. (1978) Stickstoffnachlieferung in saure Waldböden Nordwsetdeutschlands. *Scr. Geobot*, Göttingen **14**, 1–66.

Kreutzer, K. (1989) *DVWK-Mitteilungen,* **17**, 121.

Kronzucker, H.J., Siddiqi, M.Y and Glass, A.D.M. (1997) Conifer root discrimination against soil nitrate and the ecology of forest succession. *Nature* , **385**, 59–61.

Lepistö, A., Andersson, L., Arheimer, B. and Sundblad, K. (1995) Influence of catchment characteristics, forestry activities and deposition on nitrogen export from small forested catchments. *Water, Air and Soil Pollution,* **84**, 81–102.

Lamersdorf, N.P., Beier, C., Blanck, K., Bredemeier, M., Cummins, T., Farrell, E.P., Kreutzer, K., Rasmussen, L., Ryan, M., Weis, W. and Xu, Y. (1998) Effect of drought experiments using roof installations on acidifiication/nitrification of soils. *Forest Ecology and Management.* **101**, 95–110.

Lavoie, N., Vezina, L.P. and Margolis, H.A. (1992) Absorption and assimilation of nitrate and ammonium ions by jack pine seedlings. *Tree Physiology,* **11**, 171–183.

Lükewille, A. and Wright, R.F. (1997) Experimentally increased soil temperature causes release of nitrogen at a boreal forest catchment in southern Norway. *Global Change Biology*, **3**, 13–21.

Magill, A.H., Aber, J.D., Hendricks, J.J., Bowden, R.D., Melillo, J.M. and Steudler, P.A. (1997) Biogeochemical response of forest ecosystems to simulated chronic nitrogen deposition. *Ecological Applications*. **7**, 402–415.

Magill, A.H., Downs, M.R., Nadelhoffer, K.J., Hallett, R. and Aber, J.D. (1996) Forest ecosystem response to four years of chronic nitrate and sulfate additions at Bear Brooks Watershed, Maine, USA. *Forest Ecology and Management*, **84**, 29–37.

Marschner, H., Haussling, M. and George, E. (1991) Ammonium and nitrate uptake rates and rhizosphere pH in non-mycorrhizal roots of Norway spruce [*Picea abies* (L.) Karst.]. *Trees,* **5**, 14–21.

Matzner, E. and Grosholz, C (1997) Beziehung zwischen NO_3^--Austrägen, C/N-Verhältnissen ser Auglage und N-Einträgen in Fichtenwald (Picea abies Karst.)-Ökosystemen Mitteleuropas. *Forstw. Cbl.*, **116**, 39–44.

Mc Nulty, S.G., Aber, J.D. and Boone, R.D. (1991) Spatial changes in forest floor and foliar chemistry of spruce–fir forests across New England. *Biogeochemistry,* **14**, 13–29.

Mc Nulty, S.G. and Aber, J.D (1993) Effects of chronic nitrogen additions on nitrogen cycling in a high-elevation spruce–fir stand. *Canadian Journal of Forest Research*, **23**, 1252–1263.

Mc Nulty, S.G., Aber, J.D and Newman, S.D (1996) Nitrogen saturation in a high elevation New England spruce–fir stand. *Forest Ecology and Management,* **84**, 109–121.

Miller, J.D., Cooper, J.M. and Miller, H.G. (1993) A comparison of above-ground component weights and elements in four forest species at Kirkton Glen. *Journal of Hydrology,* **145**, 419–438.

Miller H.G. and Miller J.D (1988) Response to heavy nitrogen applications in fertilizer experiments in British Forest. *Environmental Pollution*, **54**, 219–232.

Mitchell, M.J., Driscoll, C.T., Kahl, J.S., Likens, G.E., Murdoch, P.S. and Pardo, L.H. (1996) Climatic control of nitrate loss from forested watersheds in the Northeast United States. *Environmental Science and Technology,* **30**, 2609–2612.

Mitchell, M.J., Raynal, D.J., White, E.H., Stehman, V.S., Driscoll, C.T., David, M.B., McHale, P.J. and Bowles, F.P. (1994) Increasing soil temperature in a northern hardwood forest; effects on elemental dynamics and primary production: *USDA Forest Service, progress report.*

Mohren, G.M.J., Van den Burg, J. and Burger, F.W. (1986) Phosphorus deficiency induced by nitrogen input in Douglas fir in the Netherlands. *Plant and Soil,* **95**, 191–200.

Moldan, F. and Wright, R.F. (In press) Changes in runoff chemistry after 5 years of N addition to a forested catchment at Gårdsjön, Sweden. *Forest Ecology and Management.*

Moldan, F., Hultberg, H., Nyström, U. and Wright, R.F (1995) Nitrogen saturation at Gårdsjön , southwest Sweden, induced by experimental addition of ammonium nitrate. *Forest Ecology and Management,* **71**, 89–97.

Nadelhoffer, K.J., Giblin, A.E., Shaver, G.R., and Linkins A.E. (1992) Microbial processes and plant nutrient availability in arctic soils. In: Chapin FS III, Jeffries, R.L., Reynolds, J.F., Shaver, G.R., Svoboda, J. (Editors). *Arctic ecosystems in a changing climate. An ecosphysiological perspective.* Academic Press, San Diego, pp 281–300.

Nadelhoffer, K.J., Downs, M.R., Fry, B., Aber, J.D., Magill, A.H. and Melillo, J.M. (1995) The fate of ^{15}N–labelled nitrate additions to a northern hardwood forest in eastern Maine, USA. *Oecologia,* **103**, 292–301.

Näsholm, T., Nordin, A., Edfast, A. and Högberg, P (1997) Identification of coniferous forests with incipient nitrogen saturation through analysis of arginine and nitrogen–15 abundance of trees. *Journal of Environmantal Quality,* **26**, 301–309.

Nihlgard, B. (1985) The ammonium hypothesis – An additional explanation to the forest dieback in Europe. *Ambio,* **14** (1), 1–8.

Nodvin, S.C., Driscoll, C.T. and Likens, G.E. (1988) Soil processes and sulphate loss at the Hubbard Brook Experimental Forest. *Biogeochemistry,* **5**, 185–199.

Nodvin, S.C., Van Miegroet, H., Lindberg, S.E., Nicholas, N.S. and Johnson, D.W (1995) Acidic deposition, ecosystem processes, and nitrogen saturation in high elevation southern Appalachian watershed. *Water, Air and Soil Pollution,* **85**, 1647–1652.

Peterjohn, W.T., Adams, M.B. and Gilliam, F.S. (1996) Symptoms of nitrogen saturation in two central Appalachian hardwood forest ecosystems. *Biogeochemistry,* **35**, 507–522.

Peterjohn, W.T., Melillo, J.M., Bowles, F.P. and Steudler, P.A. (1993) Soil warming and trace gas fluxes: experimental design and preliminary flux results. *Oecologia,* **93**, 18–24.

Postek, K.M., Driscoll, C.T., Aber, J.D. and Santore, R.C. (1995) Application of PNet–CN/CHESS to a spruce stand in Solling, Germany. *Ecological Modelling,* **83**, 163–172.

Reynolds, B., Emmett, B.A., and Woods, C. (1992) Variations in streamwater nitrate concentrations and nitrogen budgets over 10 years in a headwater catchment in mid-Wales. *Journal of Hydrology,* **136**, 155–175.

Reynolds, B., Ormerod, S.J. and Gee, A.S. (1994) Spatial patterns in stream nitrate concentrations in upland Wales in relation to catchment forest cover and forest age. *Environmental Pollution,* **84,** 27–33.

Rustad, L.E., Fernandez, I.J. and Arnold, S. (1995) Experimental soil warming effects on C, N, and major element cycling in a low elevation spruce–fir forest soils: In: *Gen. Tech. Rep. NE. Radnor* (Editors, Hom, J., Birdsey, R., O'Brien, K.), pp. 1–7. USDA Forest Service, PA.

Skeffington, R.A. and Wilson, E.J. (1988) Excess nitrogen deposition: issues for consideration. *Environmental Pollution,* **54**, 159–84.

Stark, J.M. and Hart, S.C. (1997. High rates of nitrification and nitrate turnover in undisturbed coniferous forests. *Nature,* **385**, 61–64.

Stevens, P.A., Harrison, A.F., Jones, H.E., Williams, T.G. and Hughes, S. (1993) Nitrate leaching from a Sitka spruce plantation and the effect of fertilization with phosphorus and potassium. *Forest Ecology and Management,* **58**, 233–247.

Stevens, P.A. and Hornung, M. (1990) The effect of harvest intensity and ground flora establishment on inorganic N leaching from a Sitka spruce plantation in North Wales. *Biogeochemistry,* **10**, 53–65.

Stevens, P.A., Norris, D.A., Sparks, T.H., Hidgson, A.L. (1994) The impacts of atmospheric inputs on throughfall, soil and stream water interactions for different aged forest and moorland catchments in Wales. *Water, Air and Soil Pollution*, **73**, 297–317.

Stoddard, J.L. (1994) Long term changes in watershed retention of nitrogen: its causes and aquatic consequences. In; *Environmental Chemistry of Lakes and Reservoirs*. (Ed. Baker, L.A.) American Chemical Society.

Stuanes, A.O. and Kjønaas, O.J. (1998a) Soil solution chemistry during four years of NH_4NO_3 additions to a forested catchment at Gårdsjön, Sweden. *Forest Ecology and Management*. **101**, 215–226

Stuanes, A.O., Huse, M., Kjønaas, O.J., and Nygaard, N.H., (1998b) Forest health responses in the Gårdsjön catchments. In: Hultberg, H., and Skeffington, R. eds. *Experimental Reversal of Acid Rain Effects: the Gårdsjön Roof Project*. New York: John Wiley and Sons p. 327–34.

Swank, W.T. and Vose, J.M. (In press) Long term nitrogen dynamics of Coweeta Forested Watersheds in the Southeastern USA. *Global Biogeochemistry*.

Tamm, C.O. (1991) Nitrogen in terrestrial ecosystems. *Ecological Studies* **81**, 115.

Thornley, J.H.M. and Cannell, M.G.R. (1992) Nitrogen relations in a forest plantation – soil organic matter ecosystem model. *Annals of Botany*, **70**, 137–151.

Tietema, A. (1998) Microbial carbon and nitrogen dynamics in coniferous forest floor material collected along a European nitrogen deposition gradient. *Forest Ecology and Management*. **101**, 29–36.

Tietema, A. and Beier, C. (1995) A correlative evaluation of nitrogen cycling in the forest ecosystems of the EC projects NITREX and EXMAN. *Forest Ecology and Management*, **71**, 143–152.

Tietema, A., Emmett, B.A., Gundersen, P., Kjønaas, O.J. and Koopmans, C.J. (1998) The fate of ^{15}N–labelled nitrogen deposition in coniferous forest ecosystems. *Forest Ecology and Management*. **101**, 19–28.

Tuovinen, J.P., Barrett, K. and Styve, H. (1994) transboundary acidifying pollution in Europe: calculated fields and budgets. EMEP Report: *EMEP/MSC–W Report1/94*. Publ. The Norwegian Meteorological Institute, Oslo, Norway.

Ulrich B. (1983) A concept of forest ecosystem stability and of acid deposition as driving force for destabilization. In: B. Ulrich and J. Pankrath (editors); *Effects of Accumulation of Air Pollutants in Forest Ecosystems*. Publ. D. Reidal Publishing Company, Dordrecht, Holland, pp. 1–29.

United Nations (1994) Protocol to the 1979 Convention on Long Range Transboundary Air Pollution on future reductions of S emissions. *EB.AIR/R.84*

Van Breemen, N. and Mulder, J. and Driscoll, C.T. (1983) Acidification and alkalinization of soils. *Plant and Soil*, **54**, 249–74.

Van Breemen, N. and Van Dijk, H.F.G. (1988) Ecosystem effects of atmospheric deposition of nitrogen in The Netherlands. *Environmental Pollution*, **75**, 283–308.

Van Cleve, K., Oechel, W.C. and Hom, J.K. (1990) Response of black spruce (*Picea mariana*) ecosystems to soil temperature modification in interior Alaska. *Canadian Journal of Forestry Research*, **20**, 1530–1535.

Van Den Driessche, R. and Dangerfield, J. (1975) Response of Douglas Fir seedlings to nitrate and ammonium nitrogen sources under various environmental conditions. *Plant and Soil*, **42**, 685–702.

Van Dijk, H.F.G. and Roelofs, J.G.M (1998) Effects of excessive ammonium deposition on the nutritional status and condition of pine needles. *Physiologia Plantarum*, **73**, 494–501.

Van Miegroet, H., Cole, D.W. and Foster, N.W. (1992) Nitrogen distribution and cycling. In: D.L. Johnson and S.E. Lindberg (Editors). *Atmospheric Deposition and Forest Nutrient Cycling. A Synthesis of the Integrated Forest Study*. Springer, New York, pp178–199.

Vitousek, P.M. and Reiners, W.A. (1975) Ecosystem succession and nutrient retention: a hypothesis. *Bioscience*, **25**, 376–381.

Warmerdam, B. (1992) Effects of increased nitrogen input on nitrogen transformations in European forest ecosystems. *Internal Report. Laboratory of Physical Geography and Soil Science, University of Amsterdam, Report No. 50.*

Werner, B. and Spranger, T (Eds) (1996) Manual on methodologies and criteria for mapping critical levels/loads. UN ECE Convention on Long Range Transboundary Air Pollution. Published by Federal Environmental Agency, Germany.

Whitehead, P.W., Wilson, E.J. and Butterfield, D. (In press) A semi-distributed process based nitrogen model for multiple source assessment in catchments. *Science of the Total Environment.*

Wilson, E.J and Skeffington, R.A (1994) The effects of excess nitrogen deposition on young Norway spruce trees. Part I. The Soil. *Environmental Pollution*, **86**, 141–151.

Williams, M.W., Brooks, P.D., Mosier, A. and Tonnessen, K.A. (1996) Mineral nitrogen transformations in and under seasonal snow in a high-elevation catchment in the Rocky Mountains, Unites States. *Water resources Research*, **32**, 3161–3171.

Williard, K.W.J. (1996) Indicators of nitrate export from forested watersheds of the Chesapeake Bay Region. *MS Thesis, Environmental Pollution Control, Pennsylvania State University.* 72 pp.

Wright R.F. and Van Breemen, N. (1995) The NITREX project; an introduction. *Forest Ecology and Management,* **71**, 1–6.

THE IMPACT OF ATMOSPHERIC NITROGEN DEPOSITION ON THE BEHAVIOUR OF NITROGEN IN SURFACE WATERS

P. J. CHAPMAN AND A. C. EDWARDS
Macaulay Land Use Research Institute, Craigiebuckler, Aberdeen, AB15 8QH, UK.

6.1. Introduction

There is now much evidence that NO_3 and NH_4 deposition, both wet and dry, has increased over much of Europe and North America in the last 50 years (see Chapter 2). It was originally believed that this 'extra' N would have little effect upon surface water chemistry in temperate semi-natural ecosystems. This is because vegetation and soil communities within these systems are generally regarded to be N-limited and thus any additional N would be rapidly sequestered. In addition, any N that eventually reached surface waters would similarly be expected to be utilised by aquatic plants and algae. Thus streams draining these ecosystems should contain very low concentrations of N with fluxes being small, usually less than 1–2 kg ha^{-1} yr^{-1} (Driscoll *et al.*, 1989; Reynolds and Edwards, 1995). The majority of this N is exported during the winter when runoff and NO_3 concentrations are greatest.

Over the last decade, studies in forested catchments of northern Europe and the northeastern United States have reported elevated concentrations of NO_3 in both streams and lakes that appear to be strongly correlated with a trend of increased atmospheric N deposition (Grennfelt and Hultberg, 1986; Henriksen and Brakke, 1988; Driscoll *et al.*, 1989; Hauhs *et al.*, 1989; Dise and Wright, 1995; Peterjohn *et al.*, 1996). It has been proposed that this increase in stream water NO_3 has resulted from a decrease in the capacity of terrestrial ecosystems to retain N as a result of long-term N loading from atmospheric deposition (Aber *et al.*, 1989). This decrease in capacity to retain N, generally termed N saturation, occurs when the supply of N to terrestrial ecosystems exceeds their biological demand, with excess N being removed in drainage waters (Aber *et al.*, 1989). However, it should be remembered that the biological utilisation of N is highly seasonal and also dependent upon the availability of carbon and other nutrients such as phosphorus, and to a lesser extent potassium (Stevens *et al.*, 1993a; Kaste *et al.*, 1997).

The consequences of increased N leaching can lead to; (i) acidification of surface waters, when the NO_3 is accompanied by Al or H^+ and (ii) eutrophication, where N is the limiting nutrient. However, identifying the precise impact of any increased leaching on aquatic ecosystems is complicated by the variety of chemical forms in which N may exist in atmospheric deposition, catchment waters and surface waters. For example, the proportion of N as either NO_3 or NH_4 in deposition determines the way N enters the terrestrial cycle with the majority of NH_4 being retained by the soil initially through ion exchange processes. A variety of terrestrial processes (see Chapter 3) therefore separate N deposition from its effects and reduces the ability to attribute observations of aquatic effects to known rates of N deposition (Stoddard, 1994).

In this chapter we review the extent of current knowledge on the behaviour of N in surface waters and how it may change in response to increasing N deposition. The chapter is subdivided into four separate sections describing;

(1) a brief overview of the N cycle in semi-natural ecosystems;
(2) how the dynamics of N in surface waters is likely to change in response to increasing N deposition;
(3) the consequences of increased N leaching on the functioning of aquatic ecosystems;
(4) and finally, a review of surface water N data from three different regions having significant areas of semi-natural temperate vegetation which vary in their N deposition, climatic and pedological properties.

6.2. The Nitrogen Cycle In Semi-Natural Ecosystems

Atmospherically derived N reaches surface waters by one of two pathways; (i) direct deposition to water surfaces or (ii) via drainage waters from the terrestrial part of the catchment. Since the former is only significant for catchments where water represents a large proportion of the surface area, the majority of deposited N to the catchment is routed via the soil where it undergoes a range of chemical and biological transformations (see Chapter 3). The interactions of deposited N with vegetation and soil modify the amount and form of N as it passes through the terrestrial part of the catchment. Hydrological processes regulate the degree to which these interactions occur and act as the major transfer mechanism for soluble N. If it is assumed that the direct leaching of atmospherically derived NO_3 from natural and semi-natural ecosystems only occurs during winter storms, there is considerable potential for delay in the timing between NO_3 deposition and NO_3 loss. Thus the short-term pattern of NO_3 export from the catchment may bear little relationship to the original deposition. Therefore, to determine whether increased N deposition is likely to lead to a change in either the amount, form or timing of N loss that occurs to surface waters a clear understanding of the terrestrial N cycle is required.

Figure 6.1 shows how the annual variation in atmospherically derived N inputs together with soil processes regulate the NO_3 concentration of stream water in semi-natural ecosystems. There is a pronounced seasonality to the concentration of NO_3 and NH_4 in precipitation; concentrations are generally largest in the late spring (April/May) and usually decline to a minimum in December/January for Britain (RGAR, 1990). In contrast, the quantity of precipitation is generally smallest in the late spring and greatest throughout the winter. As a result maximum wet deposition of N occurs in March and April (Figure 6.1a). In general, the seasonality of atmospheric inputs is not reflected by the pattern of variation in NO_3 concentrations in either soil or stream waters (Figure 6.1c); which provides further evidence of the importance of modifying effect of soil and plant processes.

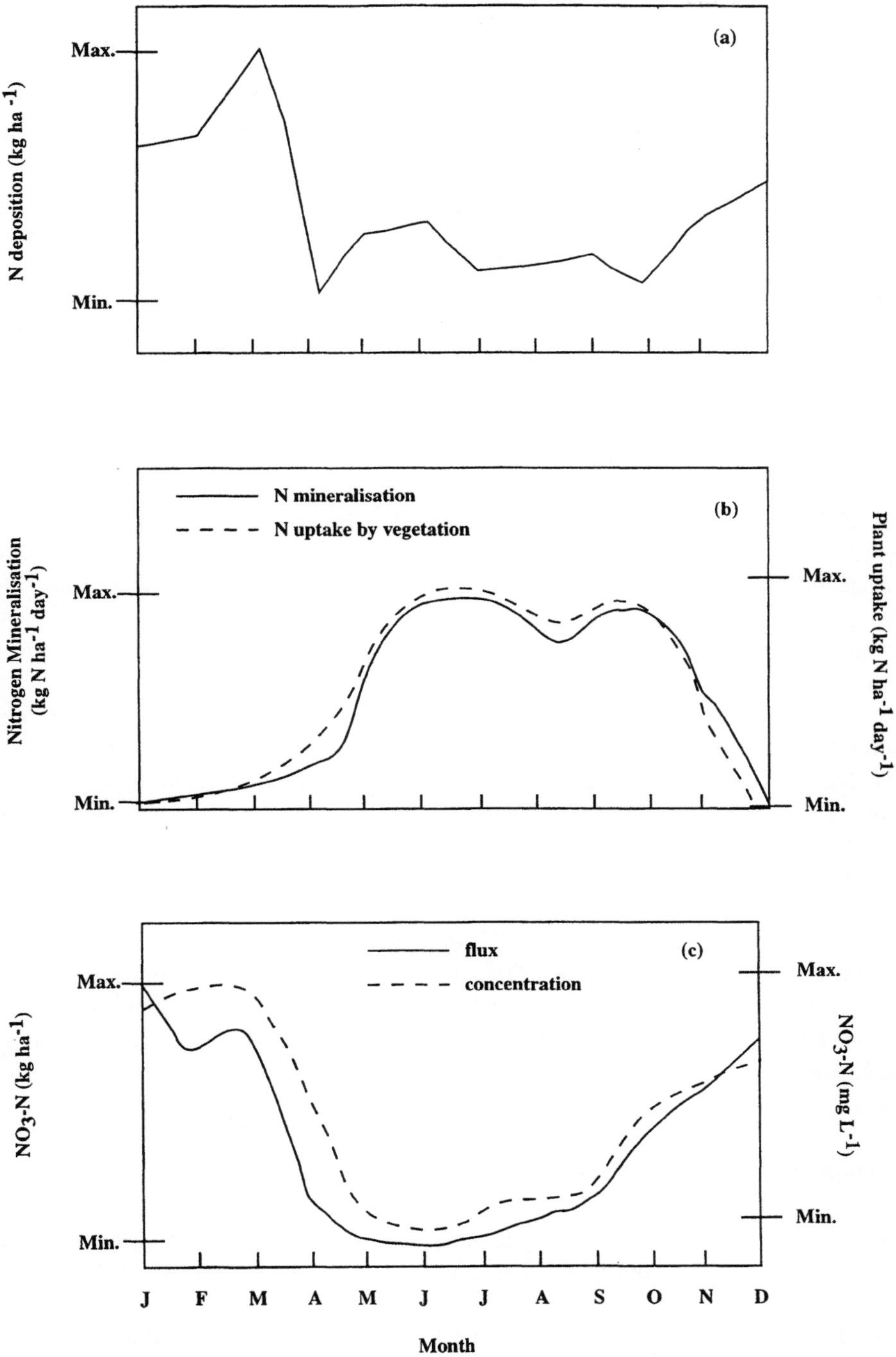

Figure 6.1: Schematic representation of the typical temporal variation in (a) N deposition, (b) N mineralisation and plant uptake rate and (c) NO_3 concentrations and fluxes in streams common to upland catchments in western Britain.

The largest amount of net mineralisation (i.e. the excess of mineralisation over immobilisation) in temperate semi-natural ecosystems typically occur in late spring/early summer with a secondary peak in the autumn (Figure 6.1b). During the winter, mineralisation is limited by low temperatures, but as soils warm up in spring the rate increases rapidly, roughly coinciding with the start of the growing season. As mineralisation may be limited by moisture stress under very dry conditions (Reddy, 1982) the process may be suppressed in late summer where well drained soils dry out. Under most summer conditions the quantity of soil NO_3 in excess of plant demand will be small. As the growth of vegetation is generally N-limited in temperate systems, atmospherically derived N will be rapidly sequestered during the growing season. It is worth noting that atmospheric N deposition can be considerably smaller than that produced by mineralisation. During the late autumn/winter period plant uptake declines, resulting in increased soil water NO_3 concentrations, and soil water content increases. This seasonality to soil NO_3 availability within the soil is partly responsible for the commonly observed pattern of larger stream and lake water NO_3 concentrations in the winter than in the summer (Figure 6.1c). Therefore soils have a much lower ability to retain atmospheric inputs of NO_3 during the winter than the summer when it is more likely to enter the terrestrial N cycle. In contrast, nearly all deposited NH_4 is initially retained by soil. This exchangeable NH_4 forms a readily available source of N for nitrification and may also be taken up directly by plants. Thus little N is lost from soil in the form of NH_4 and its concentration in surface waters is usually small (<0.01 mg N L^{-1}) and displays no obvious seasonal cycle.

Inorganic forms of N are not necessarily the only or indeed the major form of N in soils or surface waters (Roberts *et al.*, 1983; Stevens and Wannop, 1987; Chapman *et al.*, 1998). Nitrogen containing organic compounds from biological origins and generally referred to as dissolved organic N (DON), are also present in surface waters. Their concentrations may exceed the amount of inorganic N, particularly in areas dominated by organic soils (Lepisto *et al.*, 1995; Chapman *et al.*, 1998). The N can either be present as an integral part of protein molecules or as discrete compounds formed from the partial breakdown of these larger molecules, for example, peptides, urea and amino acids. Anderson *et al.* (1991) divided the DON component into three broad fractions, fulvic acids, hydrophilic acids and humic acids and found that the former two groups contributed more than half the total dissolved N lost from two forested catchments in central Scotland, while N associated with humic acids was negligible.

6.3. Symptoms Of Nitrogen Saturation In Surface Waters

Much of the debate about whether aquatic systems are being adversely affected by N deposition centres on the concept of N saturation proposed by Aber *et al.* (1989). They presented a hypothetical time course for the response of a forested catchment to increasing N additions. When N inputs from the atmosphere exceed biological demand any excess N will ultimately reach surface drainage waters.

Stoddard (1994) proposed 4 stages of N loss from terrestrial ecosystems which correspond to the stages of terrestrial N saturation described by Aber *et al.* (1989).

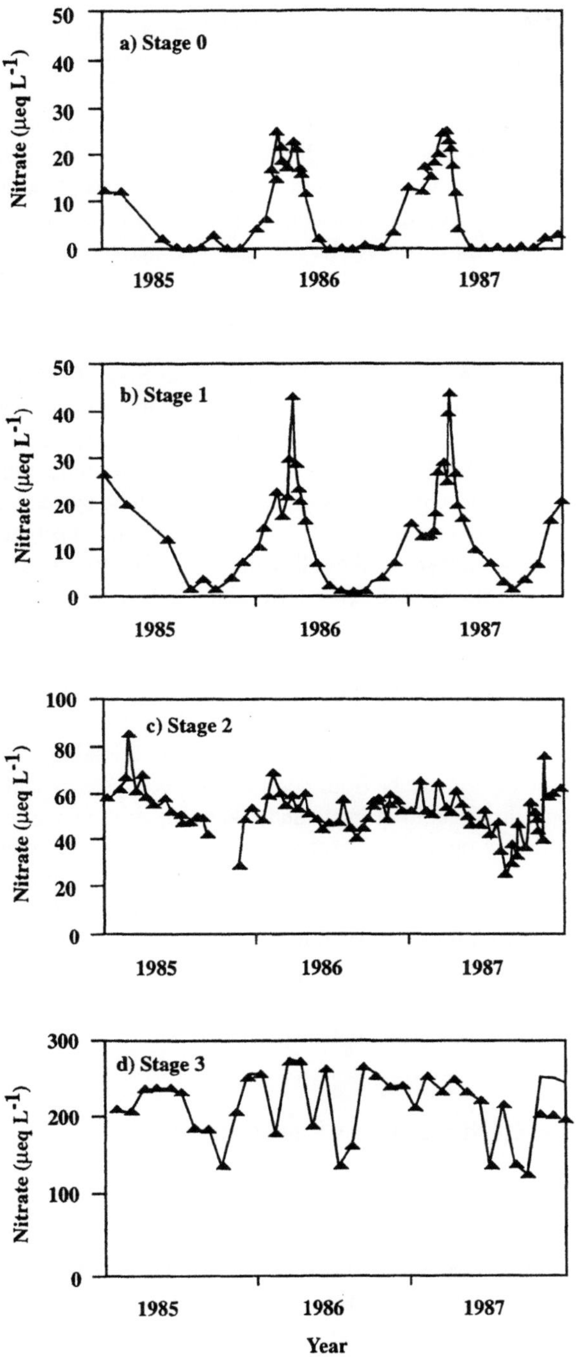

Figure 6.2: Diagrammatic representation of expected changes in the seasonal and long-term patterns of surface water NO_3 concentrations in response to the progressive development of N saturation (a) Stage 0, (b) Stage 1, (c) Stage 2 and (d) Stage 3 (from Stoddard, 1994).

Figure 6.2. illustrates the expected changes in the seasonal and long-term patterns of surface water NO_3 concentrations in response to the progressive development of N saturation. At Stage 0 the ecosystem is N-deficient and the N cycle is dominated by vegetation and microbial uptake and the demand for N has the major influence on the seasonal NO_3 pattern in surface waters (as discussed in Section 6.2). Thus at this stage, NO_3 concentrations in surface waters should be negligible except perhaps during the winter/spring period where direct losses of atmospherically derived N arising from either snowmelt or spring rains may occur. The longer the period of small stream NO_3 concentrations the greater the extent of N limitation within the ecosystem.

With the progressive development of N saturation, amplification of the seasonal NO_3 pattern in surface waters would be expected (Stage 1) as winter NO_3 concentrations continue to increase and summer values remain small compared to Stage 0. During Stage 2, the extent which seasonal cycles of surface water NO_3 concentrations occur has diminished as significant loss of NO_3 also occurs during the summer primarily because NO_3 supply exceeds biological demand. By Stage 3, the catchment has become a net source of N rather than a sink and the characteristics of surface waters in these systems is elevated NO_3 concentrations which lack any seasonal trend. Thus as a catchment becomes progressively saturated with respect to N, changes in the following would be envisaged; (i) the seasonal pattern of NO_3 concentrations and (ii) long-term trends of surface water NO_3 concentrations.

As most national monitoring programmes of N in surface waters only analyse for NO_3, and in some cases NH_4, there are few data describing either organic and total nitrogen concentrations and virtually no information on either their seasonal patterns or long-term trends. Therefore it is difficult to assess what impact increased loadings of atmospheric N have had on the export of organic N from semi-natural ecosystems and thus dissolved organic N (DON) concentrations in streams and lakes. In addition the few data that are available present conflicting opinions. Yesmin *et al.* (1995) observed enhanced leaching of DON from ombrotrophic peat microcosms collected from 9 sites along an increasing N deposition gradient (2–17 kg N ha^{-1} yr^{-1}) in Britain. In contrast, no clear relationship was observed between N deposition and DON fluxes in streams from 20 small catchments in Norway (Wright *et al.*, 1997). However, both Wright *et al.* (1997) and Yesmin *et al.* (1995) observed a significant inverse relationship between the C:N ratio of dissolved organic matter and N deposition (see Chapter 3).

6.4. The Consequences Of Nitrogen Loss From Catchments

6.4.1. SURFACE WATER ACIDIFICATION

Surface waters in many of the semi-natural areas of Europe and North America are sensitive to elevated inputs of acidic deposition. This is because many of these ecosystems are located in areas of base-poor geology with acidic soils, thus inputs of strong acid anions (SO_4, NO_3 and Cl) are only neutralised to a limited extent. Hence the strong acid anions are transported through the terrestrial system along with base cations and/or Al and H^+ to surface waters. Acidification of surface waters is dependent on the extent that SO_4, NO_3 or Cl are leached with H^+ or Al.

Acid neutralisation capacity (ANC) is commonly regarded as a better indicator of the acidic status of water than pH alone. The ANC of a lake or stream is a measure of the water's capacity to buffer acidic inputs. It results from the presence of carbonate or bicarbonate (alkalinity), aluminium and organic acids (Sullivan *et al.,* 1989). Surface waters with an ANC less than zero are often defined as being acidic. Surface water acidification can be either chronic (long-term) or episodic (event-based) in nature. Chronic acidification occurs throughout the annual cycle and is generally associated with elevated concentrations of SO_4 relative to concentrations of base cations (Driscoll and Schaefer, 1989). Episodic acidification is the process by which lakes and streams experience a short-term decrease in ANC, usually during hydrological events associated with rainfall and snowmelt (Wigington *et al.,* 1990). Acidic episodes may result from a number of different processes, such as:

(i) dilution of basic cation concentrations and/or
(ii) increase in SO_4 concentrations (e.g. Christophersen and Wright, 1981),
(iii) increase in organic acid concentrations (e.g. Grip and Bishop, 1990),
(iv) increase in NO_3 concentrations (e.g. Murdoch and Stoddard, 1992),
(v) sea-salt events (e.g. Langan, 1989).

In the past, emphasis has been placed on the role of SO_4 leaching in the acidification of surface water, largely because rates of S deposition are usually greater than those for N and because concentrations of NO_3 in surface waters are generally very small compared to SO_4. However, in recent years SO_4 deposition has decreased due to a reduction in SO_2 emissions so that wet deposition of N is now larger than S deposition on an equivalent basis in many areas of Europe (e.g. Henriksen *et al.,* 1997a). Therefore, the key question is to what extent does the leaching of NO_3 along with Al and/or H^+ from the soil contribute to the acidification of surface waters relative to other sources of acidity.

Chronic acidification

Recent studies in NE America and Europe have highlighted the increasing role of NO_3 in the acidification of lakes (Henriksen and Brakke, 1988; Kopacek *et al.,* 1995; Sullivan *et al.,* 1997). In the Adirondacks in the northeastern United States, lake water NO_3 concentrations display a pronounced geographic pattern, with large concentrations observed in the west and southwest which corresponds closely with the distribution of (i) lake water acidity and (ii) diatom-inferred acidification from pre-industrial times to the present (Sullivan *et al.,* 1997). Although lake water SO_4 concentrations are much larger than NO_3 concentrations, they display no spatial pattern. Hence, Sullivan *et al.* (1997) concluded that these patterns lend support to the hypothesis that NO_3 plays a significant role in the chronic acidification of lakes in the Adirondacks. In the 1986 regional survey of 1005 lakes in Norway, largest concentrations of NO_3 were also observed in those lakes with the lowest pH and for 300 lakes in southern Norway NO_3 concentrations had almost doubled since the previous survey in 1974 (Henriksen and Brakke, 1988). This increase in lake water NO_3 concentrations was associated with increasing concentrations of Al

(Henriksen *et al.*, 1988), which is toxic to fish (Brown, 1988). It has been suggested that NO_3 may have a greater ability to mobilise Al from the soil than does SO_4 (James and Riha, 1989).

Henriksen (1988) proposed that the ratio (in equivalents) of NO_3:(NO_3+SO_4) in surface waters be used as an index of the influence of NO_3 on chronic acidification status. The index assesses the importance of NO_3 relative to that of SO_4. If the amounts of S and N in deposition are equivalent, and both S and N are not retained in the catchment and all deposited NH_4 is converted to NO_3, the factor would approximate 0.5. Thus a value greater than 0.5 indicates that NO_3 has a greater influence on the chronic acidification of surface waters than SO_4.

Table 6.1: Concentrations (μeq L^{-1}) and ratios of NO_3 to (NO_3+SO_4) in lakes of acid-sensitive regions of Europe and North America

Region	Number of Lakes	pH	NO_3	SO_4	NO_3:(NO_3+SO_4)	Year	Reference
UK	11	5.42	13	71	0.14	1988–93	Patrick *et al.* (1995)
Galloway, Scotland	22	4.97	21	103	0.17	1979	Wright & Henricksen (1980)
Norway	1011		7	62	0.10	1986	Henricksen & Brakke (1988)
Sumava Mts., Czech Rep.	4	4.46–4.78	44	131	0.25	1993	Kopacek *et al.* (1995)
High Tatra Mts., Slovakia	53	5.92	30	79	0.27	1980s	Kopacek *et al.* (1995)
Adirondacks, USA	17	5.9	17	136	0.11	1985	Driscoll & Newson (1985)
Eastern Lake Survey, USA							
Southern Blue Ridge		6.98	3	32	0.09	1985	Stoddard (1994)
Florida		6.56	1	94	0.01		
Upper Midwest		7.09	0.7	57	0.01		
Maine		6.91	0.2	75	0		
Southern New England		6.81	0.8	141	0.01		
Central New England		6.77	0.3	101	0		
Adinordack Mountains		6.71	0.6	119	0.01		
Catskills–Poconos		7.02	0.7	159	0		

Mean concentrations and ratios of NO_3:(NO_3+SO_4) in lakes from different acid-sensitive regions of Europe and North America are presented in Table 6.1. For lake waters in Norway and Scotland, NO_3 concentrations are small and the NO_3:(NO_3+SO_4) ratio less than 0.17, indicating a strong retention of N in the catchment. In central European lakes, where NO_3 concentrations are much larger the ratio of NO_3:(NO_3+SO_4) exceeds 0.25 indicating that a high percentage of the incoming N is moving through the terrestrial system and that NO_3 is contributing significantly to lake water acidity, but SO_4 is still the major contributor. The Eastern Lake Survey data from the United States (Table 6.1), suggests that NO_3 does not contribute significantly to chronic acidification. However, this survey was based on a probability sampling of lakes during the autumn (Stoddard, 1994) when NO_3 concentrations are often at their lowest. Hence the NO_3:(NO_3+SO_4) ratio obtained from these data is a minima. For example, the Eastern Lake Survey's NO_3:(NO_3+SO_4) ratio for lakes in the Adirondack mountains was 0.01, whereas the ratio based on mean monthly samples collected over several years from 17 Adirondack Long Term Monitoring (ALTM) lakes was 0.11 (Table 6.1). Driscoll and Newton (1985) observed that NO_3 contributes up to 17% of the strong acid anions in these ALTM lakes compared to 2% in most other areas in the US. In Norway, NO_3 is also a small contributor (generally < 10%) to the total acid anions (Henriksen and Brakke, 1988), while in Central Europe Kopacek and Stuchlik (1994) observed that NO_3 represented >30% of the strong acid anions in lake waters from the High Tatra Mountains, Slovakia.

Table 6.2: Concentrations (μeq L^{-1}) and ratios of NO_3 to (NO_3+SO_4) in streams of acid-sensitive regions of Europe and North America

Region	Number of streams	pH	NO_3	SO_4^{-}	NO_3:(NO_3+SO_4)	Year	Reference
UK	11	5.5	17	71	0.13	1988–93	Patrick *et al.* (1995)
Norway	12	5.48	12	48	0.2	1986	Henriksen & Brakke (1988)
Southern Norway	10	5.06	14	37	0.29	1992–95	Henriksen *et al.* (1997a)
Catskill Mts, USA	51	6.06	29	138	0.17	1984–86	Stoddard (1994)
National Stream Survey, USA							
Catskills		6.96	6	169	0.03	1986	Stoddard (1994)
Northern Appalachians		6.60	30	171	0.14		
Valley and Ridge		7.05	10	154	0.06		
Southern Blue Ridge		6.99	8	17	0.28		
Piedmont		6.80	2	48	0.03		
Southern Appalachians		7.33	16	58	0.32		

Mean concentrations and ratios of NO_3:(NO_3+SO_4) in streams from different acid-sensitive regions of Europe and North America are presented in Table 6.2. In Norway, the NO_3:(NO_3+SO_4) ratios for 11 rivers (calculated from monthly samples collected in 1986) ranged between 0.07 and 0.34. These values compare well with the average lake ratios for the counties in which the rivers are located (Henriksen and Brakke, 1988). In contrast, the NO_3:(NO_3+SO_4) ratios for streams in different regions of the United States are considerably larger than those for lakes in the same region (compare values in Tables 6.1 and 6.2). This probably reflects the fact that the National stream survey was carried out in the spring (Kaufmann *et al.*, 1991) when NO_3 concentrations are often at their maximum, whereas the Lake Survey was carried out in autumn when NO_3 concentrations are at a minimum. Thus caution is needed when interpreting and comparing surface water data from different studies.

Overall, SO_4 is the most important strong acid anion in the acidification of surface waters. Typically <15% of the acidity in nutrient-poor lakes and streams is explained by NO_3. In addition, while surface water NO_3 concentrations generally display a seasonal trend with maximum concentrations observed in the winter/spring, SO_4 concentrations vary little on an annual basis. Thus the contribution of NO_3 to the acidification of surface waters will also vary seasonally. However, as S deposition decreases and if N deposition continues to increase and the ability of a catchment to store atmospheric N is exceeded, increased NO_3 in runoff may counterbalance any reduction in SO_4 due to reductions in emissions of SO_2 (Henriksen and Brakke, 1988, Van Miegroet, 1994), resulting in little change in the acidity of surface waters.

Episodic acidification

To determine the contribution of NO_3 to episodic acidification, the importance of short-term increases in NO_3 concentration must be assessed in relation to the other processes which contribute to acidic episodes. Except in extreme cases, N loss from catchments is more likely to be an episodic or seasonal process than a chronic one. Therefore data used to assess the contribution of NO_3 to acidification must be collected intensively during storms (Stoddard, 1994). As regional surveys tend to collect water samples on a regular basis, low water flow rates will be sampled most of the time and thus limited data exists to assess the contribution of NO_3 to episodic acidification on a regional basis.

In the Adirondacks, strong NO_3 pulses in both lakes (Galloway *et al.*, 1980; Driscoll and Schafren, 1984) and streams (Driscoll *et al.*, 1987; Sullivan *et al.*, 1997) are apparently the primary factor contributing to depressed ANC and pH during snowmelt. Schaefer *et al.* (1990) concluded that the size of episodes experienced by 11 Adirondack lakes depended strongly on their base-cation concentration. They concluded that lakes with low ANC values undergo episodes that result largely from increases in NO_3 concentrations, while lakes with high base cations and thus high ANC values undergo episodes that are largely the result of dilution by snowmelt. At intermediate ANC values, lakes are affected by both base cation dilution and NO_3 increases. Murdoch and Stoddard (1992) reported similar findings for streams in the Catskills Mountains, where NO_3 increases were the primary determinant of acidic episodes in low to moderate ANC streams during the spring. In contrast, base-cation dilution and increases in organic acids were the major factors contributing to acidic episodes in the autumn. Sullivan *et al.* (1997) presented data for Adirondack streams that showed that the relative importance of NO_3 versus SO_4 acidity increased as the ANC decreased during storm events.

In a review of acidic episodes in Europe, Davies *et al.* (1992) concluded that SO_4 appears to be the strong acid anion most strongly related to pH depression during episodes. In contrast, NO_3 are more commonly associated with acidic episodes in the USA (Wigington, 1990). This difference may result from differences in the nature of catchments, more forested catchments have been studied in the USA, or it may be a consequence of different acidic deposition histories, or simply due to the fact that compared to SO_4 few NO_3 data have been collected during storms in Europe. Relatively few NO_3 data have been reported for rainfall events in Europe (Davies *et al.,* 1992). However, studies in both Britain (e.g. Jenkins *et al.,* 1993) and Norway (e.g. Johannessen and Henriksen, 1978) have shown that NO_3 concentrations can be large in the initial melt water from the snowpack and thus can contribute to producing an acidic pulse at the catchment outflow (e.g. Davies *et al.,* 1993).

6.4.2. EUTROPHICATION

The majority of water draining natural and semi-natural regions, such as those being considered here, are oligotrophic and therefore by definition have very limited supplies of plant available nutrients. Thus a possible consequence of increased N leaching from terrestrial ecosystems is the eutrophication of lakes and streams. For an increase in terrestrially derived N to have any significant influence upon the trophic status of the aquatic system, the productivity of the system should already be N limited. While phosphorus (P) is widely regarded as the principal element limiting biological productivity, under certain circumstances N deficiency has also been identified (e.g. Axler *et al.,* 1994). The N:P ratio is commonly used to provide an indication of which nutrient may be present at limiting concentrations. Establishing meaningful guideline ratio values is difficult and often tend to be site or region specific and by far the majority of studies have concentrated upon standing waters. Generally the TN:TP ratio (by weight) is reported to be wide for oligotrophic lakes and narrow for eutrophic lakes (Downing and McCauley, 1992). In a comparison of 228 north latitude lakes, Smith (1982) confirmed that chlorophyll yield is dependent upon both the P concentration and the TN:TP ratio. Critical boundary values for the N:P ratio (by weight) have been suggested as being N deficient if narrower than 10 and P deficient if wider than 17 (Forsberg *et al.,* 1978).

In natural and semi-natural ecosystems, it is very likely that because of the small N and P concentrations that are actually present in lakes and streams, individual sites will fluctuate between N and P limitation. In a study of Mountain lakes in Colorado, Morris and Lewis (1988) suggested that at ratios (by atoms) between dissolved inorganic N and total P of less than 9, phytoplankton would be stimulated by additions of both N and P, while at ratios less than 2, there was a strict P limitation. Using this criterion, they observed that the majority of lakes in their study were limited by N alone or N and P combined. By way of contrast, in a study of 109 lakes in southern Norway and based on the criteria of Morris and Lewis (1988), only 6 lakes were limited by N alone, 17 were N and P limited and the remainder were P limited (Hessen *et al.,* 1997a). Thus increased N concentrations in lakes and streams of natural and semi-natural ecosystems may skew the N:P ratio in freshwater towards a more strict P limitation, which would have minor impact on the trophic status but may induce competitive shifts in algal communities (Faafeng and Hessen, 1993). Therefore, establishing a link between N deposition and the eutrophication of freshwater systems depends on a determination that N deposition is the main source of N to the system and that productivity of the system is limited by N availability (Stoddard, 1994).

Nitrogen uptake

The potential impacts that any increased N concentrations might have on the trophic status of freshwater is greatly influenced by the physical properties of the receiving water body. In upland ecosystems, the growth of aquatic vegetation and algal communities are generally considered to be N and P limited. Therefore any increase in bioavailable N into a nutrient deficient stream would be expected to be rapidly sequestered by plant and algae, particularly in the summer when plant growth is at a maximum and warm temperatures, light conditions and low flows enhance the growth of algae. Furthermore, the benthic microorganisms involved in depleting N may be lost from the stream bed during storm events (Williamson and Cooke, 1985). Such events are more frequent in winter, which together with low temperatures and shorter days will restrict re-colonisation during this period. Nutrient uptake by algae will therefore be proportionally greater during the summer. Therefore in rapid flowing upland streams, the timing of increased N loss with respect to the biological demand for N is critical. It is likely therefore, that if the increased loss of N from the terrestrial system is limited only to winter months then this will have little immediate impact upon the productivity of streams. This does not have to be the case for standing water bodies, which because of their greater retention/turnover time, should be less likely to be influenced by the seasonality of N loss.

The capacity of in-stream processes to influence the amount, timing and composition of N in river water can be substantial. Various studies have indicated that a combination of physical, chemical and biological mediated processes can operate simultaneously and it is within this context that the forms of N present become especially significant. With respect to inorganic nitrogen, NH_4 is generally considered to be the preferred form for plant uptake and is also likely to be involved in ion-exchange reactions with sediment. Triska *et al.* (1990) observed that a large proportion of NH_4 added to a small mountain stream was temporarily retained within the channel, probably by sediment sorption, as SO_4 (the tracer) decreased almost immediately to background levels while NH_4 attenuated over 6 days. In the same study, NO_3 concentrations increased by 12% within hours of the NH_4 addition to the stream indicating that some of the added NH_4 was oxidised. The significance of biological uptake influencing stream water NO_3 concentrations has been demonstrated by Chapman *et al.* (1996) who reported a 20% retention of NO_3 during a summer addition experiment compared to an almost complete recovery during a winter experiment. Summer losses of NO_3 in headwater streams have been attributed to uptake by algal communities (Sebetich *et al.*, 1984; Mulholland, 1992) and aquatic vegetation (e.g. Howard-Williams *et al.*, 1982), whilst denitrification within stream sediments may also be important in other systems (Hill, 1979; Swank and Caskey, 1982). The relative importance ascribed to each process will vary between sites (Munn and Meyer, 1990) and with changing conditions, such as stream discharge (Cooper, 1990), at individual sites, but all processes represent potential pathways for the transformation of N in stream water (Kaushik *et al.*, 1983).

The immobilisation or uptake capacity of the stream channel may decrease following periods of prolonged inputs and a swing towards conditions where P is the limiting nutrient may occur, particularly for streams which no longer display the typical summer depression of NO_3 concentrations, indicative of an advanced stage of N saturation (Stoddard, 1994). Following the addition of P to a small moorland catchment in southwestern Norway, NO_3 concentrations were observed to decrease and pH increase (Hessen *et al.*, 1997b). The diurnal oscillation of NO_3 after the P addition suggests that the NO_3 was utilised by epiphytes and submerged vegetation in the stream (Hessen *et al.*, 1997b). The results from this study suggest that for extreme P deficient systems an increase in N inputs from the terrestrial to aquatic system will lead to higher stream water NO_3 concentrations throughout the year than in systems where P is more readily available.

The general significance of DON as a potential source of N for macrophytes and microflora has not been fully explored, although there is increasing evidence that the N from certain compounds can be taken up either directly or after hydrolysis. Conversion of dissolved N from organic to inorganic or particulate forms or vice versa can also occur (Meyer *et al.*, 1981) and although this results in little change in the total N loading in the stream, any transformations that alter the bioavailability of N may have an important consequence for water quality.

6.5. The Effects Of Nitrogen Deposition On Surface Water Chemistry

In this section the effects of N deposition on stream water chemistry are summarised from information available in the literature. Over the last decade, research projects concerned with the impact of increased N deposition on the quality of surface waters have been undertaken in Europe and North America. In Europe a large proportion of work on semi-natural systems has been carried out in Britain and Scandinavia, while in the United States, much of that effort has focused on north-eastern lakes and streams. Thus, in the following discussion, the concentrations, fluxes, seasonal patterns, long-term trends and episodic responses of N in surface waters are presented separately for each region.

6.5.1. BRITAIN

Characteristics of the uplands

In Britain, upland and marginal upland landscapes as defined in Bunce and Howard (1992) represent 37% of the total land area and are located predominantly in the north and west (Figure 6.3). In general the uplands are defined as land over 300 m above sea level, although the definition is subject to local modification (Reynolds and Edwards, 1995). The climate is cool and wet, with annual rainfall ranging between 1000 and 3000 mm. The growing season is short, ranging from 150 to 200 days (Batey, 1982). The general large precipitation and small evaporation (<500 mm) means that the uplands are normally areas of water surplus and therefore an important source of potable water.

The dominant soils range from brown earths through gleys and podzols to peats. These soils are generally acidic, imperfectly drained, organic rich and nutrient poor, although they contain substantial amounts of organic carbon, nitrogen and phosphorus in the upper horizons. Consequently streams draining upland areas are usually acidic, organic-rich and contain small concentrations of nutrients.

Table 6.3: Distribution (%) of land cover in England, Wales, Scotland and Britain

	Semi-natural	Coniferous Plantations	Managed Grass	Arable	Urban	Other
England	17	2	33	32	10	6
Wales	40	4	38	5	3	10
Scotland	57	6	15	8	2	12
Britain	33	3	27	21	7	9

Data from Barr *et al.* (1993)

Total deposition (wet, dry and occult) of inorganic N exceeds 30 kg ha^{-1} yr^{-1} in the Pennines and Lake District, whereas in the uplands of Wales and SW Scotland total N deposition ranged between 20 and 30 kg ha^{-1} yr^{-1} (INDITE, 1994). In Scotland, N deposition decreases from a maximum of 30 kg ha^{-1} yr^{-1} in SW Scotland to less than 10 kg ha^{-1} yr^{-1} in Northern and NE Scotland (INDITE, 1994).

Nitrogen behaviour in surface waters: (a) Lakes

Concentrations. In Britain, data is mainly confined to lowland lakes and reservoirs and surprisingly little information is available for the upland regions (INDITE, 1994). Unlike Scandinavia and the United States, where long-term lake water monitoring programmes were initiated in the 1980s, no database of lake water chemistry for the whole of Britain was available prior to 1990 when a major programme of water chemistry sampling was set up to provide data suitable for Critical Loads mapping. For this programme, water samples were collected, between 1990 and 1992 during the spring and autumn, from the most sensitive (to acidification) standing water in each 10 x 10 km^2 grid square in upland areas (Kreiser *et al.,* 1995). Where no suitable standing water was available headwater streams were used. Concentrations of NO_3-N ranged from <0.014 to >0.28 mg L^{-1} with the highest concentrations observed in the uplands of Wales, the Pennines, the Lake District, Galloway and the Cairngorms (Allott *et al.,* 1995). The pattern of surface water NO_3 largely reflects that of total deposition of inorganic N, with greatest NO_3 concentrations clearly associated with large total N deposition (INDITE, 1994). However, only 45% of the variance in NO_3 concentration could be explained by N deposition (Allott *et al.*, 1995). Clearly there are other important factors which govern the concentration of NO_3 in upland surface waters.

A second source of data regarding lake water NO_3 is from the UK Acid Water Monitoring Network (UKAWMN) which was established in 1988 to assess the response of surface waters in upland areas to reductions in SO_2 emissions over the long-term (Patrick *et al.,* 1995). Eleven lakes were included in the network and mean NO_3 concentrations for the first five years are presented in Table 6.4. Concentrations are extremely variable ranging from <0.035 to 0.33 mg N L^{-1} and in general, smallest mean concentrations occur in Scotland and larger mean concentrations in Wales, the Lake District and Northern Ireland. The contribution of NO_3 to the acidity of the lakes can be assessed by calculating the ratio (in equivalents) of NO_3 to the sum of NO_3 and SO_4 (Henriksen, 1988).

Table 6.4: Concentrations (μeq L^{-1}) and ratios of NO_3 to (NO_3+SO_4) in lakes of acid-sensitive regions of the UK (data from Patrick *et al.* 1995)

Site name	Region	Total N deposition ($kg\ ha^{-1}\ yr^{-1}$)	pH	NO_3	SO_4	NO_3:(NO_3+SO_4)	No.of samples	Season -ality
Loch Coire nan Arr	N Scotland	14.3	6.39	3	41	0.06	20	yes
Lochnagar	NE Scotland	15.6	5.40	11	61	0.15	19	no
Loch Chon	NW Scotland	24.1	5.47	10	72	0.12	19	yes
Loch Tinker	NW Scotland	24.1	6.10	<2.5	53		19	no
Round Loch of Glenhead	SW Scotland	26.6	4.90	5	68	0.07	19	yes
Loch Grannoch	SW Scotland	24.1	4.65	14	98	0.13	19	yes
Scoat Tarn	Lake District	22	4.93	21	61	0.26	20	yes
Burnmoor Tarn	Lake District	22	6.50	6	81	0.07	20	yes
Llyn Llagi	N Wales	25.8	5.23	11	62	0.15	19	yes
Llyn Cwm Mynach	N Wales	22.8	5.38	22	88	0.20	19	no
Blue Lough	N Ireland	12.7	4.67	24	97	0.19	12	no

A wide range in the NO_3:(NO_3+SO_4) ratio is observed from 0.06 to 0.26 (Table 6.4), indicating that the while N is strongly retained in some catchments, in others a high percentage of incoming N is moving through the catchment.

The most recent N data for standing waters in upland Britain is available from the Northern European Lake Survey 1995 (Henriksen *et al.,* 1997b). Water samples were collected from 188 lakes (>0.04 km^2) in Scotland and Wales during December 1995 and January 1996. Nitrate concentrations ranged from 0.006 to 2.53 mg N L^{-1} with a median of 0.088 mg N L^{-1} and 90% of samples contained less than 0.42 mg N L^{-1}. The water samples were also analysed for total dissolved N (TDN); the median concentration was 0.322 mg N L^{-1}. Thus the majority of N in lake water was present as dissolved organic N (DON).

Seasonality

In the Lake District, a pronounced seasonal cycle in lake water NO_3 concentrations is observed (Heron, 1961; Sutcliffe *et al.,* 1982), with a summer minima (August/ September) and winter maxima (February/March) (Figure 6.4). This annual pattern is linked with biological activity and depletion during the summer months.

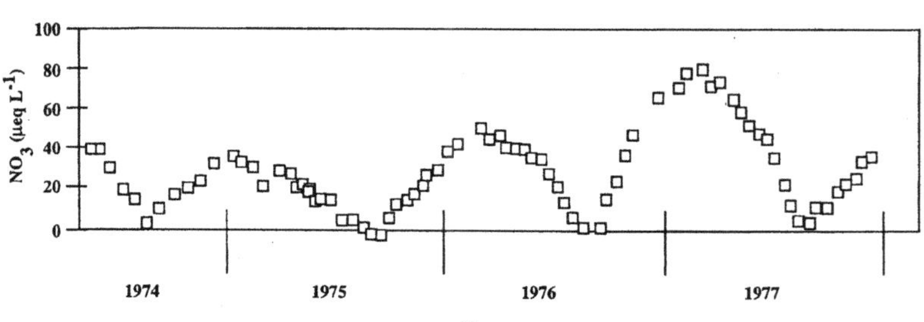

Figure 6.4: Seasonal variation in NO_3 concentration in Esthwaite Water (0–5cm), from May 1974 to December 1977 (from Sutcliffe *et al.,* 1982)

At seven of the UK AWMN lakes, NO_3 concentrations were generally larger in the winter and spring samples than the summer and autumn samples (Table 6.4). Although eight of the lake catchments received similar inputs of N deposition, the mean NO_3 concentration and seasonal NO_3 response varied between them (Table 6.4), suggesting that N deposition is not the only factor controlling rates of NO_3 leaching.

At Round Loch of Glenhead, southwest Scotland, NO_3 concentrations are significantly lower than at Scoat Tarn, Lake District throughout the year (Figure 6.5). These catchments receive similar inputs of atmospheric N, they have the same land use and are of a similar size (Table 6.5). The major differences between the catchments are altitude range and soil type. Scoat Tarn is at a higher altitude and has a considerably larger area of bare rock and much shallower soils than Round Loch of Glenhead. Thus although Round Loch of Glenhead receives slightly more annual atmospheric N than Scoat Tarn, the greater NO_3 concentrations in Scoat Tarn probably reflects the fact that the bare rock and shallow soils allow atmospheric N to pass more rapidly through the catchment and into the lake. In addition, Scoat Tarn probably receives more N from occult deposition than Round Loch of Glenhead due to its greater altitude range. From of a study of 76 lakes in the Snowdonia region, north Wales, Kernan (1997) observed that NO_3 concentrations displayed a positive correlation with altitude and area of bare rock.

Table 6.5: Nitrate concentrations and catchment characteristics of two UK AWMN sites (Patrick *et al.* 1995)

	Round loch of Glenhead SW Scotland	Scoat Tarn Lake district
Mean NO_3-N concentration (mg L^{-1})	0.07	0.3
Minimum NO_3-N concentration (mg L^{-1})	<0.035	0.13
Maximum NO_3-N concentration (mg L^{-1})	0.13	0.59
Area (ha)	95	95
Altitude range (m)	295–531	602–825
Land use	100% moorland	100% moorland
Soils	Blanket peat, peaty podzols	Shallow rankers, bare rock
N deposition (kg ha^{-1} yr^{-1})	26.6	22.2

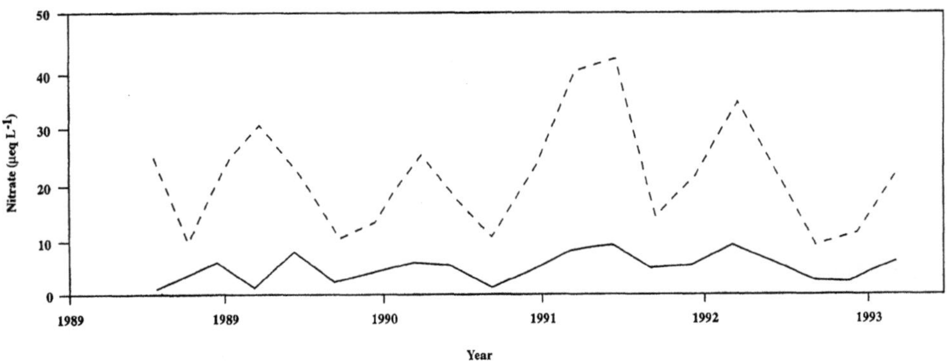

Figure 6.5: Time-series plot of lake water NO_3 concentration at Scoat Tarn and Round Loch of Glenhead (from Jenkins *et al.*, 1996)

Long-term trends

In the Lake District, NO_3 has been determined in water samples collected weekly from three lakes (Belham Tarn, Esthwaite Water and Windermere – south and north basin) between 1946 and 1980 (Sutcliffe *et al.*, 1982). As the seasonal variation in lake water NO_3 concentrations were large (Figure 6.4), comparisons between years were based on the mean of 10 consecutive determinations after NO_3 reached maximum concentrations during the winter. The timing of the winter/spring maximum varied between lakes and was reached first in the catchment with the shortest retention time: Belham Tarn<Esthwaite Water<Windermere. The data show that the winter/spring concentrations of NO_3-N have increased during the 1946–1980 period in all four lake basins (Figure 6.6). The most rapid period of increase occurred in the late 1960s and 1970s. A change in analytical methods in 1964 was responsible for a large part (32–40%) of the observed increase, nevertheless an increase is still evident. The fact that high maxima years were shared by all lakes suggests that climatic factors are important in controlling lake water NO_3 concentrations. As NO_3 data for other lakes in the Lake District indicated little change in the overall mean concentrations of NO_3 between the surveys of 1955–1956 and 1974–1978, Sutcliffe *et al.* (1982) suggested that the increase in the NO_3 content of the three lakes may result from an increase in the use of N fertilizers and sewage inputs.

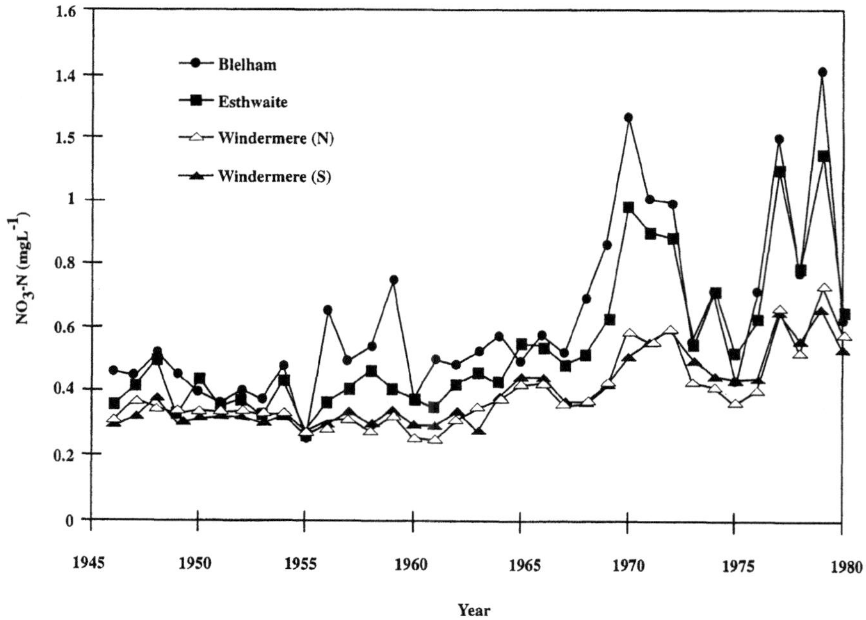

Figure 6.6: Long-term patterns of NO_3 concentrations in four lakes in the English Lake District (data are from Sutcliffe *et al.*, 1982)

Sutcliffe *et al.* (1982) found no evidence to suggest that the alkalinity or pH of Belham Tarn, Esthwaite Water and Windermere have decreased over the same period. In fact alkalinity increased significantly in the late 1960s and 1970s. The most marked rise in alkalinity coincided with the increase in winter concentrations of NO_3 and PO_4-P in the late 1960s. Sutcliffe *et al.* (1982) suggested that the correlation between nutrient enrichment and increased alkalinity may have developed partly as a by product of biological activity. For example, the reduction of NO_3 to N_2 or NH_3 by a combination of biological denitrification and algal assimilation results in the generation of alkalinity. Thus higher concentrations of NO_3 in winter may lead to an increase in lake water alkalinity.

(b) Streams

Concentrations. In Britain, the Environment Agency (EA) and Scottish Environment Protection Agency (SEPA) monitor river water quality. However, most of the sites are mainly located in the lower reaches which are invariably affected by agricultural and sewage inputs. Betton *et al.* (1991) used data collected at 743 of these monitoring sites to map mean concentrations across the country. The monitoring sites were all located near source rather than further downstream on major rivers. Rivers in upland areas were associated with mean NO_3-N concentrations < 2.5 mg N L^{-1} and within the higher massifs of southwest England, Wales, northern England and Scotland mean river NO_3-N concentrations were < 1 mg L^{-1}.

In a summer (April–August) survey of 61 upland rivers from throughout Britain, mean NO_3-N concentrations ranged from <0.005 to 3.05 mg L^{-1} with a median of 0.318 mg L^{-1} and 90% of samples contained less than 1.0 mg L^{-1} (Chapman *et al.*, 1998). In addition, NO_3 concentrations varied significantly between upland regions (Figure 6.7); the mean NO_3 concentration was over twelve times greater in rivers draining the uplands of south Wales (mean = 1.02 mg L^{-1}) than the highlands of Scotland (mean = 0.08 mg L^{-1}). This regional pattern in upland river NO_3 concentrations is consistent with that observed across the UK AWMN sites (Patrick *et al.*, 1995), where NO_3 concentrations in sites in northwest and central Scotland had small mean concentrations with small standard deviations, while sites in Wales had larger means and larger standard deviations.

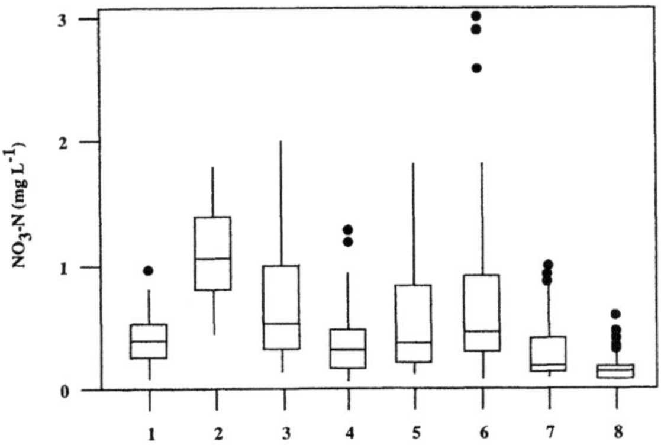

Figure 6.7: Box and whisker plots summarising concentrations of NO_3-N in samples of river water collected from different upland regions of Britain; (1) SW England, (2) S Wales, (3) N Wales, (4) Pennines, (5) Upper Tweed catchment, (6) SW Scotland, (7) NE Scotland, (8) Highlands of Scotland. In a box and whisker plot, the middle horizontal line of the box represents the median value, the ends of each box delineate the upper and lower quartiles and the whiskers extend to the 90th and 10th percentile. Outliers are represented by a closed circle (from Chapman *et al.* 1998).

This variation in stream water NO_3-N concentrations between regions may reflect atmospheric deposition patterns (Allott *et al.*, 1995), but it is also likely to reflect variability in the characteristics of upland regions.

In studies of moorland catchments which ranged in size between < 1–50 km^2 mean annual NO_3-N concentrations in streams are generally very small (Table 6.6), typically below 0.2 mg L^{-1} and NH_4 concentrations are usually below the detection limits of routinely used analytical methods. Mean NO_3 concentration for the majority of streams ranged between 0.11 and 0.21 mg N L^{-1} which is surprisingly small given the geographical spread of samples. Streams draining forested catchments may have mean annual NO_3-N concentrations which exceed 0.2 mg L^{-1} (Table 6.7) and a number of studies have shown that streams draining mature conifer plantations contain more NO_3 than streams draining adjacent moorland (Table 6.8). The reasons for this have been widely discussed (e.g. Reynolds and Edwards, 1995) and contributing factors are considered to include a combination of:

(i) Deposition effects – forests receive approximately double the atmospheric N inputs of moorland vegetation (Fowler *et al.*, 1989).
(ii) Vegetation effects – changes in N requirement with tree growth (Stevens *et al.*, 1993b).
(iii) Deposition, vegetation, soil interactions – increased interception and transpiration by forests leads to drier forest soils compared to moorland soils which results in the increased oxidation of organic N compounds (Hornung *et al.*, 1987) and reduced runoff.

Table 6.6: Annual mean dissolved inorganic N concentrations (mg N L^{-1}) and fluxes (kg N ha^{-1} yr^{-1}) for upland moorland catchments in the UK

Catchment	Region	Mean concentration NO_3-N	Mean concentration NH_4-N	Estimated flux NO_3-N	Estimated flux NH_4-N	Length of study	Number samples	Reference
Wye	mid-Wales	0.18	0.013	3.8	0.27	7/76–7/80	1460	Roberts *et al.* (1983)
Cyff	mid-Wales	0.18	<0.1	3.6		1/83–12/85	156	Reynolds *et al.*(1989)
Gwy	mid-Wales	0.21	<0.1	4.4		1/83–12/85	156	Reynolds *et al.*(1989)
C2	mid-Wales	0.15	<0.1	3.3		10 years	286	Reynolds *et a.l* (1992)
Camddwr	mid-Wales	0.17	0.1	2.5	0.19	2/81–1/83	58	Stoner *et al.* (1984)
Glendye	NE Scotland	0.19	0.13	2.2	1.5	10/77–9/78	47	Reid *et al.* (1981)
Glendye	NE Scotland	0.11	<0.01	1.4		6/84–6/85	12	Edwards *et al.* (1985)
Peatfold	NE Scotland	0.2	<0.01	1.7		6/84–6/85	12	Edwards *et al.* (1985)
Mharchaidh	NE Scotland	<0.035	<0.01	<0.33		7/88–3/93	57	Patrick *et al.* (1995)
Ciste Mhearad	NE Scotland	0.13	0.01	2.2	0.17	2/84–4/8585	60	Cooper *et al.* (1987)
Halladale	N Scotland	0.039	0.027	0.27	0.19	1/93–1/94		Miller, 1995
Dargall Lane	SW Scotland	0.2	0.008	4.9	0.19	1/87–12/89	156	Lees, 1991
Trout Beck	Pennines	0.12	0.01	1.87	0.16	1/93–12/95	156	Adamson *et al.*(1998)
Narrator	Dartmoor	0.13	<0.01	1.5		4/88–3/93	57	Patrick *et al.* (1995)
Loughgarve*	N Ireland	0.04	0.03	0.55	0.4	1/93–12/93	26	Gibson *et al.* (1995)

* Catchment includes a lake

Table 6.7: Annual mean dissolved inorganic N concentrations (mg N L^{-1}) and fluxes (kg N ha^{-1} yr^{-1}) for upland forested catchments in the UK

Catchment	Region	Mean concentration		Estimated flux		Length of study	Number of samples	% Forested (stand age)	Ref
		NO_3-N	NH_4-N	NO_3-N	NH_4-N				
Severn	mid-Wales	0.27	0.01	5.0	0.19	7/76–7/80	1460	70 (35)	1
Hafren	mid-Wales	0.39	<0.1	7.1		1/83–12/85	156	48 (35)	2
Hafren	mid-Wales			6.5	0.39	1985–1990	260	48 (35)	3
Hore	mid-Wales	0.32	<0.1	6.0		1/83–12/85	156	77 (35)	2
Hore	mid-Wales			11.1	0.39	1985–1990	260	77–30(35)	3
Upper Hore	mid-Wales			5.3	0.23	1985–1990	260		3
Tywi	mid-Wales	0.32	0.014	3.2	0.2	2/81–1/83	58	80(30)	4
Beddgelert	N Wales	0.72	<0.1	12.7		6/82–6/84	312	100(50)	5
Kelty	Scotland	0.14	0.19	2.7	3.6	10/85–5/87		65(40)	6
Chon	Scotland	0.04	0.1	0.7	1.8	10/85–5/87		65(40)	6
Green Burn	SW Scotland	0.14	0.009	3.0	0.19	1/87–12/89	156	70(20)	7
White Laggan	SW Scotland	0.18	0.025	3.9	0.54	1/87–12/89	156	30(20)	7
Kershop	NW England	0.92	0.1	6.2	0.67	4/82–4/83	100	100(35)	8

References: (1) Roberts *et al.* (1983), (2) Reynolds *et al.* (1989), (3) Durand *et al.* (1994), (4) Stoner *et al.* (1984), (5) Stevens *et al.* (1989), (6) Miller *et al.* (1990), (7) Lees (1991), (8) Stevens *et al.* (1990).

Table 6.8: Mean NO_3-N concentrations (mg L^{-1}) draining forested and moorland catchments in upland Britain

Site	Moorland	Forest[1]		Reference
Wales				
Llyn Brianne	0.17	0.23	(22–24)	Stoner *et al.* (1984)
Plynlimon	0.21	0.39	(21–40)	Reynolds *et al.* (1989)
Beddgelert	0.08	0.63	(46–54)	Hornung *et al.* (1990)
Scotland				
Loch Dee	0.10	0.08	(8–12)	Farley and Werrity (1989)
Loch Chon	0.12	0.14	(Mature)	Harriman & Morrison (1982)
Duchray	0.14	0.14	(Young)	Harriman & Morrison (1982)
SWAP sites	0.01	0.13	(Mixed ages)	Harriman *et al.* (1990)

[1]Forest age in brackets

Recent evidence from a survey of moorland and forested catchments in Wales suggests that forest age is the major control of NO_3 leaching (Emmett *et al.*, 1993; Stevens *et al.*, 1994). Streams draining moorland and afforested catchments with stand ages less than 30 years old characteristically had small NO_3 concentrations compared to stands > 30 years old which possessed characteristics similar to those described for 'N saturated' sites described by Aber *et al.* (1989), acting as net sources rather than sinks of NO_3 (Emmett *et al.*, 1993).

Upland agricultural improvement can also influence the N content of surface waters. Pulses of NH_4 were observed in stream water following the initial improvement and after remedial fertilizer application to 22 ha of moorland (Roberts *et al.*, 1989) and large quantities of NO_3 were released, 61, 22 and 14 kg N ha^{-1} in the three years following ploughing and fertilization of deep peat (Roberts *et al.*, 1986). Studies of the short-term effects of land improvement have shown that liming alone can increase NH_4 and NO_3 production in upland soils (Reynolds and Emmett, 1990), by stimulating microbial activity (Shah *et al.*, 1990). Hornung *et al.* (1985) observed that mean annual NO_3 concentrations in streams draining land that had been agriculturally improved 10 and 40–50 years previously were significantly larger than in streams draining unimproved moorland (0.28, 0.18, 0.07 mg L^{-1}, respectively).

In NE Scotland, the upland semi-natural catchments of the Rivers Dee and Don have small total N concentrations (<1.0 mg L^{-1}), with organic N being the major component (Edwards *et al.*, 1996). The concentration of NO_3 generally increases downstream (Oborne *et al.*, 1980; Talling *et al.*, 1989), as does total N, closely reflecting changes in land use (MacDonald *et al.*, 1994). Small areas of improved land can have a disproportionate effect upon stream water NO_3 concentrations. For example, the annual mean NO_3 concentration below a small area of improved land was 0.443 mg L^{-1} compared with 0.053 mg L^{-1} above it (Edwards *et al.*,1985). Thus the larger the catchment and the further downstream the monitoring site, the more likely the influence of agriculture on the NO_3 concentration of rivers.

Table 6.9: Concentrations (μeq L^{-1}) and ratios of NO_3 to (NO_3+SO_4) in upland streams of the UK

Site	Region	pH	NO_3	SO_4	NO_3:(NO_3+SO_4)	Reference
Moorland						
C2	mid-Wales	4.98	6	91	0.06	Reynolds *et al.* (1983)
Cyff	mid-Wales	5.43	13	98	0.12	Reynolds *et al.* (1989)
Gwy	mid-Wales	4.89	15	75	0.17	Reynolds *et al.* (1989)
Camddwr	mid-Wales	5.30	12	77	0.13	Stoner *et al.* (1984)
Duchray	Scotland	4.90	10	115	0.08	Harriman & Morrison (1982)
Dargall Lane	SW Scotland	5.18	12	82	0.13	Lees (1991)
Ciste Mhearad	NE Scotland	5.15	9	39	0.19	Cooper *et al.* (1987)
Forested						
Hafren	mid-Wales	4.69	28	98	0.22	Reynolds *et al.* (1989)
Hore	mid-Wales	4.80	23	117	0.16	Reynolds *et al.* (1989)
Beddgelert	N Wales	4.30	51	137	0.27	Stevens *et al.* (1989)
Tywi	mid-Wales	4.70	16	137	0.10	Stoner *et al.* (1984)
Chon	Scotland	4.61	3	93	0.03	Miller *et al.* (1990)
Kelty	Scotland	4.02	10	100	0.09	Miller *et al.* (1990)
Duchray	Scotland	4.66	10	142	0.09	Harriman & Morrison (1982)
Green Burn	SW Scotland	5.20	8	73	0.10	Lees (1991)

In general NO_3 contributes less than 5% to the strong anion charge of moorland streams (Reynolds and Edwards, 1995) and thus plays a minor role in the acidity of surface waters. In Table 6.9, the ratio of NO_3 to the sum of NO_3 and SO_4 reflects the contribution of NO_3 to the acidity of some moorland and forested streams. A value of 0.1 or less indicates a strong retention of N in the catchment and little contribution of NO_3 to stream water acidity (see Section 6.4.1). Largest values are observed for the forested catchments in Wales. However, no relationship is observed between pH and the NO_3:(NO_3+SO_4) ratio, suggesting that NO_3 is not the major control on acidity in upland streams.

Fluxes

Estimates of NO_3 losses from semi-natural and forested upland catchments are presented in Tables 6.6 and 6.7, respectively. These values were calculated from the mean annual NO_3 concentration and mean annual runoff for the catchment. Producing accurate estimates of chemical loads in rivers is problematic even where regular sampling and continuous flow monitoring are carried out (Walling and Webb, 1985; Littlewood, 1992). Regular sampling programmes tend to over emphasise base flow components which is compounded the longer the time interval between samples. Samples taken during storms, when changes in NO_3 concentration are likely to be greatest (Webb and Walling, 1985; Roberts *et al.,* 1984), are less likely to be represented in the data set. Therefore, the NO_3 fluxes presented in Tables 6.6 and 6.7 are only intended to provide a first approximation, and are likely to underestimate the actual load. For example, NO_3 fluxes calculated by Roberts *et al.* (1983) for the headwaters of the River Wye and Edwards *et al.* (1985) for Glendye, based on mean monthly concentration and flow data rather than mean annual values, were 27% and 16%, respectively, higher than calculated here. However, in the absence of more detailed approaches recommended by Littlewood (1992), these estimates still provide an insight into variations in NO_3 fluxes from upland catchments in Britain.

For semi-natural catchments losses of NO_3 tend to be larger from catchments with runoff in excess of 2000 mm yr^{-1}, such as those in Wales and SW Scotland. Therefore, while NO_3 concentrations may display no discernable geographical pattern, differences in precipitation produce a broad regional pattern in load, with greatest losses of between 4.9 and 2.5 kg N ha^{-1} yr^{-1} observed in Wales and SW Scotland and considerably smaller rates (<2.2 kg N ha^{-1} yr^{-1}) observed from catchments in the Highlands of Scotland, NE Scotland, SW England and N Ireland (Table 6.6). For those catchments where NH_4 was detected in stream water, estimates of NH_4 losses are presented in Table 6.6. In general, NH_4 fluxes from upland catchment are small, only at Glendye did NH_4 losses exceed 1 kg ha^{-1} yr^{-1}.

For forested catchments, the NO_3 flux varied from 0.7 to 12.7 kg N ha^{-1} yr^{-1} (Table 6.7). The largest flux was observed at Beddgelert, Wales. As observed for semi-natural catchments, larger NO_3 fluxes were observed from forested catchments in Wales than Scotland. Only at the two forested catchments in Central Scotland were NH_4 concentrations and loads larger than those of NO_3.

Comparisons have only been made of inorganic N losses from the catchments in Tables 6.6 and 6.7, as very few studies have attempted to quantify fluxes for organic N. In mid-Wales, Roberts *et al.* (1983) observe that DON accounted for an average of 36% of the annual TDN flux of 8.6 kg N ha^{-1} from a moorland catchment and 26% of the annual TDN flux of 9 kg ha^{-1} yr^{-1} from a forested catchment. These annual losses of DON from the headwaters of the Wye and Severn are of a similar magnitude to the annual losses of DON, 2–3 kg ha^{-1}, calculated for the eight tributaries of the Rivers Dee and Don, NE Scotland, by Edwards *et al.* (1996) using the average C:organic N ratio of 20:1 with published values of organic C losses (Hope *et al.*, 1994), which are in the range of 50 kg C ha^{-1} yr^{-1}. As DON is associated with DOC, largest losses of DON are likely to occur from catchments with the largest DOC fluxes. Hope *et al.* (1997) showed that the size of the soil carbon pool is the single most important factor in determining DOC fluxes in British rivers. Therefore, a positive relationship between DON flux and peat coverage in upland catchments may be expected.

When inorganic N inputs in bulk precipitation are plotted against inorganic outputs in stream water for the catchments in Tables 6.6 and 6.7, no discernable relationship is observed (Figure 6.8). The data do demonstrate that inorganic N inputs exceed streamflow losses of N in all catchments except two, indicating that N appears to be retained within these upland ecosystems. However this observation must be treated with some caution as all the main inputs and outputs of N have not been fully considered.

The wide range in N outputs for a given N deposition reflects the importance of site-specific factors which together determine the extent of the soil and vegetation sinks for the incoming N (Skeffington and Wilson, 1988). Thus Skeffington and Wilson (1988) suggested that a significant relationship between N inputs and outputs for semi-natural ecosystems is most unlikely unless consideration is limited to a single type of ecosystem, covering a restricted geographic region where the geology, soil types and climate of the catchments are all similar.

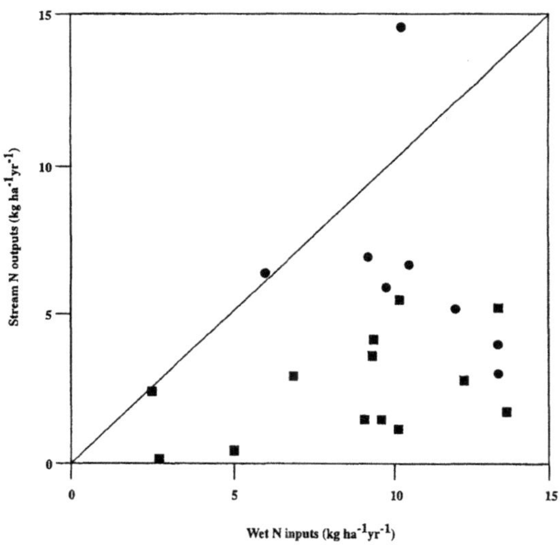

Figure 6.8: Relationship between wet deposition of N (NO_3+NH_4) and outputs of N (NO_3+NH_4) in stream water for the moorland (■) and forested (●) catchments in Tables 6.6 and 6.7. The solid line is the one-to-one line.

Seasonal variability

A pronounced seasonal cycle in NO_3 concentrations with summer minima and winter (February or March) maxima are commonly, but not exclusively, observed in upland moorland catchments (Roberts *et al.*, 1984; Edwards *et al.*, 1985; Betton *et al.*, 1991;

Reynolds *et al.*, 1992; Chapman, 1994; Patrick *et al.*, 1995) and may also be apparent for afforested catchments (Roberts *et al.*, 1984; Reynolds *et al.*, 1994; Langan *et al.*, 1997). Reynolds *et al.* (1994) determined the ratio of mean winter to mean summer NO_3 concentrations for 136 Welsh streams and observed that seasonality was significantly more pronounced at sites where the average age of conifer trees was over 30 years old, by comparison to moorland catchments (mean ratio 1.56 versus 1.25). In general the mean summer:mean winter NO_3 concentration increased with the average age of the conifer trees on the catchment indicative that these catchments may be displaying the initial signs of N saturation (e.g. Stoddard and Murdoch, 1991). Reynolds *et al.*, (1994) suggested that the difference in NO_3 seasonality between old forests and young forests/moorlands may result from that fact that:

(i) older forests accumulate larger amounts of mineral N in the soil during the summer which is subsequently flushed out during the winter.
(ii) older forests receive larger atmospheric inputs, particularly as cloud water, compared with younger trees and moorland vegetation.
(iii) the microclimate created by the forest canopy may allow soil microbial N transformations to continue for longer in the winter than on the more exposed moorland.

Stevens *et al.* (1993b) observed contrasting seasonal patterns in stream water for a forested and moorland catchment in North Wales which they attributed to a reduction in N utilisation by the 50 yr old forest stand. In the moorland stream NO_3 concentrations were generally below 0.01 mg N L^{-1} in the growing season and rarely exceeded 0.4 mg N L^{-1} in the winter (Figure 6.9) which is typical of Stage 0 N loss. In contrast, stream water NO_3 concentrations in the forest stream were normally within the range 0.4 to 1.5 mg N L^{-1}, with highest concentrations observed in the summer (Figure 6.9) suggesting that the system has reached Stage 3 of N loss, becoming a net source rather than sink of N.

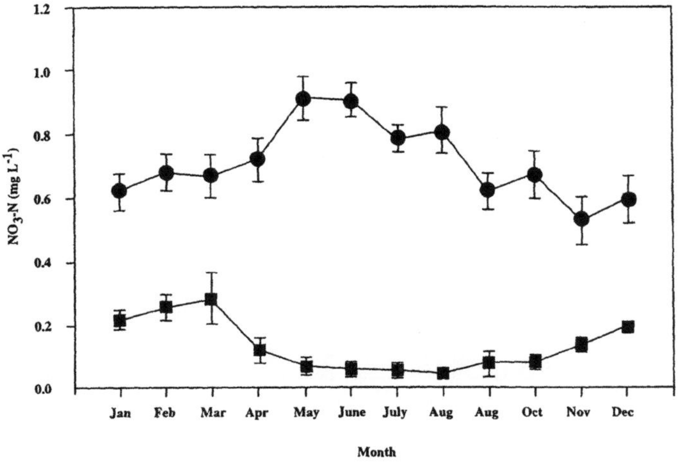

Figure 6.9: Seasonal trends in NO_3-N concentration (mean ± standard error) for the forest (●) and moorland (■) stream at Beddgelert, N Wales. Means are based on 9 years data for forest stream and 8 years data for the moorland stream (from Stevens *et al.*, 1993).

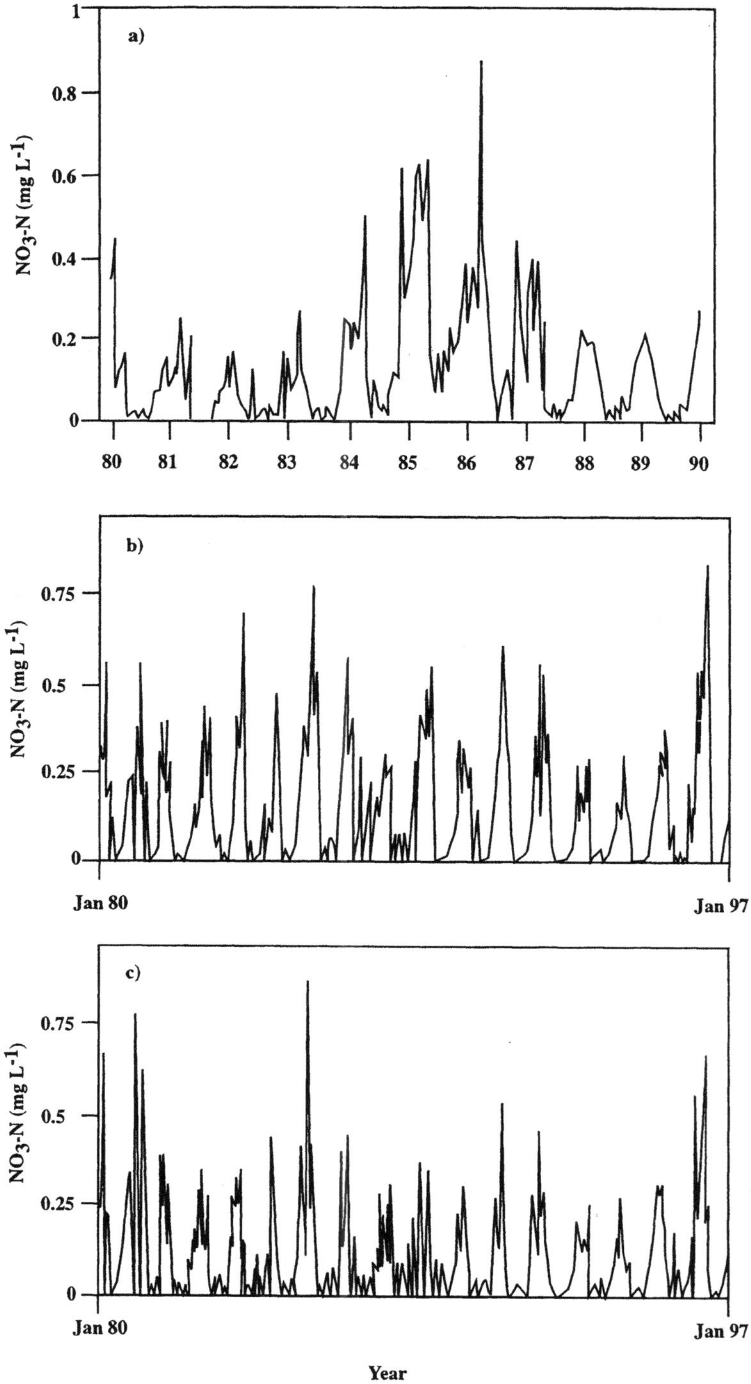

Figure 6.10: Time-series plot of stream water NO_3-N concentration at (a) the C2 catchment (moorland), Plynlimon, mid-Wales (from Reynolds *et al.,* 1992), (b) Dargall Lane (moorland), Loch Dee, and (c) Green Burn (70% forested), Loch Dee, SW Scotland (from Langan *et al.*, 1997).

While changes in seasonal variability in stream water NO_3 concentrations have been observed for forested catchments, little information exists that suggests that moorland catchments are being affected. Although in a study of seven moorland catchments in northeast Scotland, it appeared that much of the NO_3 deposited on acid peats in the summer was not retained, resulting in significant NO_3 leaching from these soils in comparison with the mineral soils lower in the catchments (Black *et al.*, 1993). By way of contrast, a three year study of a blanket peat catchment in the Pennines found that inorganic N export from the catchment in stream water was very much lower than atmospheric inputs because of immobilisation in the blanket peat (Adamson *et al.*, 1998). Denitrification may also remove NO_3 from poorly drained peats.

Long-term trends

Betton *et al.* (1991) used data collected from 743 monitoring sites within Britain to analyse the pattern of change in NO_3 concentrations over the period 1980–1986. Trend analysis of the data showed a significant increase, a significant decrease and no significant change in NO_3 concentrations at 51%, 11% and 38% of the sites, respectively. The highest rates of increase in mean concentrations were observed in the lowland areas supporting high intensity agriculture and in the winter. However, the largest (>10%) relative increase in NO_3 concentrations were observed in the upland areas of Northern England and the Scottish Highlands. As these areas receive little additional inputs of N, other than from the atmosphere, these results suggest that changes in atmospheric N deposition have resulted in the observed increase in stream water NO_3.

In contrast, no significant long-term trends in stream water NO_3 concentrations have been observed in studies of upland catchments. In a 300 ha moorland catchment in mid-Wales, no discernable annual trend in stream water NO_3 concentrations was observed over the 10 year period 1980–1990 (Figure 6.10a), although there were large annual variations attributable to climatic variations (Reynolds *et al.*, 1992). Similarly in the Cairngorm region of Scotland, one of the most remote areas of Britain, no evidence of long term increase in NO_3-N concentrations was observed between 1983 and 1992 in either the headwaters of the River Dee or a small stream draining into a large loch system (INDITE, 1994). At Loch Dee in southwest Scotland, stream water NO_3-N concentrations draining both forested and moorland catchments show no trend over a 16 year period, 1980–1996 (Figure 6.10b) (Langan *et al.*, 1997).

Response during storms

Although the seasonal variability of NO_3 concentrations in upland streams has been well documented, considerably fewer studies have been concerned with the behaviour of NO_3 during storms. In a small (0.04 km^2) headwater moorland catchment of the River Wye in mid-Wales, Chapman (1994) observed that the response of NO_3 varied between storms. During summer (June and July) storms, when baseflow NO_3 concentrations were low, a small increase in NO_3 concentrations was observed, as NO_3 concentrations in storm runoff were larger than in stream baseflow. In contrast during storms in October, November and March, when baseflow NO_3 concentrations were approximately 50% greater than in the summer, an inverse relationship between concentration and discharge was observed due to the fact that NO_3 concentration in storm runoff were lower than in stream baseflow. As the

NO_3 content of both storm runoff and stream baseflow vary seasonally, the storm response of NO_3 is dependent on the relative difference between the NO_3 content of stream baseflow and storm runoff (Chapman, 1994). For the larger Wye and Severn catchments in mid-Wales, Roberts *et al.* (1984) also observed contrasting storm response for NO_3; concentrations increased with flow in storms following warm, dry periods, whereas in winter storms concentrations varied little. In a small (2.5 km^2) forested catchment at Llyn Brianne, mid-Wales, Soulsby (1995) monitored 19 storm events over a four year period (1986–1990) and observed that NO_3 concentrations were generally small in storm runoff. Only during one event associated with substantial snowmelt did NO_3 concentrations increase, reaching a maximum of 1.4 mg N L^{-1}.

The most detailed investigation of stream NO_3 response to storm events in Britain was carried out by Webb and Walling (1985) for a grassland catchment in Devon in which they recorded nearly 600 storm events over eight years (1975–1983). Although the catchment has its source in the upland area of Dartmoor, the major land use is permanent pasture grazed by dairy cows. However, the river is free from significant inputs of domestic sewage and industrial effluent. They observed that NO_3 concentrations displayed both positive and negative relationships with flow during storms and a clear seasonal distribution in the occurrence of the two contrasting types of storm NO_3 response was evident. Dilution responses dominated the months from October to April, while increases in NO_3 concentrations were more typical of the period May to September. Webb and Walling (1985) attributed this trend to seasonal variations in the dominant sources of storm runoff.

Although acidic episodes in surface waters have been studied at many sites in Scotland and Wales (Davies *et al.,* 1992), no study reported data for NO_3 except at the sub-arctic Ciste Mhearad catchment in Scotland (Copper *et al.,* 1987), where NO_3 concentration increased from 0.168 to 0.336 mg N L^{-1} during spring in response to early snowmelt. This increase in NO_3 was accompanied by an increase in SO_4 from 1.92 to 2.34 mg S L^{-1} and a decrease in Ca concentrations and pH from 5.2 to 4.7. A later study in the same catchment (Davies *et al.,* 1993) also observed an increase in stream water NO_3 concentration from 0.2 to 2.9 mg N L^{-1} in response to snowmelt. During this event, SO_4 concentrations also increased from 0.64 to 5.44 mg S L^{-1} and pH decreased to a minimum of 4.0. At the peak of this event the NO_3:(NO_3+SO_4) ratio was 0.37, although NO_3 only contributed 29% to the total acid anion concentration (NO_3+SO_4+Cl). In contrast, in a study of snowmelt in the Cairngorm mountains of Scotland, Jenkins *et al.* (1993) observed that although NO_3 concentrations were large during the early part of snowmelt, NO_3 peaks were not observed in the stream, which they attributed to biological activity. However, while snowmelt was collected daily, stream water was only collected every two weeks. Thus any increase in stream water NO_3 concentrations may have been missed.

In mid-Wales, a depression in pH was observed during all rainfall events studied by Chapman (1994) despite the contrasting behaviour in NO_3 response; an increase in organic acid concentrations was the main process responsible for the pH depression during all events. Although limited NO_3 data are currently available for storms events, it appears that the contribution of NO_3 to the episodic acidification of streams is more important in snow-melt events than rainfall events. In addition, it is these snowmelt events that produce the most acidic episodes reported in Britain.

6.5.2. Norway, Sweden and Finland

Characteristics of natural ecosystems

The geology of Norway, Sweden and Finland is dominated by the Precambrian Fennoscandinavian or Baltic shield which consists of base poor gneisses and granitic gneisses, although glacial moraine deposits are found in southern Sweden. Soils in Finland are largely podzolic, having developed from till or glacialfluvial deposits on the granitic bedrock (Kortelainen *et al.,* 1989). The topography is flat and peatlands cover approximately a third of the land area. Forestry is the major land use (Table 6.10) and coniferous forests are predominant. At lower altitudes the major soils in Sweden and Norway are podzolic whereas in the mountains the principal soils are tundra lithosols and rankers (Papadakis, 1969). Forestry is the major land use in Sweden, whereas alpine and arctic vegetation dominate in Norway where 46% of the land lies above the tree line (1000 m.a.s.l) (Table 6.10).

Table 6.10: Distribution (%) of the main landscapes in Finland, Norway and Sweden

	Mountains	Forest	Mires	Cultivated	Water	Built-up
Finland	<1	75	5	7	10	3
Norway	48	36	7	3	5	1
Sweden	8	65	8	8	9	2

Data from Henriksen *et al.* (1997b)

The difference in soils between Norway and Finland is reflected in the total organic carbon (TOC) concentrations of the rivers and lakes. From the Norwegian '1000 lake survey' a negative relationship between altitude and TOC concentration was observed, the highest concentrations being found at low elevations in coniferous forests, where the soils were thicker and forest productivity greater (Petersen et al., 1995). However, 90% of lake samples contained less than 6 mg L^{-1} TOC and 60% had values less than 2 mg L^{-1}. Above the tree line all lakes had DOC concentrations less than 2 mg L^{-1}. In contrast, the median TOC concentration in a survey of 987 lakes in Finland was 12 mg L^{-1} and over 91% of the lakes had TOC concentrations above 5 mg L^{-1} (Kortelainen *et al.*, 1989). The low TOC concentrations in Norway reflect the dominance of shallow soils, whereas the much higher values of the Finnish lakes are due to the presence of extensive peatlands throughout the country. Concentrations of TOC in Swedish lakes are similar to those observed in Finland (Henriksen *et al.,* 1997b); the median concentration of 3075 lakes sampled in 1995 was 6.05 mg L^{-1}.

There are extreme differences in climate varying from maritime in southwest Norway and Sweden to the arctic in northern Norway. Along the south–north latitudinal gradient of Scandinavia the annual mean temperature ranges from 7.8 ^{o}C in southern Sweden to –1.5 ^{o}C in the arctic region (Petersen *et al.,* 1995). There is also a wide range in precipitation amounts, which is the consequence of distance from the sea and topography (Henriksen *et al.,* 1997b). In Norway rainfall decreases from 3000 mm yr^{-1} on the west coast to 300 mm yr^{-1} in central Norway, on the Eastern side of the

Table 6.11: The four groups of rivers in Finland, Norway and Sweden

Group	Vegetation	Location	Characteristics	Land Use
1	Former deciduous forest	South coast of Norway South coast of Sweden	low gradient	Agriculture
2	Mixed coniferous forest	Southern Norway South/Central Sweden Southeastern Finland	low gradient many lakes/ponds & wetlands high DOC concentrations	Forestry
3	Boreal forest	West coast of Norway Northern Sweden (>60°N) Northern Finland	steep gradient thin soils oligotrophic waters low conductivity waters	Commercial forestry
4	'Alpine/Arctic vegetation	Northernmost Norway Mountains of Norway & Sweden	above birch tree line shallow soils/bare rock mountainous (up to 2400 m)	Sheep and rein-deer grazing

From Petersen *et al.* (1995)

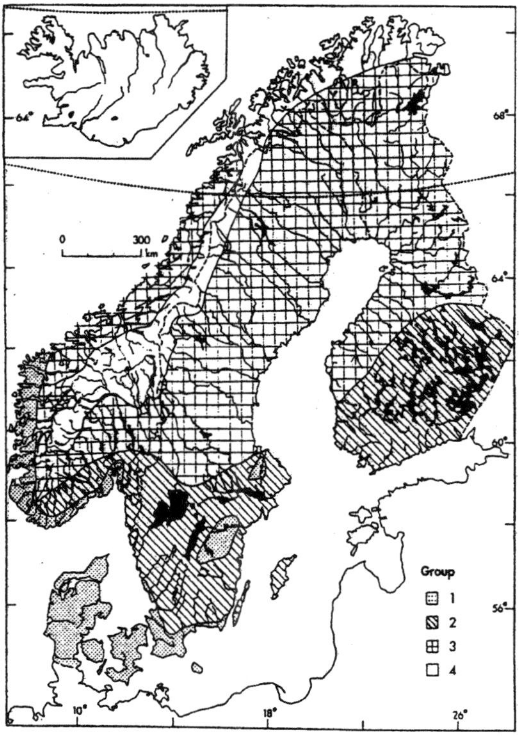

Figure 6.11: The distribution of the four Nordic River Groups in Finland, Norway and Sweden; Group 1, deciduous forest rivers; Group 2, mixed coniferous fores rivers; Group 3, boreal rivers; Group 4, arctic rivers. Major lakes shaded in black (from Petersen *et al.*, 1995).

mountain chain. For Sweden there is also a large gradient in rainfall from 1500 mm yr^{-1} close to Norway, through western Sweden (800–1000 mm yr^{-1}) to 600 mm yr^{-1} in eastern regions. The variability in rainfall is less for Finland, ranging between 600 and 800 mm yr^{-1} in the south and central areas to 500–600 mm yr^{-1} in the north.

Petersen *et al.* (1995) used the four major natural vegetation types; nemoral, boreo-nemoral, boreal and alpine to divide the rivers of Scandinavia and Finland into four groups (Figure 6.11; Table 6.11); Group 1, low-gradient streams in former deciduous forests, now extensively used for agriculture; Group 2, low-gradient streams in mixed coniferous forests, heavily interspersed with clear or humic lakes, ponds and wetlands; Group 3, high gradient streams in the coniferous and deciduous forests of the Boreal vegetation zone; and Group 4, streams of the alpine and arctic vegetation. The climatic and precipitation differences within Scandinavia and Finland are reflected in the different flow regimes of the rivers. Rivers in the south are dominated by runoff from the autumn and winter rains, while those in the north tend to be dominated by snowmelt in spring/summer with a transitional zone existing in the middle (Petersen *et al.*, 1995).

Annual total N deposition ranges from 17 kg ha^{-1} in southern Sweden to 8 kg ha^{-1} in southwestern Norway to less than 1 kg ha^{-1} in northern Norway, Sweden and Finland (Henriksen *et al.*, 1997b).

Nitrogen behaviour in surface waters: (a) Lakes

Concentrations. During the past decade extensive lake surveys have been conducted in Norway, Sweden and Finland to establish the extent of surface water acidification and to quantify long-term chemical changes (e.g. Kortelainen *et al.*, 1989; Brakke *et al.*, 1988; Henriksen *et al.*, 1988). The most recent survey was carried out during the autumn of 1995 (Henriksen *et al.*, 1997b) the results of which are summarised in Table 6.12. Although the median concentration of NO_3 was similar and small for all countries, the largest concentrations (>0.056 mg N L^{-1}) were associated with the areas receiving largest atmospheric inputs; southern and western Norway, southern Sweden and southern Finland. A decrease in NO_3 concentrations moving from south to north, was generally accompanied by a decreased N deposition and agricultural activity. For Norwegian lakes, greatest concentrations of NO_3 were observed in areas which received N deposition >7 kg ha^{-1} yr^{-1} (Wright *et al.*, 1997) although because of a wide scatter in the data only 46% of the variance in NO_3 concentration could be explained by the amount of N deposition.

Table 6.12: Median concentrations of solutes in lakes surveyed during 1995

	Finland	Norway	Sweden
Number of lakes	873	1006	3075
% of total lakes	3	2.6	7.9
Alkalinity (meq L^{-1})	111	36	119
pH	6.58	6.36	6.75
NO_3-N (mg L^{-1})	0.01	0.01	0.01
TDN (mg L^{-1})	0.4	0.14	0.4

Data from Henriksen *et al.* (1997b)

Seasonality and long-term trends

The inlet and outlet of Lake Gardsjon southwest Sweden display the typical seasonal pattern (Figure 6.12) although the summer minimum NO_3 concentration in both the inlet and outlet has increased (Grennfelt and Hultberg, 1988) which could suggest that the catchment is in a transitional phase between Stages 1 and 2 of Stoddard's N loss classification. Hornstrom and Ekstrom (1986) also observed increases in NO_3 concentrations of between 50 and 100% over the period 1971 to 1986 for deep oligotrophic lakes in southern Sweden.

In the 1986 survey of 1005 Norwegian lakes, largest NO_3 concentrations coincided with low pH values and where precipitation was most acidic and for 300 lakes in southern Norway originally surveyed in 1974, the NO_3 concentrations were almost twice those observed previously (Henriksen and Brakke, 1988). The lakes which displayed increased NO_3 were mainly in catchments having thin soil cover and sparse vegetation, while lakes surrounded by productive forest continued to have small NO_3 concentrations. Although N deposition had increased over this period, this by itself was unable to account for the higher concentrations of NO_3 in lake waters, implying that the relative uptake of N by the catchments had decreased. In 90% of the lakes that displayed increased NO_3 concentrations, Al had also increased (Henriksen *et al.,* 1988). However, since 1986 there has been no increase in the NO_3 concentration of these lakes (Hessen *et al.,* 1997c).

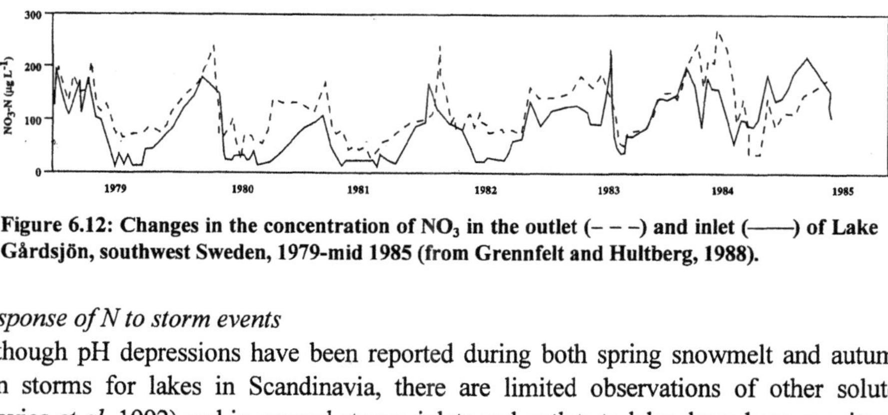

Figure 6.12: Changes in the concentration of NO_3 in the outlet (– – –) and inlet (——) of Lake Gårdsjön, southwest Sweden, 1979-mid 1985 (from Grennfelt and Hultberg, 1988).

Response of N to storm events

Although pH depressions have been reported during both spring snowmelt and autumn rain storms for lakes in Scandinavia, there are limited observations of other solutes (Davies *et al*, 1992) and in general stream inlets and outlets to lakes have been monitored during storms and not the lake itself.

(b) Streams

Concentrations and fluxes. Between 1993 and 1995 the River Bjerkreim in southwest Norway was intensively studied (Kaste *et al.,* 1997) and N budgets for 9 sub-catchments with 3 different land uses are presented in Table 6.13. Smallest NO_3 concentrations (0.08–0.16 mg N L^{-1}) and fluxes were observed in streams draining the forested catchments, while largest concentrations (0.21–0.24 mg L^{-1}) and fluxes were observed for streams draining the mountainous catchments (Table 6.13) which is perhaps in contrast to that expected in Britain. Dissolved organic nitrogen concentrations and fluxes varied little between catchments, while NH_4-N concentrations and fluxes were very small. Nitrate was the dominant N fraction, especially in the mountainous catchments where NO_3 represented 70–75% of the total N. In contrast, NO_3 represented only 60–65% of total N in the heathland catchments and 45–55% in the forested catchments.

Table 6.13: Dissolved nitrogen fluxes (kg N ha^{-1} yr^{-1}) in streams draining mountainous, forested and heathland catchments in southwest Norway.

Land use	No. catchments	Inorganic N deposition	Stream water NO_3-N	Stream water NH_4-N	Stream water DON
Forested	3	17.5	1.6	0.14	1.3
Heathland	2	16.2	2.7	0.14	2.0
Mountainous	4	19.6	5.5	0.14	2.1

Data from Kaste *et al.* (1997)

The results from the Bjerkreim study show that land use is an important factor determining N loss in an area of Norway that receives high N deposition compared to other regions. Retention of atmospheric N by the catchments in the Bjerkreim catchment decreased in the following order; forested>heathlands>mountainous. Among the forested catchments N retention increased with increasing forest coverage. Kaste *et al.* (1997) suggested that the difference in N retention between forested and mountainous sub-catchments of the Bjerkreim catchment may be due to a combination of the following factors:

(i) In mountainous areas with shallow and patchy soils, total N storage capacity may be restricted in comparison to forested catchments with deeper soils. Also, sparse vegetation, short growing season and cool temperatures lead to low biological N demand in mountainous areas compared to forested catchments.
(ii) In the mountainous catchments characterised by steep slopes and shallow soils, water movement may be fast during events of heavy rain and snowmelt so that contact time between soil and water will be short and thus N passes rapidly through the catchment and into the stream. In addition, the mountainous catchments are at a higher altitude than forested catchments and therefore receive more N as both wet and occult deposition.
(iii) Low N retention in the mountainous catchments may be due to terrestrial P-limitation.

The results from the Bjerkreim study have important implications for the interpretation of data obtained from catchments that span a N deposition gradient but have different land use. For example, atmospheric deposition and stream water are measured regularly at six catchments in Norway as part of a national program to monitor the effects of long-range transported air pollution on stream water quality (Wright *et al.*, 1997). Four of the catchments together span a gradient in N deposition from 3.6 to 17 kg N ha^{-1} yr^{-1} and fluxes of NO_3-N are larger from the catchments which receive the largest N deposition (Table 6.14). In contrast, the DON and very small NH_4-N fluxes displayed no relationship with N deposition. However, the study compared two forested catchments with two heathland catchments.

From a 10 year study of 20 forested catchments in Finland and Sweden, it was observed that 80% of the TDN was in the form of DON, with a minimum of 59% and a maximum of 96% (Lepisto *et al.*, 1995). The median DON concentration ranged between 0.2 and 1.1 mg N L^{-1}, while the median concentration of NO_3 in 17 of the 20 catchments was less than 0.1 N mg L^{-1} (Arheimer *et al.*, 1996). Nitrate fluxes varied between 0.02 and 1.33 kg N ha^{-1} yr^{-1}, having a median of 0.25 kg ha^{-1} yr^{-1} (Lepisto *et al.*, 1995).

Table 6.14: Annual dissolved nitrogen fluxes (kg N ha^{-1} yr^{-1}) for four small catchments in Norway

Catchment	Vegetation	Precipitation Inorganic N	Stream water NO_3-N	NH_4-N	DON
Birkenes	Coniferous forest	17.0	1.4	0.14	1.12
Storgama	Coniferous forest	9.5	2.0	0.14	1.26
Langtjern	Alpine heath & peatland	6.0	0.28	0.14	1.12
Karvatn	Alpine heath & peatland	3.6	0.28	0	0.14

Data from Wright *et al.* (1997)

Although atmospheric N deposition was of significant importance for the export flux of NO_3, other significant explanatory variables were temperature and high stream density. As N deposition and air temperature were strongly correlated it was difficult to distinguish individual effects. However, retention (median = 91.2%) of atmospheric N occurred in most of the catchments. Significantly lower losses of NO_3 occurred from catchments with a high percentage of organic soils.

Seasonality and long-term trends

Nitrate concentrations in Scandinavian streams display the seasonal trend typical of Stage 0 N loss, with maximum concentrations during the spring snowmelt and minimum concentrations in autumn (Hauhs *et al.*, 1989; Wiklander *et al.*, 1991). In a study of 20 forested catchments in Finland and Sweden, 14 of the catchments had significantly smaller NO_3 concentrations during the summer than the winter (Arheimer *et al.*, 1996). The most pronounced difference between NO_3 concentrations in the summer and winter was observed in catchments with a high percentage of lakes (2–3%). Those catchments that displayed no significant difference between concentrations during the growing and dormant seasons, received significantly larger inputs of N deposition (Arheimer *et al.*, 1996) which is indicative of catchments proceeding towards the chronic stages of N saturation (Stoddard, 1994).

During the three year study of the River Bjerkreim in southwestern Norway, pronounced seasonal variations in NO_3 concentrations were observed in streams draining forested catchment (Figure 6.13a) which are typical of Stage 0 of Stoddard's N loss classification (Kaste *et al.*, 1997). Similar seasonal patterns were observed in streams draining heathland catchments (Figure 6.13b), whereas seasonal NO_3 concentrations in the mountainous catchments were less clear (Figure 6.13c), with NO_3 peaks apparent during both the dormant and growing season.

Two small undisturbed forested catchments in southern Finland have shown increased

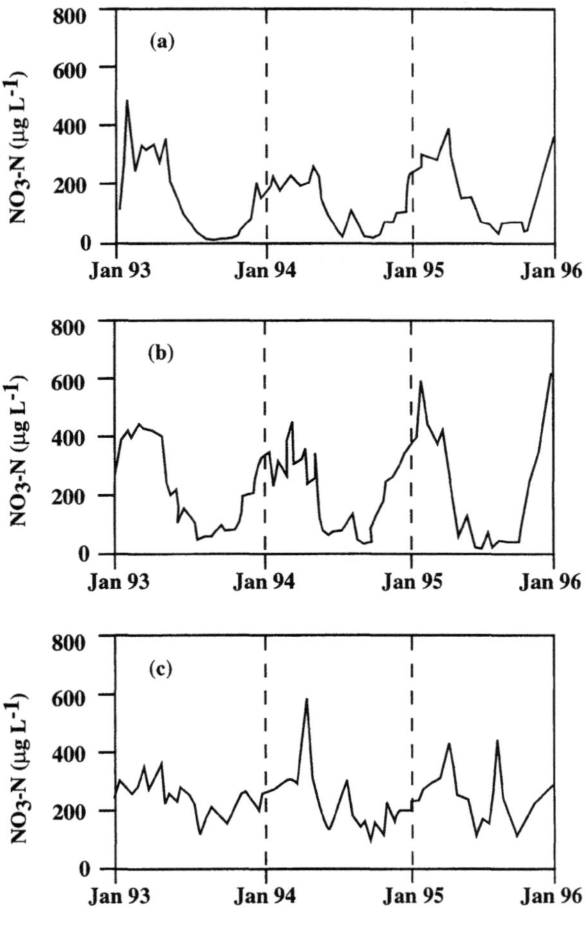

Figure 6.13: Seasonal variation in NO_3-N concentrations in (a) Svela, a forested catchments, (b) Longa, a heathland catchment and (c) a mountainous catchment of the Bjerkreim River catchment, southwestern Norway (from Kaste *et al.*, 1997).

NO_3 concentrations and fluxes during the last 20–25 years (Table 6.15, Lepisto, 1995). The greatest increase in concentration occurred during the period between the mid 1970s to the mid 1980s. The stream water flux of total N also increased and the average proportion of the NO_3 flux to the total N flux increased from 56% to 68% over the same period. During the period of this study, mean annual inorganic N deposition increased from 6.2 to 8.5 kg ha^{-1} yr^{-1}. Decreasing retention of N within the catchment was reflected by an increase in the ratio of inorganic N output compared to N inputs which increased from 12% in the 1970s to 24% in the 1980s. The seasonal pattern of NO_3 concentrations also changed between the 1970s and 1980s: concentrations were greater during all months in the 1980s (Figure 6.14). The average increase in concentration was largest (0.7–1.0 mg L^{-1}) during the dormant season, but was also clear during the growing season (0.2–0.4 mg L^{-1}). Lepisto (1995) concluded that this increase in NO_3 leaching, particularly during the growing season, provides evidence of N saturation and the inability of the biomass to utilise the available N, whereas increases during snowmelt indicates the direct impact of N deposition on NO_3 losses.

Table 6.15: Volume-weighted average NO_3 concentrations and fluxes in water draining two forested catchments in Finland

Catchment	Area (km^2)	Concentration (mg N L^{-1}) 1969–1979	Concentration (mg N L^{-1}) 1980–1990	Flux (kg N ha^{-1} yr^{-1}) 1969–1979	Flux (kg N ha^{-1} yr^{-1}) 1980–1990
Teeressuonoja	0.69	0.28	0.42	0.7	1.3
Yli–Knuutila	0.07	0.58	1.35	0.6	2.0

Data from Lepisto (1995)

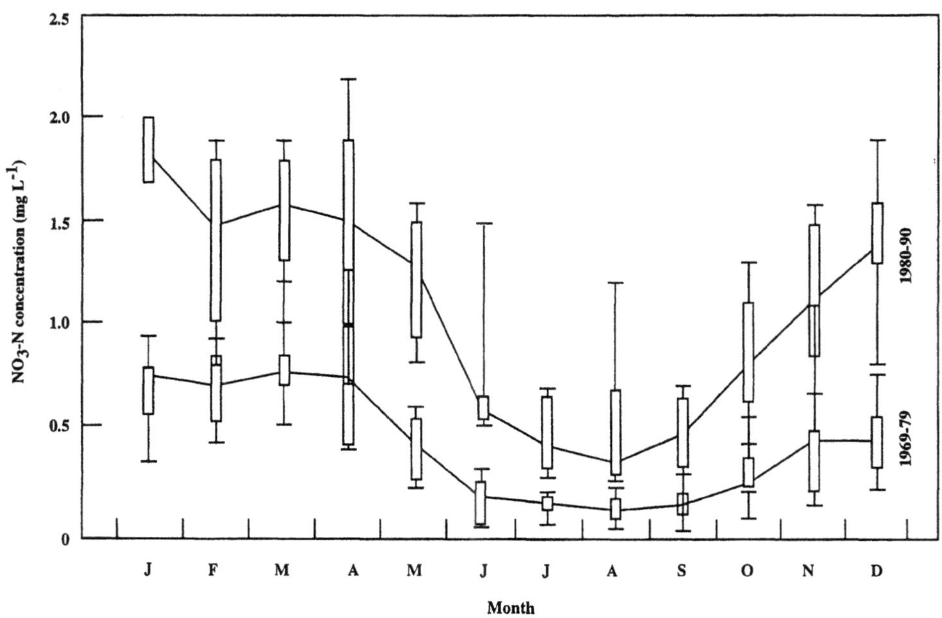

Figure 6.14: Seasonal variation of stream water NO_3 concentrations for a forested catchment, Southern Finland, shown as box and whisker plots (the medians are joined, the box defines the inter-quartile range and the whiskers define the 10 and 90% limits of the distribution). The study periods of 1969–1979 and 1980–1990 are shown separately (from Lepisto, 1995).

In three rivers monitored in southern Norway NO_3 concentrations have increased over the period 1980–1987 (Henriksen and Brakke, 1988). However the influence of changing agricultural activities cannot be ignored. By way of contrast, at Birkenes in southern Norway, where stream water NO_3 concentrations are small, there has been no significant long-term change in concentrations since 1980 although annual variations over this period have been considerable (Hessen *et al.*, 1997c). This shows that short-term observations of stream water chemistry provide a poor indicator for the effect of increased atmospheric deposition.

The response of N to storm events

In a study of 20 forested catchments in Finland and Sweden, dilution of stream water NO_3 was observed during high flows in catchments that had small annual mean NO_3 concentrations, whereas in catchments with larger mean NO_3 concentrations, concentrations increased during high flows (Arheimer *et al.*, 1996). The catchments with larger NO_3 concentrations during high flows were situated in the south and at low altitudes where mineral soils are more common and N deposition largest (Arheimer *et al.*, 1996). In 8 of the catchments, the largest increase in NO_3 was observed in the first half of the year, with most events occurring in connection with the spring flood (snowmelt) (Arheimer *et al.*, 1996). In the second half of the year, only one of these catchments displayed an increase in NO_3 with increasing flow.

In a review of acidic episodes in Europe (Davies *et al.*, 1992), NO_3 responses were mainly reported for sites in Norway, where NO_3 concentrations were observed to increase during snow events (Table 6.16). At Storgama, Upper Tovdal, Sorlandet and Lofssjon the relative increase in NO_3 was greater than the relative increase in SO_4 and in the Storgama catchments, NO_3 concentrations were 30% of SO_4 concentrations (on an equivalent basis) during the peak of the episode.

Table 6.16: Response of stream water pH, NO_3 and SO_4 to rain and snow events in Norway and Sweden

Catchment	Snow (S) or Rain (R)	pH		NO_3 (μeq L^{-1})		SO_4 (μeq L^{-1})	
		P	E	P	E	P	E
Norway							
Birkenes	R	5.3	4.2	15	30	100	240
	S	5.3	4.2	20	40	100	180
Storgama	S	4.6	4.0	15	60	90	180
Tovdal (Upper)	S	4.7	4.2	16	36	62	83
Sorlandet (1)	S	4.6	4.4	25	35	62	83
Sorlandet (2)	S	4.8	4.6	12	25	140	170
Sweden							
Lofssjon Djursvasslan	S	6.9	4.0	0	15	30	50

Data from Davies *et al.* (1992)
P=pre-episode E= peak-episode

Davies *et al.* (1992) concluded that there are relatively few NO_3 data reported for rainfall events in Europe and the data that do exist suggest that NO_3 concentrations during rainfall events vary to a lesser extent than SO_4 concentrations and NO_3 concentrations during snowmelt induced episodes. Any changes in stream water NO_3-N during rain storms tend to be more apparent after dry periods; after such an event at Birkenes, Norway, there was a quick return to pre-episodic values (Gjessing *et al.*, 1976).

6.5.3. EASTERN UNITED STATES

Characteristics of the Appalachian Highlands

The upland areas of the northeastern United States, known as the Appalachian Highlands, can be split into six physiographic provinces (Table 6.17). The New England, Adirondack and northern parts of the other Appalachian Highland provinces were glaciated (Figure 6.15) and contain a large number of lakes. The major soils of the Appalachian Highlands are brown earths and podzols (UNESCO, 1975). Natural vegetation is deciduous forest with beech and maple dominant in the Mixed Mesophytic association. Further north and at higher altitudes the Northern Hardwood association might be described as a broad transition from deciduous forest to the boreal coniferous forest (Webster *et al.*, 1995). Forestry is the major land use in this region with 58–90% of the total land area of the states covered in Forest (Table 6.18). In general forest harvesting is proceeding much more slowly than growth, consequently the forests are characterised by continuing ageing (Moore *et al.*, 1992).

Table 6.17: Physiographic provinces of the Eastern United States

Province	States	Geology	Relief	Altitude	Vegetation
New England	Maine, Vermont, New Hampshire	folded sediments with metamorphic and igneous intrusions	rolling hills and subdued mountains	300–2000	Northern Hardwood Forest
Adirondacks	New York	a precambrian dome	subdued mountains	>1629	Northern Hardwood Forest
Appalachian Plateau	Pennsylvania, West Virginia	highly variable; horizontal to gently folded strata	rugged hills and low mountains	360–1200	Mixed Mesophytic Forest
Piedmont Plateau		metamorphic gneiss and schist	rolling upland	90–460	Oak–Chestnut
Blue Ridge		complex folded strata: gneiss and sedimentary rocks	subdued mountains	900–2037	Oak–Chestnut
Valley and Ridge		folded strata of limestone sandstone and shale	valleys and ridges	600–900	Oak–Chestnut

Data from Webster *et al.* (1995)

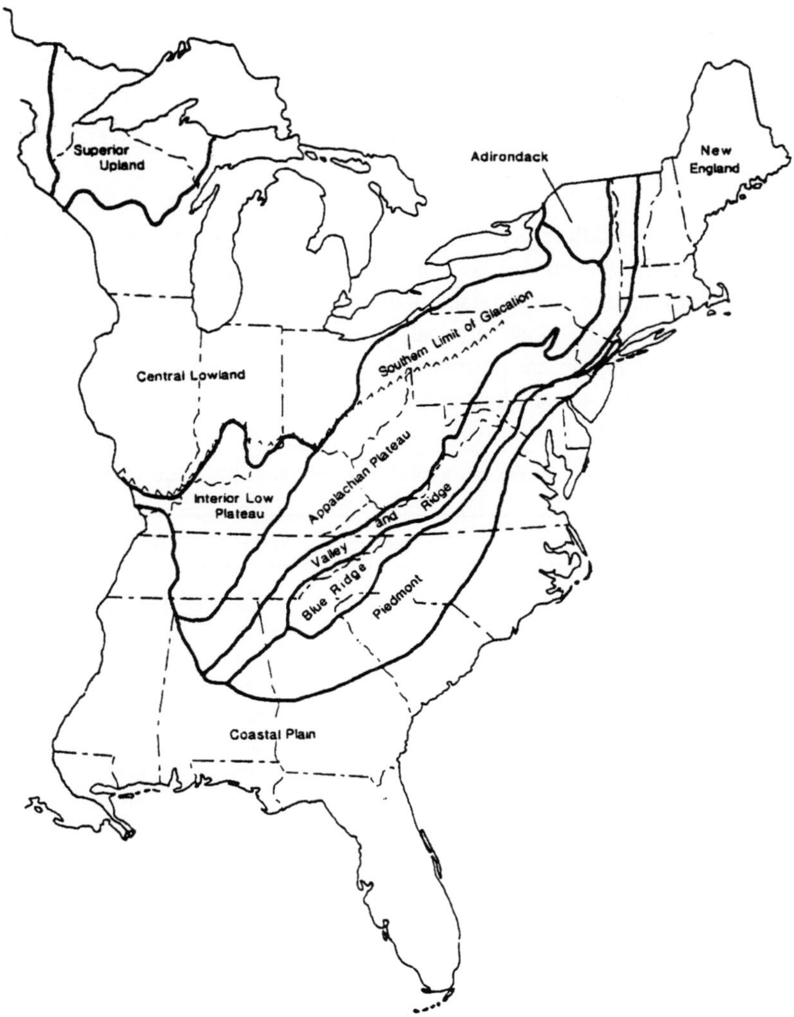

Figure 6.15: Physiographic provinces of the eastern United States (from Webster *et al.*, 1995)

Table 6.18: Percentage land cover in some eastern states of the United States in 1992

	Forested	Pasture	Crop	Other
Maine	90	1	2	7
New Hampshire	79	2	3	16
Vermont	75	6	12	7
New York	58	10	19	13
West Virginia	75	11	6	8

Data from Moore *et al.* (1992)

Nitrogen behaviour in surface waters

There are a number of comprehensive reviews that discuss the extent and patterns of N loss from natural and semi-natural ecosystems in the northeastern United States, see for example, Driscoll *et al.* (1989) and Stoddard (1994).

(a) Lakes

Concentrations. In the north-eastern United Sates, the long-term monitoring (LTM) project was initiated in 1983 to assess both short-term and long-term changes in water chemistry over gradients of acidic deposition and in different geographic regions (Ford *et al.*, 1993). Median values of various physical, chemical and deposition characteristics for the lakes in each LTM region are presented in Table 6.19. Concentrations of NO_3 are largest in Adirondack lakes and smallest in Maine lakes which reflects the gradient in NO_3 deposition across the region.

Table 6.19: Median values of various physical, chemical and deposition characteristics for lakes in each LTM region

	Maine	Vermont	Adirondacks
Number of lakes	5	24	16
Lake depth (m)	10.4	3.4	6.4
Lake surface area (ha)	4.8	14.3	25.8
In-lake retention time (yr)	1.25	0.27	0.33
Catchment area (ha)	64	108	370
ANC (meq L^{-1})	14	28	31
pH	5.79	6.08	6.03
SO_4 (mg L^{-1})	2.78	4.99	6.14
NO_3-N (mg N L^{-1})	bd	0.014	0.24
NO_3-N deposition (kg N ha^{-1} yr^{-1})	2.77	3.36	4.2

Data from Newell (1993) bd – below detection limit

Seasonality and long-term trends

In Maine, NO_3 concentrations are negligible throughout the year except during spring snowmelt; these NO_3 concentrations are characteristics of Stage 0 N loss (Stoddard, 1994). No long-term trends in NO_3 concentrations have been observed over the period 1983–1990 at the lakes in Maine (Newell, 1993). In Vermont, 6 of the 24 LTM lakes exhibit strong seasonal NO_3 patterns, with concentrations as high as 0.98 mg L^{-1} in the spring (Stoddard and Kellogg, 1993), which is typical of Stage 1 NO_3 loss. In contrast, the remaining 18 lakes display little seasonal variability and peak NO_3 concentrations are generally less than 0.28 mg L^{-1}. None of the sites display trends in NO_3 concentrations for the period 1981 to 1989 (Stoddard and Kellogg, 1993).

In lake waters from the Adirondacks, NO_3 concentrations display a pronounced geographic pattern; in the west and southwest concentrations are larger than in the east (Driscoll and Van Dreason, 1993). This pattern coincides with the gradient in N deposition across the region. Although N deposition has not increased since the late 1970s, concentrations of surface water NO_3 have increased. Thus Driscoll and Van Dreason

(1993) proposed that these patterns indicate a change in the utilisation of N by the catchment which seems to be consistent with conditions of N saturation. The geographic pattern of NO_3 concentrations also corresponds closely with the distribution of (i) lake water acidity and (ii) diatom-inferred acidification from pre-industrial times to the present (Sullivan *et al.*, 1997). Although lake water sulphate concentrations are much higher than NO_3 concentrations, they display no spatial pattern. Hence, Sullivan *et al.*, (1997) concluded that N plays a significant role in the acidification and consequent biological effects in Adirondack lakes.

Results from the Adirondack LTM lakes during the period 1982 to 1990 showed statistically significant increases in NO_3 concentrations in nine out of the sixteen lakes (Stoddard, 1994). The increase in lake water NO_3 concentration ranged from 0.0056 to 0.025 mg N L^{-1} yr^{-1}, with an average increase of 0.014 mg N L^{-1} yr^{-1}. Plots of temporal NO_3 patterns (Figure 6.16) show that the major change was in peak NO_3 concentrations during the spring, with some lakes being 1.4 mg L^{-1}, and that minimum NO_3 concentrations have changed little (Driscoll and Van Dreason, 1993). Stoddard (1994) suggested that this increasing trend in NO_3 concentrations implied that many Adirondack catchments were becoming increasingly N saturated and reached Stage 1 of N loss and that their condition is worsening. However, more recent data (post 1990) for Adirondack LTM lakes show a decline in lake water NO_3 concentrations in recent years which may reflect variations in climate (Driscoll *et al.*, 1995). These data suggest that extremely long-term records are needed to elucidate 'real' trends in lake water NO_3 concentrations due to complex control of NO_3 production and that an improved understanding of the factors that control NO_3 leaching is needed.

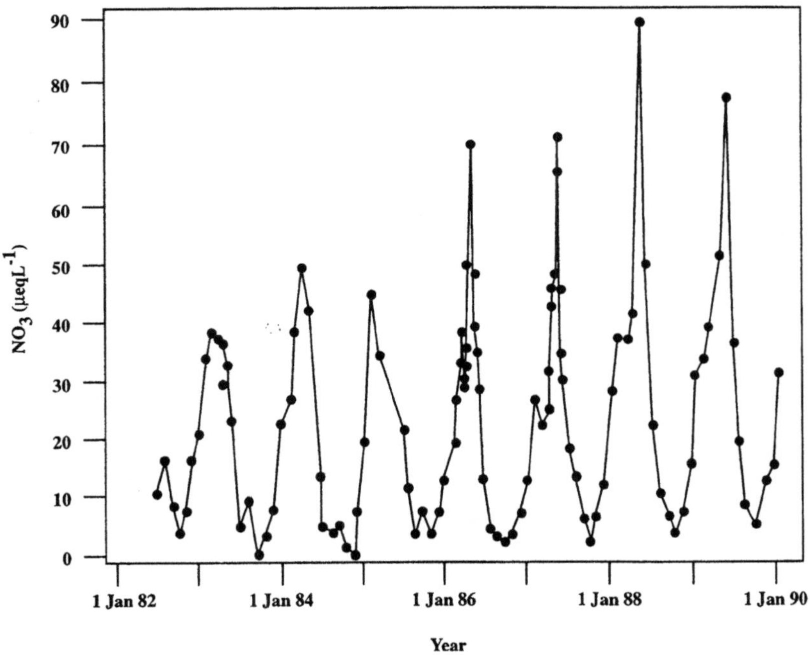

Figure 6.16: Time-series plot of NO_3 concentration at Constable Pond, Adirondack Mountains (from Sullivan *et al.*, 1997).

(b) Streams

Concentrations. In the Eastern United States the Environmental Protection Agency (EPA) conducted a National Stream Survey (NSS) of 446 streams in the spring of 1986 (Kaufmann *et al.,* 1991). Nitrogen in precipitation ranged from 0.14 to 0.42 mg L^{-1} in the region covered by the NSS, whereas subregion median stream water NO_3-N concentrations ranged from 0.006 to 0.52 mg L^{-1}. After excluding physiographic areas that were dominated by agricultural activity and sites that were not predominantly forested, a strong correlation between concentration of stream water N (NO_3 + NH_4) at spring baseflow and levels of wet inorganic N deposition in each of the NSS regions was observed. The only exception was the Catskill region where N deposition was greatest but where stream water N concentrations fell below what was expected on the basis of the data collected from the other regions. Smallest median stream water NO_3 concentrations of <0.14 mg L^{-1}, were observed where the median wet N deposition was 3.78 kg ha^{-1} yr^{-1}, whereas the median stream NO_3 concentrations >0.35 mg L^{-1} where observed where the median wet N deposition was 6.16 kg ha^{-1} yr^{-1}.

Fluxes

The balance between the input (N deposition) and output (stream water N flux) of inorganic N has been compared for a large number of forested catchments in the United States (Driscoll *et al.,* 1989; Stoddard, 1994). In general, forests that received less than 4 kg N ha^{-1} yr^{-1} of wet inorganic N deposition lost very little inorganic N in streamflow, whereas forests that received >4 kg N ha^{-1} yr^{-1} exhibited variable losses (Figure 6.17).

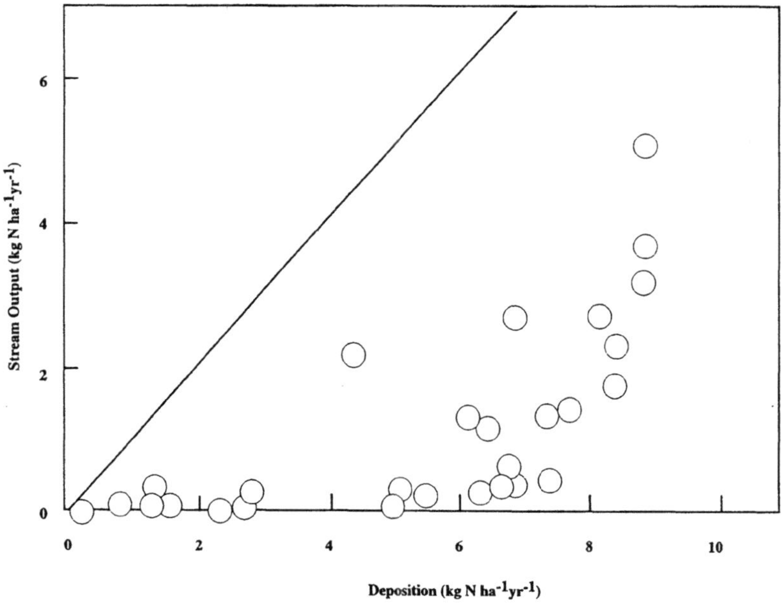

Figure 6.17: Relationship between wet deposition of N (NO_3+NH_4) and outputs of N (NO_3+NH_4) in stream water for the catchments studies throughout the Unites States (from Driscoll *et al.,* 1989)

However, Driscoll *et al.* (1989) stressed that the data in Figure 6.17 should be interpreted with caution as they were collected by using a wide range of methods and over various time scales (from 1 year to 10s of years). For example over the 10 year period 1964 to 1974, losses of NO_3 at the Hubbard Brook catchment ranged between 5.6 and 34.8 kg N ha^{-1} yr^{-1} to give an mean annual loss of 16.1 kg ha^{-1} yr^{-1}(Likens *et al.*, 1977). However, since 1974 annual losses of NO_3 have decreased substantially, while N inputs have remained similar over the whole period of record (Driscoll *et al.*,1989).

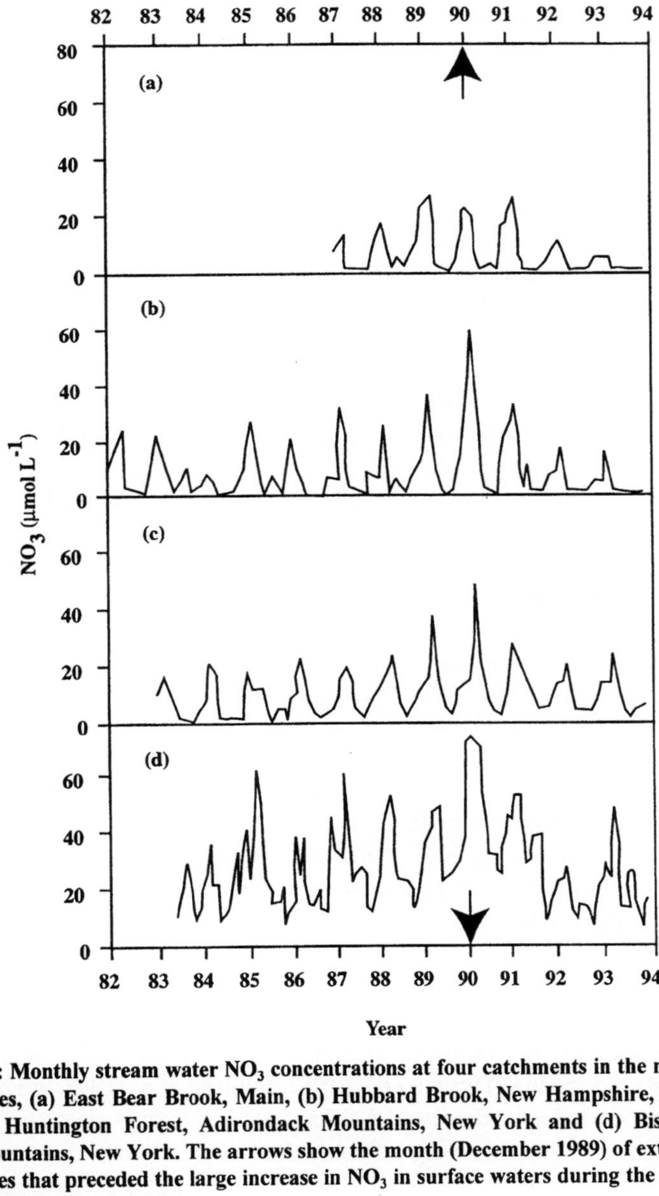

Figure 6.18: Monthly stream water NO_3 concentrations at four catchments in the northeastern United States, (a) East Bear Brook, Main, (b) Hubbard Brook, New Hampshire, (c) Arbutus catchment, Huntington Forest, Adirondack Mountains, New York and (d) Biscuit Brook, Catskill Mountains, New York. The arrows show the month (December 1989) of extremely cold temperatures that preceded the large increase in NO_3 in surface waters during the snowmelt of 1990 (from Mitchell *et al.*, 1996).

Seasonality and long-term trends

Compared to lakes, fewer long-term data sets exist for NO_3 concentrations in stream waters. In the Catskill mountains, nineteen large streams have been monitored since early in this century and trend analyses indicate that NO_3 concentrations have increased in all streams, with most of the increase occurring since 1970 (Murdoch and Stoddard, 1992). However, much of the observed increase in NO_3 concentrations occurs during high flows during the winter, whereas baseflow concentrations appear to have changed little throughout the sampling period. Murdoch and Stoddard (1992) proposed that the historical data helped to put in context the short-term NO_3 trends observed in smaller Catskill streams monitored since the early 1980s. They suggested that the historical data confirm that the short-term (1983 to 1990) trends in increasing NO_3 (0.018 to 0.042 mg L^{-1} yr^{-1}), observed in five of the eight smaller headwater streams, are part of a long-term pattern. However, as observed for lakes in the Adirondacks (Driscoll *et al.,* 1995), stream water NO_3 concentrations at Biscuit Brook, an intensively studied headwater stream in the Catskills, have decreased since 1990 (Figure 6.18a) (Mitchell *et al.*, 1996), resulting in no significant trend in stream water NO_3 for the entire period of record (1983–1994) at the site. A similar decline in stream water NO_3 concentrations since 1990 has been observed for three other streams (Figure 6.18) in the northeastern United States (Mitchell *et al.*, 1996). At all four sites, peak NO_3 concentrations occurred in 1990 and followed an anomalous cold period in December 1989. These results show that climatic variation can have a major effect on the terrestrial N cycle and thus also the temporal patterns of NO_3 in surface water.

At a small hardwood forested catchment in the Appalachians, West Virginia, twenty years of stream water chemistry data show that the annual mean NO_3 concentration has increased from 0.095 to 0.975 mg L^{-1} (Peterjohn *et al.*, 1996). Similar changes in the N content of precipitation were not observed, therefore Peterjohn *et al.* (1996) suggested that the increase in stream water NO_3 concentrations does not merely reflect changes in N inputs; rather they represent a modification to N cycling within the terrestrial ecosystem. In addition, the twenty-four year record of mean monthly NO_3 concentrations demonstrates that the seasonal variability in stream water chemistry has also changed (Figure 6.19). The data set displays three periods; an initial period (1971–1973) of small concentrations and small variability; a second period (1974–1983) of large peak concentrations and large variability, and a third period (1984–1994) of large concentrations, around 0.8 mg N L^{-1}, and small variability. This sequence of change suggests that the catchment has passed through Stages 0 and 1 to Stage 2 of N loss proposed by Stoddard (1994).

(c) The response of surface water N to storm events

In the Adirondacks, strong NO_3 pulses in both lakes (Galloway *et al.*, 1980; Driscoll and Schafren, 1984) and streams (Driscoll *et al.*, 1987; Sullivan *et al.*, 1997) are apparently the primary factor contributing to depressed ANC and pH during snowmelt. The source of this NO_3 has been attributed to the release of atmospherically derived N compounds from the snow pack (Johannessen and Henriksen, 1978).

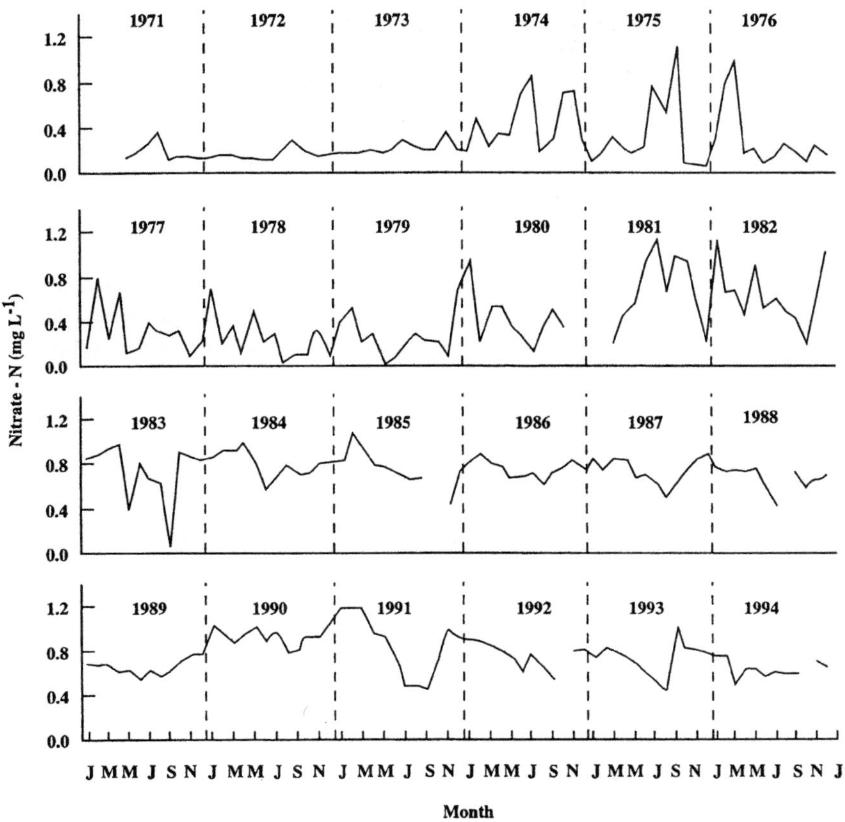

Figure 6.19: Long-term trend on the volume-weighted average monthly NO_3 concentrations for Watershed 4 at the Fernow Experimental Forest, West Virginia (from Peterjohn *et al.*, 1996).

However, it is not always possible to separate the release of NO_3 from the snow pack from that derived from the catchment soils due to freeze–thaw (Likens *et al.*, 1977; Rashcer *et al.*, 1987; Abrahams *et al.*, 1989). In the Adirondacks both nitrification of NH_4 in the snowpack and mineralisation of soil organic N contribute to NO_3 released during snowmelt (Schaefer and Driscoll, 1993).

In the Catskills, data from historical records suggest that, although episodic increases in NO_3 concentrations during snowmelt have always occurred, the magnitude of these increases has risen considerably since 1970 (Murdoch and Stoddard, 1992). The most likely causes of this increase in NO_3 concentrations are probably a change in the biological demand for atmospherically derived N, and a long-term increase in catchment N supply, from both deposition and mineralisation.

Schaefer *et al.* (1990) concluded that the magnitude of pH depression experienced by 11 Adirondack lakes during snowmelt depended strongly on their base-cation concentration. Furthermore lakes with low ANC levels undergo episodes that result largely from increases in NO_3 concentrations. In contrast, lakes with high ANC levels undergo episodes that are largely the result of dilution of base cations by snowmelt.

Murdoch and Stoddard (1992) reported similar findings for streams in the Catskill Mountains, where NO_3 increases were the primary determinant of acidic episodes in low to moderate ANC streams during the spring. In contrast, base-cation dilution and increases in organic acids were the major factors contributing to acidic episodes in the autumn. Murdoch and Stoddard (1992) also observed flushing of NO_3 during a succession of four storms; peak concentrations during the four storms were 1.1, 0.95, 0.63 and 0.53 mg l-1, respectively. This suggests that the NO_3 response during storms reflects the accumulation of NO_3 in the catchment since the last storm, as well as the magnitude and duration of the storm.

In the Catskills, although SO_4 concentrations in streams are generally higher than those of NO_3, SO_4 concentrations decrease during storms while NO_3 increases and at Biscuit Brook, NO_3 concentrations approach or exceed those of SO_4 during storms (Figure 6.20) (Murdoch and Stoddard, 1992). In addition, the ratio of NO_3:(NO_3+SO_4) increase during storms indicating that the relative importance of NO_3 to the acidity of the streams changes during storm events (Figure 6.20). Sullivan *et al.* (1997) also presented data for Adirondack streams which showed that the relative importance of NO_3 versus SO_4 acidity increased as the ANC decreased during storm events. However, in a number of Pennsylvania streams, NO_3 pulses appear to be of minimal importance and pulses of SO_4 are responsible for episodic ANC and pH depressions in streams (Herlihy *et al.,* 1993).

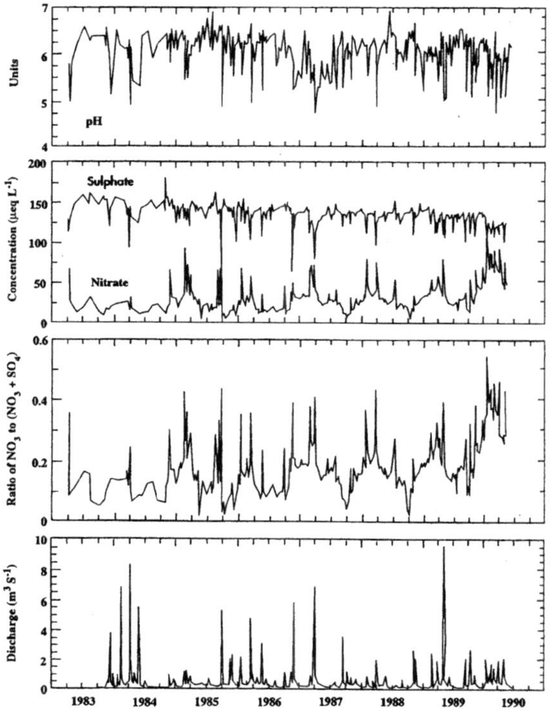

Figure 6.20: Temporal patterns of stream NO_3 and SO_4 concentrations, the ratio of NO_3:(NO_3+SO_4) and discharge in Biscuit Brook, Catskill Mountains, USA (from Murdoch and Stoddard, 1992).

6.6. Conclusions

6.6.1. CONCENTRATIONS AND FLUXES OF NITROGEN

Numerous studies of semi-natural ecosystems in Britain, Norway and the north-eastern United States have observed a positive relationship between surface water NO_3 concentrations and levels of atmospheric N deposition. However, N deposition generally only explains about 50% of the variance in surface water NO_3 concentrations (Allott *et al.*, 1995; Wright *et al.*, 1997; Reynolds, personal communication). Thus other factors such as climate, geology, topography, soil type and land use (Dillon and Molot, 1990; Lepisto *et al.*, 1995; Arheimer *et al.*, 1996; Kaste *et al.*, 1997) are also important in controlling NO_3 concentrations in surface waters. These catchment characteristics are important in determining (i) the size of vegetation and soil sinks for N (Skeffington and Wilson, 1988; Kaste *et al.*, 1997), and (ii) the hydrological dynamics which control the transport of N through the catchment (Dillon and Molot, 1990; Lepisto *et al.*, 1995). Thus the impact of atmospheric nitrogen deposition on the behaviour of nitrogen in surface waters will vary between catchments with different characteristics. In addition, relationships observed between catchment characteristics and N behaviour in surface waters in one area or country may not be the same in another. For example, in Britain a number of studies have observed larger concentrations and fluxes of NO_3 from forested catchments compared to adjacent moorland catchments, whereas in Norway larger NO_3 concentrations and fluxes have been observed from mountainous moorland catchments compared to forested catchments. The reasons for these contrasting observations may result from the following facts:

(i) in Britain, moorland and forested catchments are often found at similar altitudes, whereas in Norway moorland catchments are generally at a higher altitude than forested catchments. Therefore mountainous moorland catchments in Norway receive larger amounts of dry and occult deposition than forested catchments.

(ii) the soils of moorland catchments in Britain are deeper and more organic than the soils in mountainous catchment in Norway, which are very thin and organic poor.

(iii) as the moorland catchment of Norway are characterised by steeper slopes and shallower soils than in Britain, water movement will be faster during events of heavy rain and snowmelt so that contact time between soil and water will be short and thus N passes more rapidly through the catchment and into the stream.

The uplands of Britain are dominated by heather and acid-grassland vegetation with areas of coniferous plantation, yet considerably more research has been carried out on forested catchments (see Chapter 4). Thus there is need for a better understanding of the N dynamics in moorland catchments. Similarly in Norway, where 46% of the natural vegetation is above the tree line more information in needed on the N dynamics of these systems which are characterised by steep slopes and thin soils. In contrast, Sweden and Finland have large areas of coniferous forest which have deep organic soils. Many studies have observed a strong negative relationship between area of peat within a catchment and surface water NO_3 concentration (Lepisto *et al.*, 1995; Stevens *et al.*, 1997; Harriman *et al.*, 1998). A large proportion of atmospheric N deposition appears to be retained within peat soils. Although the natural and semi-natural areas of the northeastern United States are dominated by old forests (generally >90 years) they are usually associated with better drained soils, such as humus–iron podzols and brown forest soils, than forests in Britain and Scandinavia.

Overall, the dominant spatial trends in NO_3 concentrations and fluxes evident from the data discussed in this chapter are:

(i) in Britain, mature forested catchments with high N deposition rates have larger concentrations and fluxes of NO_3 than moorland and young forested catchments, although high altitude moorland catchments with thin soils and bare rocks also have larger concentrations and fluxes than moorland catchments with deeper soils.
(ii) in Norway, largest NO_3 concentrations and fluxes are observed from mountainous catchments in the southwest where N deposition is largest.
(iii) in the northeastern United States, forested sites further south and with high N deposition rates tend to have the largest concentrations and fluxes of NO_3.

As stressed by others (e.g. Driscoll *et al.*, 1989) comparison of data between studies must be carried out with caution as data from different studies have been collected by a variety of methods and cover various time scales. For example, comparison between the NO_3:(NO_3+SO_4) ratios for lakes (Table 6.1) and streams (Table 6.2) in the USA is not possible as the lake survey was carried out in autumn when NO_3 concentrations are normally at a minima. However, the stream survey was carried out in spring when NO_3 concentrations are often at a maxima, especially in streams receiving snow melt. In Section 6.5.1, mean NO_3 concentrations and fluxes for moorland (Table 6.6) and forested (Table 6.7) catchments in Britain are compared. These mean NO_3 concentrations were calculated from between 12 and 1460 samples and the length of study ranged from one to ten years during the period 1976 to 1995. In addition, estimates of NO_3 fluxes from the catchments in Tables 6.5 and 6.6 are probably underestimated as losses during storms have not been quantified. Nitrate concentrations are also known to display considerable annual variation, particularly between dry and wet years. For example, at Beddgelert in N Wales, annual NO_3 concentrations and fluxes between 1982 and 1990 ranged between 0.62 and 1.03 mg L^{-1} and 10.8 and 19.4 kg N ha^{-1}, respectively (Table 6.20).

Compared to inorganic forms of N, relatively few data are available for DON in upland streams, even though it has been shown to contribute significantly to total dissolved N in soil solutions and surface runoff from organic soils (Stevens and Wannop, 1987; Qualls *et al.*, 1991; Yesmin *et al.*, 1995). In those studies that have determined DON in stream water, it is usually a significant proportion of the total dissolved N present (Lepisto *et al.*, 1995; Chapman *et al.*, 1998) and may therefore be an important but often unaccounted component of total dissolved N in upland streams.

The remaining fraction of stream N losses is exported as particulate matter. Again few data are available. Particulate N is predominantly organic in nature and is usually assumed to be low compared to dissolved inorganic forms, although Crisp (1966) showed that loss of N in particulates derived from eroding peat was nearly five times that in solution. However, losses of suspended particulate are very site specific with greater losses occurring from; (i) catchments underlain with shales and slate than granite (Rigler, 1979); (ii) forested catchments than moorland catchments (Kirby *et al.*, 1991) and (iii) catchments with larger areas of agricultural land (MacDonald *et al.*, 1994).

6.6.2. SEASONAL PATTERNS OF NITROGEN

As discussed in Section 6.2, NO_3 concentrations in surface waters usually display a pronounced seasonal pattern with larger concentrations observed in the winter than summer. However, as a catchment becomes progressively saturated with respect to N, changes in this seasonal pattern would be expected as discussed in Section 6.3. It is possible to place the sites discussed in this chapter for which seasonal data are available into the Stages of N loss described for aquatic ecosystems by Stoddard (1994). Sites that fall into the later stages of N loss, Stages 2 and 3, have large NO_3 concentrations throughout the year and in some extreme cases concentrations may be larger in the summer than the winter (e.g. Stevens *et al.*, 1993b). These sites tend to be forested and occur in areas receiving large amounts of atmospheric N deposition, but substantial variability exists. For example in Britain and the United States forest stand age appears to be important; catchments with stands greater than 30 years 'leaked' more NO_3 in studies in Wales (Stevens *et al.*, 1994; Emmett *et al.*, 1993) and the northeastern United States (Stoddard, 1993), whereas in Norway, old spruce stands with well-developed undercover vegetation retained almost all N during the growing season (Mulder *et al.*, 1997).

In Britain, moorland catchments and young forested sites generally fall into Stages 0 and 1 of Stoddard's stages of N loss to aquatic systems. At these sites, NO_3 concent-rations display the typical seasonal pattern with larger concentrations observed in the winter/spring. The maximum NO_3 concentrations may become elevated in spring in those catchments in which snowmelt is important. In contrast, moorland, mountainous catchments in Norway retain less N than forested catchments and display signs of increasing N saturation. This clearly illustrates the importance of soil, slope and the influence of hydrological dynamics on the behaviour of NO_3 in surface waters.

Few studies have investigated the seasonal behaviour of DON in surface waters, although recent data from a study of upland streams and rivers in Britain suggest that DON concentrations are larger in the summer than the winter (P.J. Chapman, unpublished data) and can represent >75% of TDN between June and October inclusive. However, the significance of DON as a potential bioavailable source of N for macrophytes and microflora has not yet been quantified.

6.6.3. LONG TERM PATTERNS IN NITROGEN DATA

Some long-term data sets of NO_3 concentrations discussed in this chapter suggest that surface water NO_3 concentrations have increased in the last 10 years, while others provide no evidence of any long-term change in NO_3 concentrations. A number of studies have highlighted that caution is needed in the interpretation of 'long-term' stream and lake water chemistry and that misleading conclusions may be drawn from data sets of 10 years or less (e.g. Burt *et al.*, 1988; Driscoll *et al.*, 1989). This is especially important for N as many of the processes controlling catchment N dynamics are biologically controlled and are consequently sensitive to changes in climate as well atmospheric deposition. For example, analysis of stream water NO_3 concentrations from the Hubbard Brook Experimental catchment, northeast United States, for the period 1964–1974 suggested a strong increasing trend (Likens *et al.,* 1977). However, examination of the data for the period 1964–1987 shows no long-term trend (Driscoll *et al.,* 1989). In addition, studies that report an increasing trend in peak NO_3 concentrations during the winter/spring should also be treated with caution as the seasonal pattern in NO_3 concentrations can be severely disrupted by drought (Burt *et al.*, 1988; Reynolds *et al.*, 1992) and extreme cold periods (Mitchell *et al.,* 1997). Although some of the data reported here for the Catskill Mountains can be considered long-term (up to 65 years of records), data for other areas of the United States and Europe generally span less than 10 years and should therefore be interpreted with caution. Thus if the effects of changing atmospheric N inputs on surface water chemistry are to be seen against a background of considerable climatic driven variation extremely long-term records of surface water N concentrations are needed. At present such data sets are scarce, thus the importance of maintaining long-term (>10 yrs) surface water chemistry studies is evident.

The lack of data on long term trends in TDN and DON means that where increases in surface water NO_3 concentrations are observed it is not possible to elucidate whether this is the result of an absolute increase in the NO_3 concentration or a relative change in the proportions of the various N forms present. Assessment of the impacts of increased N deposition on N losses from semi-natural ecosystems need to consider not only the dissolved inorganic fraction but also the contribution from dissolved organic forms and that associated with particulates. Without this data it is not possible to evaluate fully the impact of increased atmospheric N deposition on the functioning of aquatic ecosystems. For example, while organic and particulate forms of N may not be directly 'available' to aquatic plants and organisms, they may become available downstream as a result of nutrient transformations within the stream channel. In addition, the transformation of DON to NO_3 via the processes of mineralisation and nitrification is a net source of acidity to stream water.

6.6.4. RESPONSE OF NITROGEN TO STORM EVENTS

Stoddard (1994) suggested that early signs of catchment N saturation are episodic increases in NO_3, especially during snowmelt. Many studies in the Adirondack and Catskill mountains show that pulses of NO_3 in both lakes and streams are the primary factor contributing to depressed ANC and pH during snowmelt. Although data from historical records in the Catskills suggest that episodic increases in NO_3 concentrations during snowmelt have always occurred, the magnitude of these increases has risen (Murdoch and Stoddard, 1992), probably in response to the long-term increase in catchment N supply from atmospheric deposition. However, considerably less information is available for rainfall events. Data from a study in the Catskills suggests that the increase in stream water NO_3 during storms decreases in progressive storms after snowmelt (Murdoch and Stoddard, 1992) which may reflect a decrease in NO_3 availability within the catchment drainage waters during the summer.

In Europe, considerably less information on the response of surface water N to storms exists and is mainly for snowmelt events where NO_3 concentrations have been observed to increase. There are relatively few NO_3 data reported for rainfall events in Europe, although a few studies in Britain have shown that the response of NO_3 during storms may display both positive and negative relationships with flow; dilution of NO_3 occurred predominantly during winter storms and positive responses occurred mainly in summer storms (Roberts *et al.*, 1983; Webb and Walling, 1985; Chapman, 1994). Hence a simple relationship between NO_3 concentration and stream discharge is not observed.

Losses of DON and particulate N during storms are rarely reported, although a study in Wales observed that DON increased during storms, particularly following long dry periods (Roberts *et al.*, 1983). Losses of particulate N are likely to be large during storms as concentrations of particulate organic matter increase markedly during storm events but tend to peak prior to maximum discharge (Crisp and Robson, 1979). Storm events may therefore represent a significant, but usually unaccounted, proportion of total N export from a catchment and total N losses from natural and semi-natural ecosystems may currently be severely underestimated. Therefore a better understanding of the behaviour of all N forms during both rainfall and snowmelt events is needed.

6.6.5. ACIDIFICATION OF SURFACE WATERS

It is known that deposition of both S and N can, given certain geological and soil conditions, cause acidification of surface waters, although our ability to differentiate between the relative role of N versus S dynamics in surface water acidification is limited. With declining S deposition, due to the reduction in SO_2 emission, and increasing N emissions, the importance of N as a potentially acidifying input should increase. Indeed, lake and stream survey data from Europe and the northeastern United States indicate significant NO_3 losses from some catchments, especially in areas where N deposition are large (Henriksen and Brakke, 1988; Stevens *et al.*, 1993b; Stoddard, 1994) and in some cases NO_3 concentrations approach 20–30% of the SO_4 concentration (see Tables 6.1 and 6.2). The NO_3:(NO_3+SO_4) ratio, however, should be used with caution as the calculation is often based on annual or multi-annual averages of a various number of samples and only really gives an indication of the contribution of NO_3 to chronic acidification, whereas NO_3 loss from the majority of catchments is seasonal and/or episodic.

Studies in the northeastern United States have observed that NO_3 pulses during both snow and rainfall events are the primary factor contributing to episodic depressions of surface water ANC and pH (Driscoll *et al.*, 1987; Murdoch and Stoddard, 1992; Sullivan *et al.*, 1997). In Europe, organic acids play an important role in episodes but SO_4 is the strong acid anion most frequently associated with acid episodes (Davies *et al.*,1992). This difference has not been investigated in detail, but may result from differences in the characteristics of the catchments being studied, more forested catchments having been studies in the USA, or may be a consequence of different acidic deposition histories, or simply due to the fact that compared to SO_4 few NO_3 data have been collected for rainfall and snow events in Europe. The data that are available, however, suggest that the contribution of NO_3 to episodic acidification is more important in snow-melt than rainfall events.

Despite the fact that S emissions and deposition have declined in Europe over the last decade, evidence of recovery in formerly acidified surface waters or a change in the importance of NO_3 relative to that of SO_4 remains scarce. In 1995 the Welsh Acid Waters Survey was undertaken to assess whether any changes in the chemistry of 77 acid sensitive headwater streams sampled in 1984 had occurred (Stevens *et al.,* 1997). The results showed that SO_4 concentrations have decreased significantly over the last ten years from an average of 6.2 mg L^{-1} in 1984 to 5.0 mg L^{-1} in 1995. This decrease was associated with a rise in stream water pH from 6.05 to 6.12. Nitrate concentrations were also lower in 1995 than 1984 and thus the contribution of NO_3 to acidity over the last ten years has changed little; the NO_3:(NO_3+SO_4) ratio decreased from 0.099 in 1984 to 0.096 in 1995.

6.6.6. EUTROPHICATION OF SURFACE WATERS

Increased N concentrations may also lead to freshwater eutrophication. However, few studies have shown that increased N alone results in eutrophication of surface waters in semi-natural ecosystems. Recent studies in Southern Norway have shown that the majority of freshwaters are strongly P limited (Faafeng and Hessen, 1993) and thus the impact of excess inorganic N on the trophic status of surface waters is considered minor (Hessen *et al.,* 1997a). The exception may be lentic systems where an increased growth in macrophytes and benthic algae could be linked to increased N deposition (Hessen *et al.,* 1997c). In Britain, a recent report on the impacts of nitrogen deposition in terrestrial ecosystems concluded that no biological impacts, directly attributable to N deposition, have been identified for lakes and that the impact of N deposition on the eutrophication of streams is even more difficult to quantify. Consequently, little information is available to make any assessment (INDITE, 1994). Stoddard (1994) also concluded from his review of N in surface waters that there is no evidence to suggest that N deposition contributes to the long-term eutrophication of lakes in the northeastern United States.

The impact of the interaction between acidification and eutrophication on aquatic processes and biota is unclear (INDITE, 1994), although it has be hypothesised by Hessen *et al.* (1997c) that N deposition may contribute to the oligotrophication of freshwaters as acidification, in general, tends to decrease the mobility of P (Kopecek *et al.,* 1995). However, this is a matter for debate and further research.

Acknowledgments

We would like to thank Simon Langan for his comments on various iterations of this chapter and for his patience thereafter. We are also grateful to Chris Soulsby for his review of the chapter. The preparation of this chapter was funded by the Natural Environment Research Agency (grant GTS/96/2/FS) and the Scottish Office Agriculture and Fisheries Department.

References

Aber, J.D., Nadelhoffer, K.J, Steudler, P. and Melillo, J.M. (1989) Nitrogen saturation in northern forest systems. *Bioscience*, **39**, 378–386.

Abrehams, P.W., Tranter, M., Davies, T.D. and Blackwood (1989) Geochemical studies in a remote Scottish upland catchment: II Streamwater chemistry during snow-melt. *Water, Air and Soil Pollution*, **43**, 231–248.

Adamson, J.K., Scott, W.A. and Rowland, A.P. (1998) The dynamics of dissolved nitrogen in a blanket peat dominated catchment. *Environmental Pollution*, **99**, 69–77

Allott, T.E.H., Curtis., C.J., Hall. J., Harriman, R. and Battarbee, R.W. (1995) The impact of nitrogen deposition on upland surface waters in Great Britain: A regional assessment of nitrate leaching. *Water, Air and Soil Pollution*, **85**, 297–302.

Anderson, H.A., Stewart, M., Miller, J.D. and Hepburn, A. (1991) Organic nitrogen in soils and associated surface waters, in *Advances in soil organic matter research: The impact of agriculture and the environment*, (ed W.S. Wilson), The Royal Society of Chemistry, Cambridge, pp. 97–106.

Arheimer, B., Andersson, L. and Lepisto, A. (1996) Variation of nitrogen concentrations in forest streams – influences of flow, seasonality and catchment characteristics. *Journal of Hydrology*, **176**, 281–304.

Axler, R.P., Rose, C. and Tikkanen, C.A. (1994) Phytoplankton nutrient deficiency as related to atmospheric nitrogen deposition in Northern Minnesota acid-sensitive lakes. *Canadian Journal of Fish Aquatic Science*, **51**, 1281–1296.

Barr, C.J., Bunce, R.G.H., Clarke, R.T., Fuller, R.M., Furse, M.T., Gillespi, M.K., Groom, G.B., Hallam, C.J., Hornung, M., Howard, D.C. and Ness, M.J. (1993) *Countryside Survey 1990 Main Report.* Report to the DOE, Institutes of Terrestrial Ecology and Freshwater Ecology, UK.

Batey, T. (1982) Nitrogen cycling in upland pastures of the UK. *Philosophical Transactions of the Royal Socirty of London*, **B296**, 551–556.

Betton, C., Webb, B.W. and Walling, D.E. (1991) Recent trends in NO_3-N concentration and loads in British rivers, in *Sediment and Stream Water Quality in a Changing Environment: Trends and Explanation* (Proc. Vienna Symposium, 1991), IAHS Publ. no. **203**, 169–180.

Black, K.E., Lave, J.A., Billett, M.F. and Cresser, M. (1993) Observations on the changes in nitrate concentrations along streams in seven upland moorland catchments in northwest Scotland. *Water Research,* **27**, 1195–1199.

Brakke, D.F., Landers, D.H. and Eilers, J.M. (1988) Chemical and physical characteristics of lakes in northeastern United States. *Environment, Science and Technology*, **22**, 155–163.

Brown, D.J.A. (1988) Effects of atmospheric N deposition on surface water chemistry and the implications for fisheries. *Environmental Pollution.*, **54**, 275–284.

Bunce, R.G.H. and Howard, D.C. (1992) *The distribution and aggregation of ITE Land Classes*. Report to the Department of Environment, Institute of Terrestrial Ecology, Grange-over -Sands, UK.

Burt, T.P., Arkell, B.P., Tudgill, S.T. and Walling, D.E. (1988) Stream nitrate levels in a small catchment in Southwest England over a period of 15 years (1970–1985). *Hydrological Processes*, **2**, 267–284.

Chapman, P.J. (1994) *Hydrochemical processes influencing episodic stream water chemistry in a small headwater catchment, Plynlimon, Mid-Wales*. Ph.D Thesis, University of London. pp 416.

Chapman, P.J., Reynolds, B. and Wheater, H.S. (1996) Experimental investigation of potassium and nitrate dynamics in a headwater stream in Mid-Wales. *Chemistry and Ecology*, **13**, 1–19.

Chapman, P.J., Edwards, A.C., Reynolds, B, Cresser, M and Neal, C. (1998) The nitrogen content of upland rivers in Britain: The significance of organic nitrogen. IAHS Publ., **248**, 443–450

Christopherson, N. and Wright, R.F. (1981) Sulphate budget and a model for sulphate concentrations in stream water at Birkenes, a small forested catchment in Southernmost Norway. *Water Resources Research,* **17**, 377–389.

Cooper, A.B. (1990) Nitrate depletion in the riparian zone and stream channel of a small headwater catchment. *Hydrobiologia*, **202**, 13–26.

Cooper, D.M., Morris, E.M. and Smith, C.J. (1987) Precipitation and stream water chemistry in a subarctic Scottish catchment. *Journal of Hydrology*, **93**, 221–240.

Crisp, D.T. (1966) Input and output of minerals for an area of Pennine moorland: The importance of precipitation, drainage, peat erosion and animals. *Journal of Applied Ecology*, **3**, 327–348.

Crisp, D.T. and Robson S. (1979) Some effects of discharge upon the transport of animals and peat in a north Pennine headstream. *Journal of Applied Ecology*, **16**, 721–736.

Davies, T.D., Tranter, M., Wigington, Jr., P.J. and Eshleman, K.N. (1992) Acidic episodes in surface waters in Europe. *Journal of Hydrology*, **132**, 25–69.

Davies, T.D., Tranter, M., Blackwood, I.L. and Abrehams, P.W. (1993) The character and causes of a pronounced snowmelt-induced 'acidic episode' in a stream in a Scottish subarctic catchment. *Journal of Hydrology*, **146**, 267–300.

Dillon, P.J. and Molot, L.A. (1990) The role of ammonium and nitrate retention in the acidification of lakes and forested catchments. *Biogeochemistry*, **11**, 23–43.

Dise, N.B. and Wright, R.F. (1995) Nitrogen leaching from European forests in relation to nitrogen deposition. *Forest Ecology and Management*, **71**, 153–161.

Downing, J.A. and McCauley, E. (1992) The nitrogen:phosphorus relationship in lakes. *Limnology and Oceanography*, **37**, 936–945.

Driscoll, C.T. and Schafren, G.C. (1984) Short-term changes in the base neutralizing capacity of an acidic Adirondack New York lake. *Nature*, **310**, 308–310.

Driscoll, C.T. and Newton, R.M. (1985) Chemical characteristics of Adirondack lakes *Environment Science and Technology*, **19**, 1018–1024.

Driscoll, C.T. and Schaefer, D.A. (1989) Background on nitrogen processes, in *The Role of Nitrogen in the Acidification of Soil and Surface Waters* (eds J.L. Malanchuk and J. Nilsson), Nordic Council of Ministers.

Driscoll, C.T. and Van Dreason, R. (1993) Seasonal and long-term temporal patterns in the chemistry of Adirondack lakes. *Water, Air and Soil Pollution*, **67**, 319–344.

Driscoll, C.T., Wyskowski, B.J., Cosentini., C.C. and Smith, W.E. (1987) Processes regulating temporal and longitudinal trends in the chemistry of a low-order woodland stream in the Adirondack region of New York. *Biogeochemistry*, **3**, 225–241.

Driscoll, C.T., Schaefer, D.A., Molot, L.A. and Dillon, P.J. (1989) Summary of North America data, in *The Role of Nitrogen in the Acidification of Soil and Surface Waters* (eds J.L. Malanchuk and J. Nilsson), Nordic Council of Ministers.

Driscoll, C.T., Postek, K.M., Kretser, W. and Raynal, J.D. (1995) Long-term trends in the chemistry of precipitation and lake water in the Adirondack region of New York. *Water, Air and Soil Pollution*, **85**, 583–588.

Durand, P, Neal, C., Jeffery, H.A., Ryland, G.P., Neal, M. (1994) Major, minor and trace element budgets in the Plynlimon afforested catchments (Wales): general trends, and effects of felling and climate variation. *Journal of Hydrology*, **157**, 139–156.

Edwards, A.C., Creasey, J. and Cresser, M.S. (1985) Factors influencing nitrogen inputs and outputs in two Scottish upland catchments. *Soil Use and Management,* **1**, 83–87.

Edwards, A.C., Ron Vaz, M.D., Porter, S. and Ebbs, S. (1996) Nutrient cycling in upland catchment areas: The significance of organic forms of N and P, in *Advances in Hillslope Processes* (eds M.G. Anderson and S.M. Brookes), Wiley, New York, pp. 253–262.

Emmett, B.A., Reynolds, B., Stevens, P.A., Norris, D.A., Hughes, S., Gorres, J. and Lubrecht, I. (1993) Nitrate leaching from afforested Welsh catchments – Interactions between stand age and nitrogen deposition. *Ambio*, **22**, 386–394.

Faafeng, B and Hessen, D.O. (1993) Nitrogen and phosphorus concentrations and N:P ratios in Norwegian lakes: perspective on nutrient limitation. *Verh. Int. Verein. Limnol.*, **25**, 505–507.

Farley, D.A. and Werrity, A. (1989) Hydrochemical budgets for the Loch Dee experimental catchments, southwest Scotland. *Journal of Hydrology*, **109**, 351–368.

Ford, J., Stoddard, J.L. and Powers, C.F. (1993) Perspectives on environmental monitoring: An introduction to the US EPA long-term Monitoring (LTM) project. *Water, Air and Soil Pollution*, **67**, 247–256.

Forsberg, C., Ryding, S.O., Claesson, A. and Forsberg, A. (1978) Water chemical analyses and algal assay? Sewage effluent and polluted lake water studies. *Mitt. Int. Ver. Thesr. Angew. Limnol.*, **21**, 352–363.

Fowler, D., Cape, J.N. and Unsworth, M.H. (1989) Deposition of atmospheric pollutants on forests. *Philosophical Transactions of the Royal Society of London*, **B324**, 247–265.

Galloway, J.N., Schofield, C.L., Hendrey, G.R., Peters, N.E. and Johannes, A.H. (1980) *Ecological Impacts of Acid Precipitation*, SNSF-project, Oslo, Norway, pp.264–265.

Gibson, C.E., Wu, Y. and Pinkerton, D. (1995) Substance budgets of an upland catchment: the significance of atmospheric phosphorus inputs. *Freshwater Biology*, **33**, 385–392.

Gjessing, E.T., Henriksen, A., Johannessen, M. and Wright, R.F. (1976) Effects of acid precipitation on freshwater chemistry, in *Impact of acid precipitation on forest and freshwater ecosystems in Norway*, SNSF project FR6/76, As, Oslo, pp. 65–86.

Grennfelt, P. and Hultberg, H. (1986) Effects of nitrogen deposition on the acidification of terrestrial and aquatic ecosystems. *Water, Air and Soil Pollution*, **30**, 945–963.

Grip, H. and Bishop, K.H. (1990) Chemical dynamics of an acid stream rich in dissolved organics, in *The Surface Waters Acidification Programme* (ed. B.J. Mason), Cambridge University Press, Cambridge, pp. 75–84.

Harriman, R. and Morrison, B.R.S. (1982). Ecology of streams draining forested and non-forested catchments in an area of Central Scotland subject to acid precipitation. *Hydrobiologia*, **88**, 251–63.

Harriman, R., Ferrier, R.C., Jenkins, A. and Miller, J.D. (1990) Long – and short-term hydrochemical budgets in Scottish catchments, in *The Surface Water Acidification Programme* (ed B. J. Mason), Cambridge University Press, Cambridge, pp. 31–43.

Harriman, R., Curtis, C. and Edwards, A.C. (1998) An empirical method of predicting nitrate leaching from upland catchments in the United Kingdom using deposition and runoff chemistry. *Water, Air and Soil Pollution*, **105**, 193–203

Hauhs, M., Rost-Siebert, K., Raben, G., Paces, T. and Vigerust, B. (1989) Summary of European data, in *The Role of Nitrogen in the Acidification of Soil and Surface Waters* (eds J.L. Malanchuk and J. Nillson), Nordic Council of Ministers.

Henriksen, A. (1988) *Critical Loads for Sulphur and Nitrogen* (eds. J. Nilsson and P. Grennfelt), Nordic Council of Ministers, Copenhagen, Denmark. Vol. 15, pp. 385–412.

Henriksen, A.and Brakke, D.F. (1988) Increasing contribution of nitrogen to the acidity of surface waters in Norway. *Water, Air and Soil Pollution*, **42**, 183–201.

Henriksen, A., Lien, L., Traan, T.S., Sevaldrud, I.S. and Brakke, D.F. (1988) Lake acidification in Norway – Present and predicted chemical status. *Ambio*, **17**, 259–266.

Henriksen, A., Hindar, A., Hessen, D.O. and Kaste, O. (1997a) Contribution of nitrogen to acidity in the Bjerkreim River in southwestern Norway. *Ambio*, **26**, 304–311.

Henriksen, A. *et al.* (1997b) *Results of national lake survey (1995) in Finland, Norway, Sweden, Denmark, Russian Kola, Russian Karelia, Scotland and Wales*. Report 3645–97, Norwegian Institute for Water Research, Oslo.

Herlihy, A.T., Kaufmann, P.R., Church, M.R., Wigington, Jr., P.J., Webb, J.R. and Sale, M.J. (1993) The effects of acidic deposition on streams in the Appalachian Mountain and Piedmont region of the Mid-Atlantic United States. *Water Resources Research*, **29**, 2687–2703.

Heron, J. (1961) The seasonal variation of phosphate, silicate and nitrate in waters of the English Lake District. *Limnology and Oceanography*, **6**, 338–346.

Hessen, D.O., Hindar, A. and Holtan, G. (1997a) The significance of nitrogen runoff for eutrophication of freshwater and marine recipients. *Ambio*, **26**, 312–320.

Hessen, D.O., Henriksen, A. and Smelhus, A.M. (1997b) Seasonal fluctuations and diurnal oscillations in nitrate of a heathland brook. *Water Research*, **31**, 1813–1817.

Hessen, D.O., Henriksen, A., Hindar, A., Mulder, J., Torseth, K. and Vagstad, N. (1997c) Human impacts on the nitrogen cycle: A global problem judged from a local perspective. *Ambio*, **26**, 321–325.

Hill, A.R. (1979) Denitrification in the nitrogen budget of a river ecosystem. *Nature*, **281**, 291–292.

Hornstrom, E. and Esktrom, C. (1986). *Acidification and liming effects on Phyto- and Zooplankton in some Swedish west coast lakes*. SNV Report 1864, pp. 110.

Hornung, M., Reynolds, B. and Hatton, A.A. (1985) Land management, geological and soil effects on streamwater chemistry in upland mid-wales. *Applied Geography*, **5**, 71–80.

Hornung, M., Reynolds, B., Stevens, P.A. and Neal, C. (1987) Stream acidification resulting from afforestation in the UK: evaluation of causes and possible ameliorative measures, in *Forest Hydrology and Watershed management*, IAHS–AISH Publ. No. **167**, pp. 65–74.

Hornung, M., Rodda, F. and Langan, S.J. (1990) *A review of small catchment studies in western Europe producing hydrochemical budgets.* Air Pollution Research Report No. 28, CEC, Brussels.

Hope, D., Billett, M.F. and Cresser, M.S. (1994) A review of the export of carbon in river water: Fluxes and Processes. *Environmental Pollution,* **84**, 301–324.

Hope, D., Billett, M., Milne, R and Brown, T.A.W. (1997) Exports of organic carbon in British rivers. *Hydrological Processes*, **11**, 325–344.

Howard-Williams, C., Davies, J and Pickmere, S. (1982) The dynamics of growth, the effects of changing area and nitrate uptake by water cress *Nasturtium officinale* R. BR. in a New Zealand stream. *Journal of Applied Ecolog.,* **19**, 589–601.

INDITE (1994) *Impacts of Nitrogen Deposition in Terrestrial Ecosystems.* UK Review Group on Impacts of Atmospheric Nitrogen, Department of the Environment.

James, B.R. and Riha, S.J. (1989) Aluminium leaching by mineral acids in forest soils: 1. Nitric sulphuric-acid differences. *Soil Science Society of America Journal*, **53**, 259–264

Jenkins, A., Ferrier, R. and Waters, D. (1993) Melt water chemistry and its impact on stream water quality. *Hydrological Processes*, 7, 193–203.

Jenkins, A., Boorman, D. and Renshaw, M. (1996) The UK acid Water Monitoring Network: an assessment of chemistry data, 1983–93. *Freshwater Biology*, **36**, 169–178.

Johannessen, M. and Henriksen, A. (1978) Chemistry of snow meltwater: changes in concentration during melting. *Water Resources Research,* **14**, 615–19.

Kaste, O., Henriksen, A and Hindar, A. (1997) Retention of atmospherically-derived nitrogen in subcatchments of the Bjerkreim river in southwestern Norway. *Ambio*, **26**, 296–303.

Kaufmann, P.R., Herlihy, A.T., Mitch, M.E., Messer, J.J. and Overton, W.S. (1991) Stream chemistry in the eastern United States: Synoptic survey design, acid-base status and regional patterns. *Water Resources Research*, **27**, 611–627.

Kaushik, N.K., Robinson, J.B., Stammers, W.N. and Whitely, H.R. (1983) Aspects of nitrogen transport and transformation in headwater streams. In M.A. Lock and D.D. Williams (eds), *Perspectives in Running Water Ecology.* Plenum, New York, pp. 110–117.

Kernan, M. (1997) Nitrate variation in Snowdon lakes. In: *Surface water acidification: I. the increasing importance of nitrogen*, Abstracts of one day symposium, University College London. p. 11.

Kirby, C., Newson, M.D. and Gilman, K. (1991) *Plynlimon research: The first two decades.* Institute of Hydrology Report No. 109.

Kopácek, J. and Stuchlík, E. (1994) Chemical characteristics of lakes in the High Tatra Mountains, Slovakia. *Hydrolobiologia*, **274**, 49–56.

Kopácek, J., Procházkovà, L., Stuchlík, E., Blacka, P. (1995) The nitrogen-phosphorus relationship in mountain lakes: Influence of atmospheric input, watershed, and pH. *Limnology and Oceanography*, **40**, 930–937.

Kortelainen, P., Mannio, J., Forsius, M., Kamari, J. and Verta, M. (1989) Finnish Lake Survey: The role of organic and anthropogenic acidity. *Water Air Soil Pollution*, **46**, 235–249.

Kreiser, A.M, Patrick, S.T., Batterbee, R.W., Hall, J. and Harriman, R. (1995) In Critical Loads of Acid Deposition for UK Freshwaters, Critical Loads Advisory Group, Institute of Freshwater Ecology, Penicuik, pp. 15–18.

Langan, S.J. (1989) Sea-salt induced stream water acidification. *Hydrological Processes,* **3**, 25–41.

Langan, S., Hirst, D., Helliwell, R., Ferrier, B. (1997) A scientific review of the Loch Dee water quality and quantity data sets. Report no. SR97(01)F.

Lees, F.M. (1991) *A summery report of water chemistry and hydrology of Loch Dee, 1987–1989.* Loch Dee working paper June 1991. Solway Purification Board, Dumfries, Scotland.

Lepisto, A. (1995) Increased leaching of nitrate at two forested catchments in Finland over a period of 25 years. *Journal of Hydrology*, **171**, 103–123.

Lepisto, A., Andersson, L., Arheimer, B. and Sundblad, K. (1995) Influences of catchment characteristics, forestry activities and deposition on nitrogen export from small forested catchment. *Water, Air and Soil Pollution,* **84**, 81–102.

Likens, G.E., Bormann, F.H., Pierce, R.S. Eaton, J.S. and Johnson, N.M. (1977) *Biogeochemistry of a forested ecosystem,* Springer, New York.

Littlewood, I.G. (1992) *Estimating contaminant loads in rivers: a review.* Institute of Hydrology Report No. 117. Institute of Hydrology, Wallingford, UK.

MacDonald, A.M., Edwards, A.C., Pugh, K.B and Balls, P.W. (1994) The impact of land use on nutrient transport into and through three rivers in north east Scotland, in *Integrated River Basin Development* (eds C. Kirby and W.R. While), Wiley, pp. 210–214.

Meyer, J.L., Likens, G.E. and Sloane, J. (1981) Phosphorus, nitrogen and organic carbon flux in a headwater stream. *Archives in Hydrobiology*, **91**, 28–44.

Miller, J. D. (1995) Base-line water studies at Halladale, Caithness, 1993–1994. In *Final Report on Research at Halladale, February 1993–April 1994.* Report prepared for the Forestry Commission. Macaulay Land Use Research Institute, Aberdeen, UK.

Miller J.D., Anderson, H.A., Ferrier, R.C. and Walker, T.A.B. (1990) Hydrochemical fluxes and their effects on stream acidity in two forested catchments in central Scotland. *Forestry*, **63**, 311–331.

Mitchell, M.J., Driscoll, C.T., Kahl, J.S., Likens, G.E., Murdoch, P.S. and Pardo, L.H. (1996) Climatic control of nitrate loss from forested watersheds in the northeast United States, *Environment, Science. Technology*, **30**, 2609–2612.

Moore M.V., Pace, M.L., Mather, J.R., Murdoch, P.S., Howarth, R.W., Folt, C., Chen, C.Y., Hemond, H.F., Flebbe, P.A. and Driscoll, C.T. (1997) Potential effects of climate change on freshwater ecosystems of the New England/Mid-Atlantic Region. *Hydrological Processes*, **11**, 925–947.

Morris, D.P. and Lewis, W.M. Jr. (1988) Phytoplankton nutrient limitation in Colorado mountain lakes. *Freshwater Biology*, **20**, 315–327.

Mulder, J., Nilsen, P., Stuanes, A.O. and Huse, M. (1997) Nitrogen pools and transformations in Norwegian forest ecosystems with different atmospheric inputs. *Ambio*, **26**, 273–281.

Mulholland, P.J. (1992) Regulation of nutrient concentrations in a temperate forest stream: role of upland, riparian and in-stream processes. *Limnology and Oceanography*, **37**, 1512–1526.

Munn, N.L. and Meyer, J.L. (1990) Habitat-specific solute retention in two small streams: an intersite comparison. *Ecology,* **71**, 2069–2082.

Murdoch, P.S. and Stoddard, J.L. (1992) The role of nitrate in the acidification of streams in the Catskill Mountains of New York. *Water Resources Research,* **28**, 2707–2720.

Newbould, P. (1985) Improvement of native grassland in the uplands. *Soil Use and Management* **1**, 43–49.

Newell, A.D. (1993) Inter-regional comparison of patterns and trends in surface water acidification across the United States. *Water, Air and Soil Pollution*, **67**, 257–280.

Oborne, A.C., Brooker, M.P. and Edwards, R.W. (1980) The chemistry of the River Wye. *Journal of Hydrology*, **45**, 233–252.

Ollinger, S.V., Aber, J.D., Lovett, G.M., Millham, S.E., Lathorp, R.G. and Ellis, J.M. (1993) A spatial model of atmospheric deposition for the northeastern US. *Ecological Applications*, **3**, 459–472.

Papadakis, J. (1969) *Soils of the world*, Elsevier, Amsterdam.

Patrick, S., Monteith, D.T. and Jenkins, A. (eds) (1995). *UK Acid Waters Monitoring Network: Analysis and Interpretation of Results, April 1988–March 1993,* ENSIS, London.

Peterjohn, W.T., Adams, M.B., Gilliam, F.S. (1996) Symptoms of nitrogen saturation in two central Appalachian hardwood forest ecosystems. *Biogeochemistry*, **35**, 507–522.

Petersen, Jr., R.C., Gislason, G.M., Vought, L. B.M. (1995) Rivers of the Nordic countries, in *River and Stream Ecosystems* (eds C.E. Cushing *et al.*), Elsevier, Amsterdam, pp. 295–341.

Qualls, R.G., Haines, B.L. and Swank, W.E. (1991) Fluxes of dissolved organic nutrients and humic substances in deciduous forest. *Ecology*, **72**, 254–266.

Rascher, C.M., Driscoll, C.T. and Peters, N.E. (1987) Concentrations and flux of solutes from snow and forest floor during snowmelt in the West-Central Adirondack region of New York. *Biogeochemistry,* **3**, 209–22.

Reddy, K.R. (1982). Mineralization of nitrogen in organic soils. *Soil Science Society of America Journal*, **46**, 561–566.

Reid, J.M., Macleod, D.A. and Cresser, M.S. (1981) Factors affecting the chemistry of precipitation and river water in an upland catchment. *Journal of Hydrology*, **50,** 129–145.

Reynolds, B. and Emmett, B.A. (1990) *Annual report on the soil and soil water chemistry of selected sites at Llyn Brianne*. Report to Department of the Environment, Institute of Terrestrial Ecology, Bangor, UK.

Reynolds, B. and Edwards, A.C (1995) Factors influencing dissolved nitrogen concentrations and loadings in upland streams of the UK. *Agricultural Water Management,* **27**, 181–202.

Reynolds, B., Hornung, M. and Hughes, S. (1983) Some factors controlling variations in chemistry of an upland stream in mid-Wales. *Cambria*, **10**, 130–145.

Reynolds, B., Hornung, M. and Hughes, S. (1989) Chemistry of streams draining grassland and forest catchments at Plynlimon, Mid-Wales. *Hydrolgical Sciences Journal*, **34**, 667–686.

Reynolds, B., Emmett, B.A. and Woods, C. (1992) Variations in streamwater nitrate concentrations and nitrogen budgets over ten years in a headwater catchment in Mid-Wales. *Journal of Hydrology*, **136**, 155–175.

Reynolds, B., Ormerod, S.J. and Gee, A.S. (1994) Spatial patterns in stream nitrate concentrations in upland Wales in relation to catchment forest cover and forest age. *Environmental Pollution*, **84**, 27–33.

RGAR (1990) *Acid deposition in the United Kingdom 1986–1988*. Third report of the United Kingdom review Group on Acid Rain, Department of the Environment, London.

Rigler F.H. (1979) The export of phosphorus from Dartmoor catchments: A model to explain variations of phosphorus concentrations in streamwater. *Journal of Marine Biology Ass*ociation, **59**, 327–348.

Roberts, G., Hudson, J. and Blackie, J.R. (1983) *Nutrient cycling in the Wye and Severn at Plynlimon,* Institute of Hydrology Report No. 86. Institute of Hydrology, Wallingford, UK.

Roberts, G., Hudson, J. and Blackie, J.R. (1984) Nutrient inputs and outputs in a forested and grassland catchment at Plynlimon, Mid-Wales. *Agricultural Water Management,* **9**, 177–191.

Roberts, G., Hudson, J. and Blackie, J.R. (1986) Effect of upland pasture improvement on nutrient release in flows from a natural lysimeter and field drain. *Agricultural Water Management* , **11**, 231–245.

Roberts, G., Reynolds, B. and Talling, J. (1989) *Upland Management and Water Resources*. Report to DOE and Welsh Office, Institute of Hydrology, Wallingford, UK.

Schaefer, D.A. and Driscoll, C.T. (1993) Identifying sources of snowmelt acidification with a watershed mixing model. *Water, Air and Soil Pollution*, **67**, 345–365.

Schaefer, D.A., Driscoll, C.T., Jr., Van Dreason, R. and Yatsko, C.P. (1990) The episodic acidification of Adirondack lakes during snowmelt. *Water Resources Research*, **26**, 1639–1648.

Sebetich, M.J., Kennedy, V.C., Zand, S.M., Avanzio, R.J. and Zellweger, G.W. (1984) Dynamics of added nitrate and phosphorus compared in a northern California woodland stream. *Water Resources Bulletin*, **20**, 93–101.

Shah, Z, Adams, W.A. and Haven, C.D.V. (1990) Composition and activity of the microbial population in an acidic upland soil and effect of liming. *Soil Biology and Biochemistry*, **22**, 257–263.

Skeffington, R.A. and Wilson, E.J. (1988). Excess nitrogen deposition: issues for consideration. *Environmental Pollution.*, **54**, 159–184.

Smith, V.H. (1982) The nitrogen and phosphorus dependence of algal biomass in lakes: An empirical and theoretical analysis. *Limnology and Oceanography*, **27**, 1101–1112.

Soulsby, C. (1995) Contrast in storm event hydrochemistry in an acidic afforested catchment in upland Wales. *Journal of Hydrology*, **170**, 159–179.

Stevens, P.A. and Wannop, C.P. (1987) Dissolved organic nitrogen and nitrate in an acid forest soil. *Plant and Soil*, **102**, 137–139.

Stevens, P.A., Hornung, M. and Hughes, S. (1989) Solute concentrations, fluxes and major nutrients cycles in mature Sitka-spruce plantation in Beddgelert Forest, North Wales. *Forest Ecology and Management*, **27**, 1–20.

Stevens, P.A., Adamson, J.K., Reynolds, B and Hornung, M. (1990) Dissolved nitrogen concentrations and fluxes in three British Sitka Spruce plantations. *Plant and Soil*, **128**, 103–108.

Stevens, P.A., Harrison, A.F., Jones, H.E., Williams, T.G. and Hughes, S. (1993a) Nitrate leaching from a Sitka spruce plantation and the effects of fertilization with phosphorus and potassium. *Forest Ecology and Management*, **58**, 233–247.

Stevens, P.A., Williams, T.G., Norris, D.A. and Rowland, A.P. (1993b) Dissolved inorganic nitrogen budgets for a forested catchment at Beddgelert, North Wales. *Environmental Pollution*, **80**, 1–8.

Stevens, P.A., Norris, D.A., Sparks, T.H. and Hodgson, A.L. (1994) The impacts of atmospheric N inputs on throughfall, soil and stream water interactions for different aged forest and moorland catchments in Wales. *Water, Air and Soil Pollution*, **73**, 297–317.

Stevens, P.A., Ormerod, S.J. and Reynolds, B (1997) *Final report on the acid waters survey of Wales: Volume I Main text.* Institute of Terrestrial Ecology, Bangor, N. Wales.

Stoddard, J.L. (1994) Long-term changes in watershed retention of nitrogen, in *Environmental chemistry of lakes and reservoirs* (ed L.A. Baker), American Chemical Society, Washington DC. pp. 223–284.

Stoddard J.L. and Kellogg, J.H. (1993) Trends and patterns in lake acidification in the State of Vermont: evidence from the long-term monitoring project. *Water, Air and Soil Pollution*, **67**, 301–318.

Stoner, J.H., Gee, A.S. and Wade, L.R. (1984) The effects of acidification on the ecology of streams in the upper Tywi catchment in West Wales. *Environmental Pollution.* A, **35**, 125–157.

Sullivan, T.J., Driscoll, C.T., Gherini, S.K., Munson, R.K., Cooke, R.R., Charles, D.F. and Yatsko, C.P. (1989) Influence of aqueous aluminium and organic acids on measurements of acid neutralisation capacity in surface waters. *Nature*, **338**, 408–410.

Sullivan, T.J., Eilers, J.M., Cosby, B.J. and Vache, K.B. (1997) Increasing role of nitrogen in the acidification of surface waters in the Adirondack mountains, New York. *Water, Air and Soil Pollution*, **95**, 313–336.

Sutcliffe, D.W., Carrick, T.R., Heron, J., Rigg, E., Talling, J.F., Woof, C. and Lund, J.W.G. (1982) Long-term and seasonal changes in the chemical composition of precipitation and surface waters of lakes and tarns in the English Lake District. *Freshwater Biology*, **12**, 451–506.

Swank, W.T. and Caskey, W.H., 1982. Nitrate depletion in a second order mountain stream. *Journal of Environment Quality*, **11**, 581–584.

Talling, J.F., Carrick, T.R., Lishman, J.P. (1989). A survey of the chemical composition of upland runoff in Pennine river systems, in *Upland Management and Water Resources* (eds G. Roberts, B. Reynolds and J.F. Talling). Report to Department of Environment and Welsh Office, Institute of Hydrology, Wallingford, UK.

Triska, F.J., Duff, J.H. and Avanzino, R.J. (1990) Influence of exchange flow between the channel and hyporheic zone on nitrate production in a small mountain stream. *Canadian Journal of Fish Aquatic. Science*, **47**, 2099–2111.

UNESCO (1975) *Soil map of the world.* Volume II North America.

Van Miegroet, H. (1994) The relative importance of sulpher and nitrogen compounds in the acidification of fresh water, in *Acidification of Freshwater Ecosystems: Implications for the Future* (eds C.E.W. Steinberg and R.F. Wright), John Wiley and Sons Ltd., pp. 33–49.

Walling, D.E. and Webb, B.W. (1985) Estimating the discharge of contaminants to coastal waters: some cautionary comments. *Marine Polluion Bulletin*, **16**, 488–492.

Webb, B.W. and Walling, D.E. (1985) Nitrate behaviour in streamflow from a grassland catchment in Devon, UK. *Water Research*, **19**, 1005–1016.

Webster, J.R., Wallace, J.B. and Benfield, E.F. (1995) Organic processes in streams of the eastern United States, in *River and Stream Ecosystems* (eds C.E. Cushing *et al.*), Elsevier, Amsterdam, pp. 117–187.

Wigington, P. J., Davies, T. D., Tranter, M. and Eshleman, K. N. (1990) Episodic acidification of surface waters due to acid precipitation, in *National Acid precipitation Assessment Precipitation Assessment Programm*, Washington, D.C., State of Science/Technology. Rep No. 12.

Wiklander, G., Norlander, G and Andersson, R. (1991) Leaching of nitrogen from a forest catchment at Soderasen in southern Sweden. *Water, Air and Soil Pollution*, **55**, 263–282

Williamson, R.B. and Cooke, J.G., 1985. The effect of storms on nitrification rates in a small stream. *Water Research*, **19**, 435–440

Wright, R.L. and Henriksen, A. (1980) *Regional survey of lakes and streams in Southwestern Scotland, April 1979*. SNSF-project, Oslo, Norway, Report IR 72/80.

Wright, R.L., Raastad, I. A. and Kaste, O. (1997) *Atmospheric deposition of nitrogen, runoff of organic nitrogen and critical loads for soils and waters*. Report 3592–97, Norwegian Institute for Water Research, Oslo.

Yesmin, L., Gammack, S.M., Sanger, L.J. and Cresser, M.S. (1995) Impact of atmospheric N deposition on inorganic and organic-N outputs in water draining from peat. *Science Total Environment,* **166**, 201–209.

DERIVATION OF CRITICAL LOADS BY STEADY-STATE AND DYNAMIC SOIL MODELS

M. POSCH[1] AND W. DE VRIES[2]

[1]*Coordination Center for Effects (CCE)*
National Institute of Public Health and the Environment (RIVM)
P.O.Box 1, NL-3720 BA Bilthoven, The Netherlands
[2]*DLO Winand Staring Centre for Integrated Land, Soil and Water Research (SC-DLO)*
P.O.Box 125, NL-6700 AC Wageningen, The Netherlands

7.1 Introduction

7.1.1 HISTORICAL BACKGROUND

Critical loads are now widely accepted in Europe as a basis for negotiating control strategies for transboundary air pollution as evidenced by the signing of the Second Sulphur Protocol (UN/ECE 1994) in Oslo in June 1994 and their use in formulating an Acidification Strategy for the European Union (Amann *et al.* 1996). The United Nations Economic Commission for Europe's (UN/ECE) Executive Body of the Convention on Long-range Transboundary Air Pollution (LRTAP) has set up a Task Force on Mapping Critical Levels/Loads under its Working Group on Effects. Under this Task Force, critical load data from individual countries are collected, collated and mapped by the Coordination Center for Effects (see Posch *et al.* 1995) and provided to the relevant UN/ECE bodies under the LRTAP Convention. This information is used in formulating emission reduction strategies in Europe, such as the current negotiations on a revision of the 1988 Nitrogen Protocol.

The first workshop on critical loads held under the auspices of the UN/ECE was organized in 1988 by the Nordic Council of Ministers at Skokloster (Sweden) providing the still valid definition of a critical load as 'the quantitative estimate of an exposure to one or more pollutants below which significant harmful effects on specified sensitive elements of the environment do not occur according to present knowledge' (Nilsson and Grennfelt 1988). As the role of nitrogen in the acidification of soils and surface waters gained increasing attention at the end of the 1980s in both the scientific and policy arena, a workshop was organized by the Nordic Council of Ministers and the U.S. Environmental Protection Agency on that topic in Copenhagen in 1988 (Malanchuk and Nilsson 1989). The purpose of that workshop was to review the state of science on the role of nitrogen in the acidification of the environment. The foundation for the mapping of critical loads in the ECE countries was laid in a UN/ECE workshop held in 1989 in Bad Harzburg

(Germany), resulting in a manual for mapping critical levels and loads, which has been updated recently (UBA 1996). Furthermore, in a workshop on critical loads for nitrogen organised by the Nordic Council of Ministers in Lökeberg (Sweden) in 1992 recommendations for deriving critical loads of nitrogen and their exceedances were elaborated (Grennfelt and Thörnelöf 1992). Remaining open questions were discussed at a UN/ECE workshop in Grange-over-Sands (England) in 1994, organised by the British Department of the Environment (Hornung *et al.* 1995a).

7.1.2 USE OF MODELS TO DERIVE CRITICAL LOADS:

Soils, especially forest soils, are the most common 'sensitive element of the environment' for which critical loads have been calculated and mapped by European countries. Especially in the Nordic countries, also critical loads for surface waters (lakes) have been extensively studied. The methods for computing critical loads for lakes are similar to those for soils (soils in the terrestrial catchment influence lake water chemistry), augmented by simple formulations of in-lake processes such as sulphur and nitrogen retention (Posch *et al.* 1997a). In addition, critical loads for (semi-)natural ecosystems, such as heathlands and peatlands, have been calculated and mapped in several European countries, mostly employing empirical methods (see, e.g., Hornung *et al.* 1995b with empirical critical load maps for Great Britain). Tables with empirical critical load values and estimates of their reliability have been compiled over the years, both for acidity critical loads and for nitrogen as a nutrient (e.g. in UN/ECE 1995, UBA 1996).

Critical loads can be calculated either by steady-state models or by dynamic models with different degrees of complexity. In the case of forest ecosystems not only direct effects play a role, but also soil-mediated effects of N and S deposition (N accumulation and acidification). Roberts *et al.* (1989) concluded that spruce decline in Central Europe mainly results from foliar Mg deficiency due to (i) an increased Mg demand induced by an increased growth in response to elevated N inputs, and (ii) inhibition of Mg uptake caused by ammonium accumulation and aluminium mobilization. Other effects include vegetation changes due to increased growth in response to increased N availability and increased susceptibility of conifers to frost and diseases related to high N contents in needles (De Vries 1993). To date mostly soil chemical criteria (e.g. critical aluminium or nitrogen concentrations, critical Al/Ca ratios) have been used to derive critical loads with simple steady-state models. The largest uncertainty in these calculations is the relation between the critical chemical values and the 'harmful effects'.

Steady-state models calculate deposition levels which avoid the violation of a chosen soil chemical criterion in a steady-state situation. Therefore, effects with a finite time scale, such as cation exchange and sulphate adsorption, are not included. The standard model is the so-called Simple Mass Balance (SMB) model, which is based on the charge balance of the major ions in the soil leachate, the concentrations of which are determined by the deposition of the various components (sulphate, nitrate, base cations) and the sources and sinks of those ions within the rooting zone (e.g., weathering and uptake). Critical loads, i.e. maximum values of sulphur and nitrogen deposition, are then derived by setting a limit to the leaching of ANC, e.g., by requiring that the molar ratio of aluminium to base cations in the soil solution stays below a predefined value. The SMB model has been and is widely used to produce maps of critical loads of S and N on a European scale (Posch *et al.* 1995), and also in this chapter we concentrate on that model to discuss the critical load concept.

Dynamic soil models simulate the same processes as steady-state soil models, while including additional processes that play a role on a finite time scale. These models can also be used to calculate critical loads by running the model until a steady state is reached. By trial and error the (constant) deposition level is calculated that fulfils a chosen chemical criterium (see, e.g., Warfvinge and Sverdrup, 1995, where critical loads for Sweden are derived with the dynamic model SAFE). Critical loads calculated in this way are equal to those derived by steady-state models, if the processes in both models are simulated in the same way.

Dynamic soil models can furthermore be used to derive so-called target loads by considering a finite time period (e.g. one rotation period) in which the system is allowed (or has to) reach a chemical criterium. Unlike the critical load, the target load is influenced by the present acidification status of the soil system and also the magnitude of time-limited processes.

Finally, dynamic soil models can also be used to derive the time period before the system reaches a chosen soil chemical criterium for a given deposition scenario. Depending on the present soil status, the model thus calculates the time period before risk increases or before the system starts to recover. Dynamic models are most commonly used in this context (see Cosby *et al.* 1989, De Vries *et al.* 1994)

Some of the more widely used simple dynamic soil models are MAGIC (Cosby *et al.* 1985), SAFE (Warfvinge *et al.* 1993) and SMART (De Vries *et al.* 1989). Besides many other applications, all three models have been used to simulate the past and future time development of the soils at different integrated monitoring sites (Forsius *et al.* 1996). Here we focus on the basic principles of and results from the SMART model, since it is completely compatible with the steady-state SMB model (as illustrated in Posch *et al.* 1997b).

7.1.3 AIM OF THIS CHAPTER:

In this chapter we first describe the basic assumptions and equations underlying both the steady-state and dynamic models for (forest) soils. We proceed by describing the steady-state SMB model which is used for the derivation of critical loads for nitrogen, both as a nutrient and as an acidifying compound. In this context we also discuss the various critical chemical values used for setting critical loads and derive their relationship to the base saturation in the soil, a connection to which little attention has yet been paid in the critical loads discussion. This leads to the discussion of dynamic models in the subsequent section. We focus on the very simple formulation of nitrogen processes in the three dynamic models mentioned above, but also discuss their recent modifications and extensions which try to remedy these shortcomings. Finally, we show by way of examples how dynamic models can be used to derive target loads and to estimate time horizons for reaching critical values, thus emphasising their potential role in formulating emission reduction policies.

7.2 Assumptions And Basic Equations In Steady-State And Dynamic Soil Models

As the receptor for the deposition of atmospheric pollutants we consider non-calcareous forest soils. Since we are primarily interested in applications on a large regional scale, on which data is scarce, we assume that the soil is a single homogeneous compartment and its depth is equal to the root zone. This implies that internal soil processes (such as weathering and uptake) are evenly distributed over the soil profile, and all physico-chemical constants are assumed uniform in the whole profile. Furthermore we assume the simplest possible hydrology: The water leaving the root zone is equal to precipitation minus evapotranspiration; more precisely, all evapotranspiration occurs at the top of the soil profile, and percolation is constant through the soil profile and occurs only vertically.

7.2.1 MASS AND CHARGE BALANCE EQUATIONS

Irrespective of any simplifying assumptions, every model is based on the mass balance for each element (ion) considered, and the concentrations of these ions have to fulfil the charge balance equation. In addition, equilibrium equations determine the interaction of the soil solution with the air in the soil and the soil matrix.

Changes in the concentration of ion X in the soil compartment, and thus in the leaching, are determined by the mass balance for that ion:

$$\frac{d}{dt}X_{tot} = X_{dep} + X_{int} - X_{le} \tag{1}$$

i.e. the change of the total amount, X_{tot}, in the soil profile over time is given by the deposition flux X_{dep} plus the net internal production X_{int} (sources minus sinks) minus the amount leached, X_{le}. All quantities on the right-hand side of eq.1 are expressed in equivalents (moles of charge) per unit area and unit time (e.g. eq/m^2/a). The leaching terms are given by $X_{le}=Q[X]$ where $[X]$ is the soil solution concentration (eq/m^3) of ion X and Q (m/a) is the water percolating from the root zone. For convenience we suppress the charges from the element symbols. In dynamic models changes in the concentration of ion X in the soil compartment (both in solution and the solid phases) are included, whereas steady-state models consider these changes to be zero (see below). While the deposition of certain ions, such as aluminium and organic anions, is considered zero, the deposition of sulphur, nitrogen species, chloride and base cations is either computed by (long-range) transport and deposition models or interpolated from measurements. If the deposition is taken from large scale models one has to bear in mind that it is modified locally by the type of vegetation growing on the soil at the site under consideration.

The choice of processes to be considered determines the equations for the internal fluxes. The internally generated fluxes of sulphate (such as uptake, immobilization, reduction and precipitation) and chloride are neglected in the SMB model and in most dynamic models:

$$SO_{4,int} = 0 \quad \text{and} \quad Cl_{int} = 0 \tag{2}$$

i.e. they act as tracer. For base cations the internal flux is given by ($BC=Bc+Na$, $Bc=Ca+Mg+K$):

$$BC_{int} = BC_w - Bc_u \quad \text{and} \quad Al_{int} = Al_w \tag{3}$$

where BC_w is the weathering of base cations, Bc_u the net growth uptake (Na is not taken up by plants in significant amounts), and Al_w is the weathering of aluminium in the congruent weathering of silicates. Note that adsorption and exchange reactions are not included in the internal flux descriptions. They are included in X_{tot} and described by equilibrium reactions (see below). The internal fluxes of nitrate and ammonium are calculated according to:

$$NO_{3,int} = NH_{4,ni} - NO_{3,i} - NO_{3,u} - NO_{3,de} \tag{4}$$

and

$$NH_{4,int} = -NH_{4,ni} - NH_{4,i} - NH_{4,u} \tag{5}$$

where the subscripts *ni*, *i*, *u* and *de* stand for nitrification, immobilization, net growth uptake and denitrification, respectively. Several nitrogen processes, such as fixation, volatilization, adsorption and the loss of nitrogen due to fire and erosion, are neglected in the SMB model and in most dynamic models. In the case of complete nitrification eqs.4 and 5 simplify to:

$$N_{int} = -N_i - N_u - N_{de} \tag{6}$$

In addition to individual mass balances, all ions in the soil water leaving the root zone have to fulfil charge balance:

$$H_{le} + Al_{le} + BC_{le} + NH_{4,le} = SO_{4,le} + NO_{3,le} + Cl_{le} + HCO_{3,le} + RCOO_{le} \tag{7}$$

where *RCOO* stands for the sum of organic anions.

7.2.2 EQUILIBRIUM REACTIONS

Equilibrium equations describe the interaction of the soil solution with air and with the soil matrix. The dissociation of CO_2 is calculated according to:

$$[HCO_3][H] = K_{CO2}\, p_{CO2} \tag{8}$$

where K_{CO2} is the dissociation constant and p_{CO2} the partial pressure of CO_2. Models of different complexity have been suggested for the dissociation of organic anions (see, e.g., Driscoll *et al.* 1994), but in all of them the concentration of the various organic anions depends on the pH alone (in addition to one or more equilibrium constants). The dissolution of aluminium (hydr)oxides in its simplest form is modelled by a gibbsite equilibrium:

$$[Al] = K_{gibb}\,[H]^3 \tag{9}$$

This equation is used in the critical load calculations and in the SMART model. In the SAFE model the dissolution of two aluminium hydroxides is included, whereas the MAGIC model also includes the complexation of Al with fluoride and sulphate.

Exchange reactions are often described by Gaines-Thomas equations. Lumping all base cations, the following equations describe the Al-BC-H exchange in the SMART model:

$$\frac{f_{Al}^2}{f_{BC}^3} = K_{AlBC}\,\frac{[Al]^2}{[BC]^3} \tag{10}$$

$$\frac{f_H^2}{f_{BC}} = K_{HBC}\,\frac{[H]^2}{[BC]} \tag{11}$$

where f_X is the exchangeable fraction of ion X. Furthermore, charge balance requires that

$$f_{Al} + f_{BC} + f_H = 1 \tag{12}$$

In the MAGIC model the exchange of each base cation (and chloride) is modelled by a separate equation, but the proton exchange is neglected, whereas in the SAFE model H-BC exchange is modelled by a Gapon equation.

Both in the MAGIC and SMART model the equilibrium between dissolved and adsorbed sulphate in the soil-soilwater system is described by a Langmuir isotherm (see Cosby *et al.* 1986):

$$SO_{4,ad} = \frac{[SO_4]}{S_{1/2} + [SO_4]}\,S_{max} \tag{13}$$

where $SO_{4,ad}$ (meq/kg) is the amount of sulphate adsorbed, S_{max} is the maximum adsorption capacity of the soil and $S_{1/2}$ is the 'half saturation' constant.

In the following section we will simplify the above set of equations to derive critical loads of nitrogen as a nutrient and for acidifying nitrogen and sulphur.

7.3 Critical Loads

Critical loads are – by definition – deposition fluxes which do not change over time, i.e. they are calculated for a steady-state situation. Steady state means there is no change over time in the total amount of the ions involved (cf. eq.1):

$$\frac{d}{dt}X_{tot} = 0 \quad \text{and thus} \quad X_{le} = X_{dep} + X_{int} \tag{14}$$

i.e. the amount of ion X leached is equal to the amount deposited plus the net amount generated internally.

7.3.1 THE CRITICAL LOAD OF NUTRIENT NITROGEN

A critical load of nutrient nitrogen, $CL_{nut}(N)$, is obtained by defining an acceptable leaching of nitrogen, $N_{le,acc}$. Assuming complete nitrification, it suffices to consider total nitrogen ($NH_{4,le}=0$, $NO_{3,le}=N_{le}$) and one obtains (see eqs.6 and 14):

$$CL_{nut}(N) = N_i + N_u + N_{de} + N_{le,acc} \tag{15}$$

When defining the critical load via eq.15 it is implicitly assumed that all terms on the right-hand side do not depend on the deposition of nitrogen. This is unlikely to be the case and thus all quantities should be taken 'at critical load'. However, to compute 'denitrification at critical load' one needs to know the critical load, the very quantity one tries to compute. The only way to avoid this circular reasoning is to establish a functional relationship between deposition and the sink of N, insert this function into eq.14 and solve for the deposition (to obtain the critical load). This has been done for denitrification: in the simplest case denitrification is linearly related to the net input of N by (UBA 1996):

$$N_{de} = \begin{cases} f_{de}(N_{dep}-N_i-N_u) & \text{if } N_{dep}>N_i+N_u \\ 0 & \text{otherwise} \end{cases} \tag{16}$$

where f_{de} ($0 \leq f_{de} \leq 1$) is the so-called denitrification fraction, which has been formulated as a function of soil type (De Vries *et al.* 1993). This formulation implicitly assumes that immobilization and growth uptake are faster processes than denitrification. Inserting eq.16 into eq.14 and solving for the deposition leads to the following critical load equation:

$$CL_{nut}(N) = N_i + N_u + \frac{N_{le,acc}}{1-f_{de}} \tag{17}$$

A non-linear deposition dependence of denitrification, based on a Michaelis-Menten reaction mechanism, has been suggested by Sverdrup and Ineson (1993, unpublished manuscript; see also Posch *et al.* 1995).

The nitrogen uptake in the critical load equation is the net growth uptake, i.e. the net uptake by vegetation that is needed for long-term average growth. Nitrogen input by litterfall and nitrogen removal by maintenance uptake (needed to re-supply nitrogen to leaves) is not considered here, assuming that both fluxes are equal in a steady-state situation. Thus the net uptake is equal to the annual average removal in harvested biomass. This can be estimated from the nitrogen content in stems (and branches and leaves, if they are removed, too) in the removed biomass. Dividing that biomass (estimated, e.g. from yield tables) by the rotation time gives the annual net uptake flux of nitrogen. Care should be taken not to use biomass data from sites with luxurious growth due to increased nitrogen deposition.

As with uptake, N_i denotes the long-term net immobilization (accumulation) of nitrogen in the soil, i.e. only the continuous build-up of stable C/N-compounds in forest soils. Using data from Swedish forest soil plots, Rosén *et al.* (1992) estimated the annual nitrogen immobilization since the last glaciation at 0.2 to 0.5 kgN/ha/a. Values between 0.5 and 1.0 kgN/ha/a are currently recommended for the critical loads work under the LRTAP Convention (UBA 1996).

Once the terms on the right-hand side of eq.15 or 17 are specified, critical loads of nutrient nitrogen can be determined. Maps of $CL_{nut}(N)$ for Europe are, e.g., presented in Posch *et al.* (1995). The comparison with the present or future nitrogen depositions is done by computing the so-called exceedance of the critical load:

$$Ex_{nut}(N) = N_{dep} - CL_{nut}(N) \tag{18}$$

Regionalized exceedance values are used in integrated assessment models to determine (cost-optimal) reduction strategies for nitrogen deposition.

7.3.2 CRITICAL LOADS OF ACIDITY

A critical load of acidity is derived from the charge balance equation by specifying a critical leaching of ANC. Acid neutralization capacity (ANC) is defined as:

$$ANC_{le} = -H_{le} - Al_{le} + HCO_{3,le} + RCOO_{le} \tag{19}$$

Thus we get from the charge balance (see eq.7):

$$ANC_{le} = BC_{le} + NH_{4,le} - SO_{4,le} - NO_{3,le} - Cl_{le} \tag{20}$$

which shows that ANC is independent from the partial pressure of CO_2, one of the reasons why ANC is considered a useful quantity for characterizing the chemical composition of the soil solution. Assuming complete nitrification, the charge balance for the steady-state situation becomes (see eqs. 2, 3, 6 and 14):

$$ANC_{le} = BC_{dep} - Cl_{dep} + BC_w - Bc_u + N_i + N_u + N_{de} - S_{dep} - N_{dep} \tag{21}$$

Note, that BC_{dep} and BC_w include all four base cations ($BC=Ca+Mg+K+Na$), whereas sodium is not taken up by vegetation ($Bc=BC\text{-}Na$). Knowledge of the deposition terms, weathering and net uptake of base cations as well as nitrogen uptake, immobilization and denitrification allows to calculate the ANC leaching, and thus assess the acidification status of the soil. Conversely, critical loads of S and N are computed by defining a critical (or acceptable) ANC leaching, which is set to avoid 'harmful effects' on the 'specified sensitive element of the environment' (e.g. damage to fine roots of trees). Inserting also the equation for the deposition dependent denitrification (eq.16), one obtains for the critical loads of sulphur, $CL(S)$, and acidifying nitrogen, $CL(N)$:

$$CL(S)+(1-f_{de})CL(N)=BC_{dep}-Cl_{dep}+BC_w-Bc_u+(1-f_{de})(N_i+N_u)-ANC_{le(crit)} \quad (22)$$

When comparing S and N deposition to critical loads one has to bear in mind that the nitrogen sinks cannot compensate incoming sulphur acidity, i.e. the maximum critical load of sulphur is given by:

$$CL_{max}(S)=BC_{dep}-Cl_{dep}+BC_w-Bc_u-ANC_{le(crit)} \quad (23)$$

This expression has also been termed critical load (or deposition) of acidity; and it has been used to derive the critical deposition of S - used in the negotiations of the Second Sulphur Protocol - by multiplying it by the so-called sulphur fraction (see, e.g., Hettelingh *et al.* 1995). Furthermore, if

$$N_{dep}\le N_i+N_u=CL_{min}(N) \quad (24)$$

all deposited N is consumed by uptake and immobilization, and sulphur can be considered alone. The maximum amount of allowable N deposition (in case of zero S deposition) is given by (see eqs.22-24):

$$CL_{max}(N)=CL_{min}(N)+\frac{CL_{max}(S)}{1-f_{de}} \quad (25)$$

The three quantities, $CL_{max}(S)$, $CL_{min}(N)$ and $CL_{max}(N)$, define the so-called critical load function as depicted in Figure 7.1. For every pair of deposition (N_{dep},S_{dep}) lying on the function or below it (grey-shaded area in Figure 7.1) there is non-exceedance of the critical loads of nitrogen and sulphur.

Since it is impossible to define unique critical loads of N and S, it is also impossible to define a unique exceedance in the sense of quantifying the amount of S and N to be reduced. This is illustrated by the example in Figure 7.1: Let the point E1 denote the (current) deposition of N and S. Reducing N_{dep} substantially, one reaches the point Z1 and thus non-exceedance without reducing S_{dep}; on the other hand one can reach non-exceedance by only reducing S_{dep} (by a smaller amount) until reaching Z3; finally, with a smaller reduction of both N_{dep} and S_{dep} one can reach non-exceedance as well (e.g. point Z2). In practice external factors, such as the costs of emission reduction measures, will determine the path to be followed to reach zero exceedance.

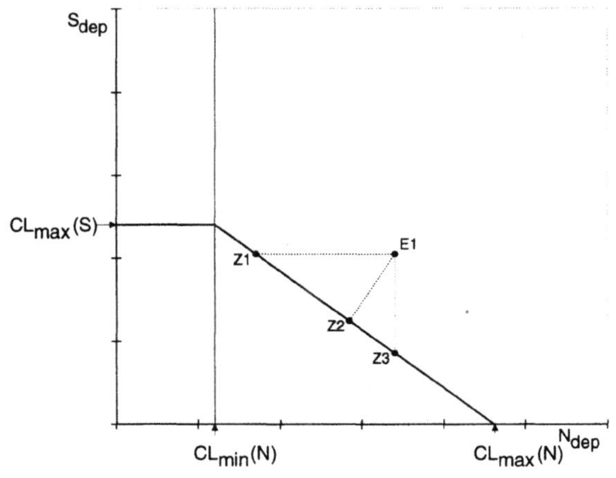

Figure 7.1 Critical load function of sulphur and acidifying nitrogen, defined by the three quantities $CL_{max}(S)$, $CL_{min}(N)$ and $CL_{max}(N)$.

To calculate critical loads the individual terms in eq.22 have to be specified. The nitrogen terms (N_i, N_u and f_{de}) are the same as in the critical load of nutrient N and have been discussed above. The deposition of base cations and chloride is assumed not to change over time and thus only the natural (non-anthropogenic) base cation deposition should be used. In addition, to account only for anthropogenic S and N depositions, the contribution of sea salts to S, Cl and base cations is usually factored out. This does not mean that sea salts play no role in soil chemistry (see, e.g., White *et al.* 1996), but critical loads are primarily used for assisting emission reduction policies. Therefore, depositions of base cations, sulphur and chloride are corrected by assuming that either all sodium or all chloride is derived from sea salts, using the following formula:

$$X^*_{dep} = X_{dep} - r_{XY} Y_{dep} \tag{26}$$

where $X = Ca, Mg, K, Na, Cl$ or SO_4, $Y = Na$ or Cl and r_{XY} is the ratio of ions X and Y in sea water (see Table 7.1); the star denotes the sea-salt corrected deposition.

Table 7.1: Ion ratios $r_{XY}=[X]/[Y]$ (in eq/eq) in sea water used for correcting base cation, chloride and sulphate deposition (see eq.26 and UBA 1996).

	X					
Y	Ca	Mg	K	Na	Cl	SO_4
Na	0.044	0.227	0.021	1	1.164	0.120
Cl	0.037	0.198	0.018	0.858	1	0.103

In the first case one has $Na_{dep}^*=0$, in second $Cl_{dep}^*=0$. In the following we assume sea-salt corrected quantities (including sulphur deposition), but we suppress the star for convenience.

The uptake of base cations (Ca, Mg and K) refers again to the net growth uptake and can be derived in the same way as nitrogen uptake (see above) by estimating the element contents in the harvested biomass. Note, that for a given species the ratios of the different elements are fairly constant in the various tree compartments and vary only mildly with location (climate, site quality).

The weathering of base cations, BC_w, is derived from input-output budgets (current weathering rates), long-term depletion studies (historical weathering rates) or computed by models such as PROFILE (Sverdrup and Warfvinge 1992, Warfvinge and Sverdrup 1992), which uses mineralogy and other physical and chemical soil parameters as input. For a review of published weathering rates see Langan *et al.* (1995).

7.3.3 CRITICAL/ACCEPTABLE CHEMICAL VALUES

Defining a critical chemical value is the most crucial step in calculating critical loads, since this quantity links deposition levels to a 'harmful effect'. It is determined by selecting a 'sensitive element of the environment' and finding a (chemical) variable which links it to soil solution chemistry.

The critical/acceptable N leaching for calculating the critical load of nutrient nitrogen can, e.g., be specified to avoid nutrient imbalances in soils. Increased amounts of N in the soil solution may lead to changes in the species composition of the ground vegetation below forests with N-tolerant species gaining ground over less tolerant species. Currently recommended values of $N_{le,acc}$ for European forests lie in the range of 0.5-4 kgN/ha/a (UBA 1996). Instead of specifying the N-leaching directly one can specify a maximum acceptable N concentration in the soil solution, $[N]_{acc}$, resulting in $N_{le,acc}=Q[N]_{acc}$. A critical N leaching could also be derived with the objective to avoid nitrogen pollution of groundwaters. Internationally agreed values, such as 25 and 50 mgN/l (EC target and limit value, resp.) could be used for this purpose.

For acidity, the critical leaching of ANC has to be specified. As stated above, all constituents of the ANC leaching (see eq.19) can be expressed in terms of $[H]$ (see eqs.8 and 9). Thus, specifying a critical pH (e.g., pH_{crit}=4.2) the critical ANC leaching can be easily computed. Alternatively, specifying a critical aluminium concentration (e.g., $[Al]_{crit}$=0.2eq/m^3), $ANC_{le,crit}$ can be computed using the gibbsite equilibrium (eq.9). The concentrations of bicarbonate and organic anions are generally neglected in calculating critical loads.

The most common critical chemical value used in the European critical loads work is the molar ratio of base cations to aluminium in soil solution, a quantity which is linked to the risk of growth reductions (see Sverdrup and Warfvinge 1993). The critical (Ca+Mg+K)/Al=Bc/Al ratio is used to calculate a critical aluminium leaching via:

$$Al_{le(crit)} = 1.5\frac{Bc_{le}}{(Bc/Al)_{crit}} \quad \text{with} \quad Bc_{le} = Bc_{dep} + Bc_w - Bc_u \tag{27}$$

where the factor 1.5 arises from the conversion of the molar Bc/Al ratio to equivalents.

Note that weathering generally includes Na, but this is taken care of by correction factors, e.g., Bc_w=0.85BC_w. Multiplying by Q and using eq.9 again, the critical ANC leaching is obtained. The same procedure can be used to compute $ANC_{le,crit}$ from a critical Ca/Al or (Ca+Mg)/Al ratio. The standard value used is $(Bc/Al)_{crit}$=1, but in Sverdrup and Warfvinge (1993) values for many different receptors (plant species) are compiled.

Although the critical base cation to aluminium ratio is the most widely used critical chemical value in the ongoing critical loads work, it is not undisputed (see Løkke *et al.* 1996, for a critical review see also Cronan and Grigal 1995), and to strengthen the link between soil chemistry and detrimental effects on ecosystems requires further research.

Base saturation, i.e the fraction of the cation exchange capacity (CEC) where base cations can be exchanged for aluminium or hydrogen ions, is not considered in the present critical load formulation, the reasoning being that CEC constitutes only a finite pool which eventually will be depleted of base cations. Base saturation, however, is an indicator of the acidity status of a soil and low values could be considered as 'critical'. In the following we show how to link base saturation with other critical values, thus allowing the specification of a critical base saturation (e.g., $f_{BC,crit}$ = 5%) for critical load calculations.

From the equations for Al-BC and H-BC exchange (eqs.10,11) and the gibbsite equilibrium (eq.9) one can derive the following equation:

$$f_{Al} = Kf_H^3 \quad \text{with} \quad K = K_{gibb}\sqrt{K_{AlBC} / K_{HBC}^3} \tag{28}$$

which allows to express the base saturation as function of f_H alone (see eq.12):

$$f_{BC} = 1 - f_H - Kf_H^3 \tag{29}$$

Using eqs.10 and 28 to replace f_{BC} in this equation yields a relationship between the Bc/Al ratio and the exchangeable fraction of H (with $[Bc]$ as additional parameter; for simplicity we assume no Na-exchange, i.e. BC=Bc):

$$Kf_H^3 + (K(Bc/Al))^{2/3}([Bc]/K_{AlBc})^{1/3} f_H^2 + f_H = 1 \tag{30}$$

The last two third-order equations establish a relationship between base saturation and Bc/Al ratio. In Figure 7.2 base saturation is shown as a function of the base cation concentration in soil solution $[Bc]=(Bc_{dep}+Bc_w-Bc_u)/Q$ for molar Bc/Al ratios of 0.25, 1 and 4, resp., setting K_{AlBc}=1 mol/l and K=3.16 (derived from $\log_{10}(K_{gibb})$=8, $\log_{10}(K_{HBC})$=5 and eq.28). It can be seen that base saturation varies widely for a fixed Bc/Al ratio as a function of net base cation input and percolation flux Q.

We have derived the connection between base saturation and ANC leaching for a simple model of cation exchange (Al-BC-H exchange only, as modelled in SMART) leading to third-order equations. No difficulty arises in principle when including more detailed descriptions of the exchange reactions, such as used in the MAGIC model. The above equations can be used to calculate the critical ANC leaching for any selected critical base saturation, thus allowing to calculate critical loads corresponding to a desired minimum base cation status of the soil.

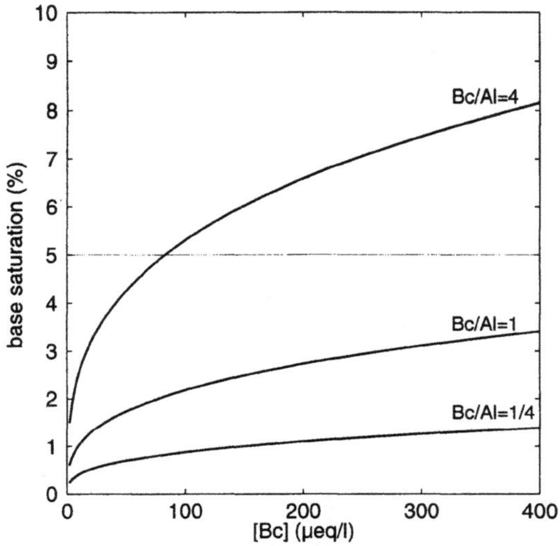

Figure 7.2 Percent base saturation as function of base cation concentration in the soil solution for three different values of molar Bc/Al ratio.

The inclusion of cation exchange reactions in steady-state models is only needed when using a fixed base saturation as a soil chemical criterion. Consideration of the time development of the base saturation, however, leads to dynamic models, which are the subject of the next section.

7.4 Dynamic Soil Models

7.4.1 BASIC PRINCIPLES

Critical loads are computed for a steady state, that is conditions in which the soil processes are in equilibrium with time-independent inputs (deposition). To answer questions about the transient behaviour of the soil variables, i.e. the behaviour under changing deposition (scenarios), dynamic models are needed. It is beyond the scope of this chapter to review even a subset of the existing dynamic models and their design philosophies. A discussion of the basic principles of frequently used dynamic models can be found in Warfvinge (1995). Here we restrict ourselves to the discussion of dynamic models which can be (and have been) applied on a regional scale and are used in connection with the computation of critical loads, i.e. MAGIC, SAFE and SMART. All three models have been applied individually on various sites in Europe and they have also been compared with each other, using the same input data sets and deposition scenarios for several intensively monitored sites (Wright *et al.* 1991, Forsius *et al.* 1996).

All three dynamic models are based on the principles described in Section 2. They describe the change over time of the amount of each element in the various phases (soil matrix, soil solution), i.e. X_{tot} in eq.1 has to be specified. For nitrogen and chloride the amount is simply given by the amount of solution in the soil profile:

$$X_{tot} = \Theta z[X], \quad X = NO_3, NH_4, Cl \tag{31}$$

where Θ is the volumetric water content at field capacity (m/m) and z is the soil depth. The total amount of sulphate per unit area is given by

$$SO_{4,tot} = \Theta z[SO_4] + \rho\, z\, SO_{4,ad} \tag{32}$$

where ρ is the bulk density and $SO_{4,ad}$ the amount of sulphate adsorbed (see eq.13). For base cations and aluminium also cation exchange is taken into account:

$$X_{tot} = \Theta z[X] + \rho \quad z f_X CEC, \quad X = Ca, Mg, K, Na, Al \tag{33}$$

where *CEC* is the total cation exchange capacity (meq/kg) and f_X is the fraction of ion X at the exchange complex ($\Sigma f_X = 1 - f_H$). Together with equilibrium reactions (see eqs.8-13) and proper initial conditions the resulting set of (differential) equations is solved to obtain the time development of the ion concentrations and other variables (such as the f_X's).

Even though the basic principles of MAGIC, SAFE and SMART are similar, they differ in many details: MAGIC and SMART are single-layer models, whereas SAFE is a multi-layer model; weathering is input to MAGIC and SMART, whereas SAFE uses the PROFILE model to calculate (pH-dependent) weathering rates. The description of the aluminium chemistry is of varying complexity: SMART uses only the simple gibbsite equilibrium (eq.9), SAFE includes the dissociation of two Al-hydroxides and MAGIC adds numerous complexation reactions of aluminium with fluoride and sulphate. Further details of the models can be found in Cosby *et al.* (1985; for MAGIC), in Warfvinge *et al.* (1993; for SAFE), and in de Vries *et al.* (1989) and Posch *et al.* (1993; for SMART).

None of the three models is very sophisticated about nitrogen processes. This has two reasons: a historic one and one related to (non-)existing knowledge (and data). Historically, the interest has been on modelling the effects of acidification, and in the early 1980s, when the first models for applications on a regional scale and with long time horizons were developed, sulphur deposition was the dominant driving force. In cases where sulphate adsorption was negligible, sulphur could be treated as tracer, and this simplified model conceptualization considerably. Nitrogen processes were treated in a very simple manner: Uptake and fixation were treated as external input, nitrification was generally assumed complete (owing to the fact that ammonium leaching is rarely observed) and denitrification was either ignored or modelled in a simple way (e.g., eq.16). Only net fluxes of nitrogen are considered and the internal cycling of N (and base cations) is generally ignored. This is not only due to the historical reasons mentioned above (emphasis on sulphur), but is also due to the lack of reliable input data, especially on a regional scale.

7.4.2 NITROGEN PROCESSES

Since the European-wide decline in sulphur depositions during the 1990s, nitrogen has become more and more the focus of attention. And in recent years models have been extended or new models have been developed which try to model nitrogen processes in more detail or consider certain processes for the first time. This is not to say that nitrogen processes have not been studied earlier, but most of previously developed models (and many present ones) focus on the simulation of forest productivity, forest vitality or ecosystem succession, with emphasis on the carbon and nitrogen pools and partitioning in soils and biomass (see, e.g., Ågren *et al.* 1991). Of course, those models serve also as starting points and inspiration for the new models developed for regional applications in the context of acidification.

As an example we consider nitrogen immobilization. In the current version of the SMART model (Posch *et al.* 1993) the net amount of N immobilized at every time step is calculated from the increase in the N-content of organic matter in the top soil. Between a critical, C/N_{crit}, and a minimal C:N-ratio, C/N_{min}, N_i depends linearly on the prevailing C:N ratio:

$$N_i = \begin{cases} 0 & \text{for } C/N \le C/N_{min} \\ \dfrac{C/N - C/N_{min}}{C/N_{crit} - C/N_{min}} N_{net} & \text{for } C/N_{min} < C/N < C/N_{crit} \\ N_{net} & \text{for } C/N \ge C/N_{crit} \end{cases} \tag{34}$$

where N_{net} is the net input of N (deposition minus uptake). The amount of immobilized N is added to the amount of N in the top soil, which in turn is used to update the C:N-ratio (assuming a constant carbon content of the organic matter). Using this formulation, Figure 7.3 shows that the amount of N immobilized over time decreases for a constant N deposition until the C:N ratio reaches its minimum value. This type of dynamic behaviour makes it difficult to derive a single value of N_i to be used for calculating critical loads. At present, due to high N-deposition and a nitrogen deficiency in many forest soils, N immobilization may be high in many parts of Europe, but in the long run this elevated input of N might lead to nitrogen saturation and consequent leaching of N to surface and ground waters. Therefore, fairly low values of N_i (0.5-1 kgN/ha/a) are recommended for computing critical loads.

The above equation for N immobilization is just one example of how to describe nitrogen processes in dynamic models, and it has been formulated as an add-on to an existing simple model (SMART). Recently SMART has been extended with a description of the complete nutrient cycle (see Kros *et al.* 1995). Using a logistic function for tree growth, SMART2 (as the model is called) includes equations for litterfall and root decay, mineralization and uptake as well as a more detailed description of the hydrology (including seepage).

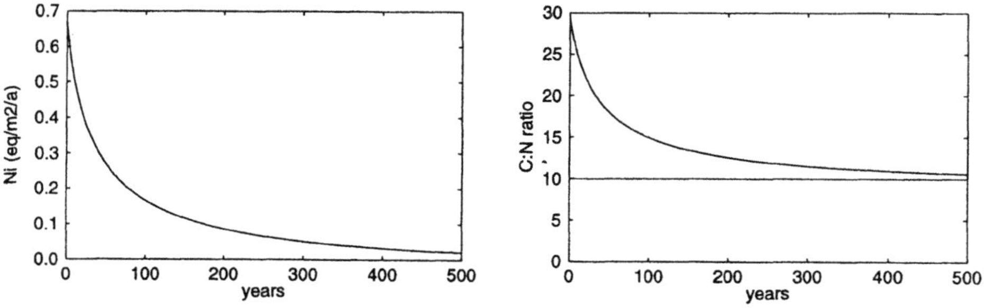

Figure 7.3 Amount of nitrogen immobilized (left) and resulting C:N ratio in the topsoil (right) for a constant net input of N of 1 eq/m^2 (20cm topsoil with 5% C-content and ρ=1.3g/cm^3). The amount of N immobilized approaches zero when the C:N ratio reaches C/N_{min}=10.

Coupled with a multiple stress model for vegetation (MOVE) the model has been used to evaluate the change in vegetation in the Netherlands due to changes in atmospheric deposition. To study organic and inorganic soil processes in more detail, a complex model called NUCSAM (Groenenberg *et al.* 1995) has been developed by the same group which uses a daily time step and several soil layers. This model requires a much larger number of input parameters to be specified, making it less suitable for regional application. A comparison with simpler models using data from an intensively monitored site in Solling (Germany) shows, however, that the long-term annual average behaviour of NUCSAM and SMART is comparable (Van der Salm *et al.* 1995), thus increasing the confidence in the use of simple models for long-term predictions.

Also the MAGIC model has been extended to include key nitrogen processes. This model, called MAGIC-WAND (MAGIC With Added Nitrogen Dynamics; Jenkins *et al.* 1997), models mineralisation, nitrification, denitrification and uptake, keeping track both of nitrate and ammonium in the soil solution. Nitrification of ammonium is represented by a first order reaction and uptake for both N species is modelled by a Michaelis-Menten equation. The model has been applied to 25 acid sensitive lochs in Scotland, examining the interaction of afforestation and changes in nitrogen deposition, assuming recently agreed reductions in sulphur deposition. The same group has also recently completed a catchment-scale mass-balance model called MERLIN (Model of Ecosystem Retention and Loss of Inorganic Nitrogen; Cosby *et al.* 1997) linking biotic and abiotic processes affecting the cycling and storage of nitrogen. The structure of the model includes the inorganic soil, a plant compartment and two soil organic compartments. Besides by deposition and hydrological discharge, fluxes in and out of the ecosystem and between compartments are regulated by plant uptake, litter production, wood production, microbial immobilization, mineralization, nitrification, and denitrification. Nitrogen fluxes are controlled by carbon productivity, the C:N-ratios of organic compartments and inorganic nitrogen in soil solution. One aim of this model (and certainly others as well) is to simulate the magnitude and timing of N saturation which has been observed empirically (see, e.g., Dise and Wright 1995).

7.4.3 TARGET LOADS AND DAMAGE DELAY/RECOVERY TIMES

As stated in the introduction, dynamic models can be used to derive target loads which consider a finite time period in which the soil is allowed to, or has to, reach a selected soil chemical criterion. Unlike critical loads, target loads are thus influenced by the present soil status. If the present soil status, e.g. base saturation, is below the critical value (i.e. the system is already at risk), the target load is less than the critical load, whereas the reverse is true, when base saturation is initially above the critical value. In the latter case dynamic soil processes such as cation exchange, lead to an additional buffering, thus causing the target load to be higher than the critical load (see also Warfvinge *et al.* 1992).

This is illustrated with the aid of the SMART model (neglecting sulphate adsorption and nitrogen processes). Using the same input parameters as for Figure 7.2 (z=0.5m, ρ =1.3g/cm^3, CEC=200meq/kg, Bc_{le}=25meq/m^2/a, Q=0.3m/a, K_{AlBC}=1mol/l and K=3.16) and selecting a critical base saturation of 5% gives a critical load of 48.2meq/m^2/a. With the chosen parameters $f_{Bc,crit}$=5% corresponds to a critical Bc/Al ratio of 4 (see Figure 7.2), and the critical load of 48.2meq/m^2/a can also be derived from eq.22 assuming Cl_{dep}=0. Figure 7.4 shows the relationship between initial (present) base saturation ($f_{Bc,0}$), deposition (target load), and the time needed to reach the critical chemical value.

The selected critical base saturation and the resulting critical load divide the plane in four quadrants. When a combination of deposition and initial base saturation lies in the upper left or lower right quadrant (grey shaded areas in Figure 7.4), the critical base saturation cannot be reached. In the first case no recovery is possible (deposition above critical load and $f_{Bc,0}$<$f_{Bc,crit}$), whereas in the second case the deposition below the critical load will never allow the initial base saturation to drop to $f_{Bc,crit}$. In the lower left and upper right quadrant, however, there exists for every finite time horizon and every initial base saturation a corresponding deposition (target load); and the isochrones (in years) shown in Figure 7.4 allow the determination of the approximate value of such a target load. Note, that for certain combinations of $f_{Bc,0}$ and time horizon there exists no target load. For example, if $f_{Bc,0}$<1.2%, it is impossible to reach $f_{Bc,crit}$=5% within 100 years, since the net base cation input by deposition and weathering is not sufficient to replenish the base cations at the exchange sites.

As mentioned before, dynamic models can also be used to calculate damage delay or recovery times of an ecosystem. Depending on the present state, answers to the following questions can be of interest (see also Warfvinge *et al.* 1992 for a more detailed discussion):

(1) The present load is greater than the critical load: How long does it take to reach a selected critical value, i.e. when does the risk of damage strongly increase? This damage delay time information can be important for the timing of mitigation measures.

(2) The system is already at risk: For a given deposition below the critical load, how long does it take for the system to recover? Again, the knowledge of this damage recovery time can help in the assessment of the effectiveness of mitigation policies.

The example in Figure 7.4 illustrates both cases: the isochrones in the upper quadrant show the damage delay time for a (still) 'healthy' system, whereas the lower left quadrant indicates the damage recovery times for a system already at risk.

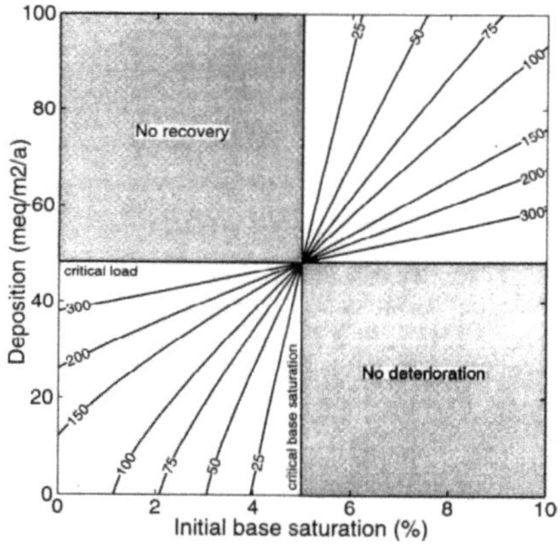

Figure 7.4 Nomogram, derived with the SMART model, for determining the target load of sulphur (net acidity) in order to reach a critical base saturation of 5% in a pre-defined time period as function of the initial (present) base saturation. Alternatively, by fixing the deposition level the damage delay/recovery time can be read from the graph. The thick curves are isochrones (in years) of damage delay/recovery times.

The values derived in the above example depend on the selected input parameters; and we again employ the SMART model to illustrate the dependence of the damage delay time on one of those parameters, the cation exchange capacity. Using the same input parameters as for Figure 7.4 and selecting an initial base saturation of 10%, we ensure that the initial Bc/Al ratio is above the critical value of 4 (corresponding to $f_{Bc,crit} = 5\%$; see above).

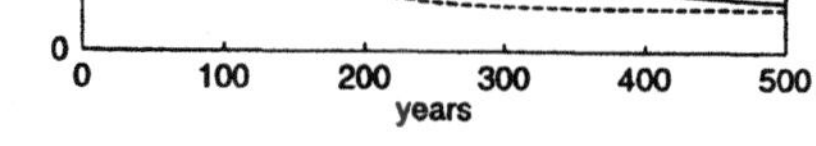

Figure 7.5 Molar Bc/Al ratio as a function of time for a constant acid load (about twice the critical load) for three values of the cation exchange capacity: *CEC*=100 (dashed line), *CEC*=200 (solid line) and *CEC*=300 meq/kg (dotted line) as simulated with the SMART model (z=0.5m, ρ =1.3g/cm^3; the horizontal line indicates the critical value of Bc/Al).

Figure 7.5 shows the result of three SMART simulations with a constant acid deposition of 100meq/m^2 (about twice the critical load) for three different values of the cation exchange capacity (*CEC*=100, 200 and 300meq/kg, resp.), keeping all other parameters constant. As can be seen, the cation exchange capacity (in fact, $\rho z CEC$)

strongly influences the time before reaching the critical value of *Bc/Al*=4 (the damage delay time), and thus the time left for taking remedial actions (55 years for *CEC*=100 to 168 years for *CEC*=300meq/kg).

Whereas the number of exchange sites strongly influences the time needed to reach a critical chemical value, it can be shown that other soil parameters have less influence. For example, varying the Al-Bc exchange constant over two orders of magnitude hardly changes the damage delay times. Although much work remains to be done, this strongly supports the view that a good knowledge of the cation exchange capacity, together with the initial base saturation, are a prerequisite for meaningful forecasts with dynamic models.

7.5 Summary and Conclusions

Critical loads are now widely accepted in sulphur and nitrogen emission reduction negotiations in Europe. Critical loads of acidifying nitrogen and nitrogen as a nutrient are derived from steady-state soil models. Since sulphur is also an acidifying anion, acidifying N and S have to be treated simultaneously, giving rise to the so-called critical load function. The most widely used critical chemical value, which links deposition to harmful effects, is the ratio of base cations to aluminium in the soil solution. In this chapter we also derived relationships which allow to use base saturation as a critical chemical value by incorporating cation exchange into the steady-state equations. The investigation of the *change* in base saturation requires dynamic models. The most commonly used dynamic models are all sulphur oriented, and nitrogen processes are modelled fairly simplistically. Only recently more realistic descriptions of nitrogen processes are being incorporated. Dynamic models are also needed to derive so-called target loads, i.e. deposition levels which allow to reach the chosen critical chemical value within a pre-specified time horizon. A target load can be larger or smaller than the steady-state critical load, depending on the initial state of the system. Finally, dynamic models can be used to estimate so-called damage delay and recovery times, i.e. the time needed to reach a chosen critical chemical value for a given deposition. Using the simple dynamic model SMART – which is completely compatible with the steady-state models underlying critical load calculations – we presented examples of target load and damage delay time calculations. These examples show that the cation exchange capacity and the initial base saturation, i.e. the remaining number of exchange sites, are the most important quantities to make credible estimates of time horizons of soil deterioration and recovery under different deposition scenarios.

In order to maintain and enhance the credibility of the critical loads concept, which it currently enjoys in the environmental policy making community in Europe, it is necessary to undertake research and development in two major directions: (1) strengthening the link between direct and indirect biological effects (damage) due to air pollutants and the (soil) chemical variables used in critical load formulations, and (2) improving the data bases needed to perform critical load calculations on a national and European level. These 'intensive' and 'extensive' aspects of steady-state modelling have to be complemented by further developments in the field of dynamic modelling, especially with respect to the role of nitrogen, to enlarge the set of tools necessary to make informed decisions in effect-oriented air pollution abatement policies.

References

Ågren, G.I., R.E. McMurtie, W.J. Parton, J. Pastor and H.H. Shugart, 1991. State-of-the-art models of production–decomposition linkages in conifer and grassland ecosystems. *Ecological Applications* **1**:118-138.

Amann, M., I. Bertok, J. Cofala, F. Gyarfas, C. Heyes, Z. Klimont and W. Schöpp, 1996. Cost-effective control of acidification and ground-level ozone. *Second Interim Report to the European Commission*, DG-XI, International Institute for Applied Systems Analysis, Laxenburg, Austria, 111 pp.

Cosby, B.J., G.M. Hornberger, J.N. Galloway and R.F. Wright, 1985. Modeling the effects of acid deposition: Assessment of a lumped parameter model of soil water and streamwater chemistry. *Water Resources Research* **21**:51-63.

Cosby, B.J., G.M. Hornberger, R.F. Wright and J.N. Galloway, 1986. Modeling the effects of acid deposition: Control of long-term sulfate dynamics by soil sulfate adsorption. *Water Resources Research* **22**:1283-1291.

Cosby, B.J., G.M. Hornberger and R.F. Wright, 1989. Estimating time delays and extent of regional de-acidification in southern Norway in response to several deposition scenarios. In: J. Kämäri, D.F. Brakke, A. Jenkins, S.A. Norton and R.F. Wright (eds): *Regional Acidification Models: Geographic Extent and Time Development*, Springer, Berlin, Heidelberg, pp. 151-166.

Cosby, B.J., R.C. Ferrier, A. Jenkins, B.A. Emmett, R.F. Wright and A. Tietema, 1997. Modelling the ecosystem effects of nitrogen deposition: Model of Ecosystem Retention and Loss of Inorganic Nitrogen (MERLIN). *Hydrology and Earth System Sciences* **1**:137-158.

Cronan, C.S. and D.F. Grigal, 1995. Use of calcium/aluminium ratios as indicators of stress in forest ecosystems. *Journal of Environmental Quality* **24**:209-226.

De Vries, W., M. Posch and J. Kämäri, 1989. Simulation of the long-term soil response to acid deposition in various buffer ranges. *Water, Air, and Soil Pollution* **48**:349-390.

De Vries, W., 1993. Average critical loads for nitrogen and sulfur and its use in acidification abatement policy in the Netherlands. *Water, Air, and Soil Pollution* **68**:399-434.

De Vries, W., M. Posch, G.J. Reinds and J. Kämäri, 1993. *Critical loads and their exceedance on forest soils in Europe*. The Winand Staring Centre for Integrated Land, Soil and Water Research, Report 58 (revised version), Wageningen, The Netherlands, 116 pp.

De Vries, W., 1994. *Soil response to acid deposition at different regional scales: Field and laboratory data, critical loads and model predictions*. Thesis, Agricultural University, Wageningen, The Netherlands, 487 pp.

Dise, N.B. and R.F. Wright, 1995. Nitrogen leaching from European forests in relation to nitrogen deposition. *Forest Ecology and Management* **71**:153-161.

Driscoll, C.T., M.D. Lehtinen and T.J. Sullivan, 1994. Modeling the acid-base chemistry of organic solutes in Adirondack, New York, lakes. *Water Resources Research* **30**:297-306.

Forsius, M., M. Alveteg, A. Jenkins, M. Johansson, S. Kleemola, A. Lükewille, M. Posch, H. Sverdrup, S. Syri and C. Walse, 1996. Dynamic model applications at selected ICP IM sites. In: S. Kleemola and M. Forsius (eds): *Fifth Annual Report* 1996. UN ECE ICP Integrated Monitoring. The Finnish Environment 27. Finnish Environment Institute, Helsinki, Finland. pp 10-24.

Grennfelt, P. and E. Thörnelöv (eds), 1992. *Critical Loads for Nitrogen*, Nord 1992:41, Nordic Council of Ministers, Copenhagen, Denmark, 428 pp.

Groenenberg, B.-J., H. Kros, C. van der Salm and W. de Vries, 1995. Application of the model NUCSAM to the Solling spruce site. *Ecological Modelling* **83**:97-107.

Hettelingh, J.-P., M. Posch, P.A.M. de Smet and R.J. Downing, 1995. The use of critical loads in emission reduction agreements in Europe. *Water, Air, and Soil Pollution* **85**:2381-2388.

Hornung, M., M.A. Sutton and R.B. Wilson (eds), 1995a. *Mapping and Modelling of Critical Loads for Nitrogen: A Workshop Report. Institute of Terrestrial Ecology*, United Kingdom, 207 pp.

Hornung, M., K.R. Bull, M. Cresser, J. Hall, S.J. Langan, P. Loveland and C. Smith, 1995b. An empirical map of critical loads of acidity for soils in Great Britain. *Environmental Pollution* **90**:301-310.

Jenkins, A., R.C. Ferrier and B.J. Cosby, 1997. A dynamic model for assessing the impact of coupled sulphur and nitrogen deposition scenarios on surface water acidification. *Journal of Hydrology* **197**:111-127.

Kros, J., G.J. Reinds, W. de Vries, J.B. Latour and M.J.S. Bollen, 1995. *Modelling of soil acidity and nitrogen availability in natural ecosystems in response to changes in acid deposition and hydrology*. The Winand Staring Centre for Integrated Land, Soil and Water Research, Report 95, Wageningen, The Netherlands, 90 pp.

Langan, S.J., M.E. Hodson, D.C. Bain, R.A. Skeffington and M.J. Wilson, 1995. A preliminary review of weathering rates in relation to their method of calculation for acid sensitive parent materials. *Water, Air, and Soil Pollution* **85**:1075-1081.

Løkke, H., J. Bak, U. Falkengren-Grerup, R.D. Finlay, H. Ilvesniemi, P.H. Nygaard and M. Starr, 1996. Critical loads of acidic deposition for Forest soils: Is the current approach adequate? *Ambio* **25**:510-516.

Malanchuk, J.L. and J. Nilsson (eds), 1989. *The role of nitrogen in the acidification of soils and surface waters*. Nord 1989:92, Nordic Council of Ministers, Copenhagen, Denmark.

Nilsson, J. and P. Grennfelt (eds), 1988. *Critical Loads for Sulphur and Nitrogen*. Nord 1988:97, Nordic Council of Ministers, Copenhagen, Denmark, 418 pp.

Posch, M., G.J. Reinds and W. de Vries, 1993. SMART – *A Simulation Model for Acidification's Regional Trends: Model description and user manual*. Mimeograph Series of the National Board of Waters and the Environment 477, Helsinki, Finland, 43 pp.

Posch, M., P.A.M. de Smet, J.-P. Hettelingh and R.J. Downing (eds), 1995. *Calculation and Mapping of Critical Thresholds in Europe*. Status Report 1995, Coordination Center for Effects, National Institute of Public Health and the Environment (RIVM), Bilthoven, The Netherlands, 198 pp.

Posch, M., J. Kämäri, M. Forsius, A. Henriksen and A. Wilander, 1997a. Exceedance of critical loads for lakes in Finland, Norway and Sweden: Reduction requirements for acidifying nitrogen and sulfur deposition. *Environmental Management* **21**:291–304.

Posch, M., M. Johansson and M. Forsius, 1997b. Critical loads and dynamic models. In: S. Kleemola and M. Forsius (eds): Sixth Annual Report 1997. UN/ECE ICP Integrated Monitoring. The Finnish Environment. Finnish Environment Institute, Helsinki, Finland. **116**: 13–23

Roberts, T.M., R.A. Skeffington and L.W. Blank, 1989. Causes of Type 1 spruce decline in Europe. *Forestry* **62**:179–222.

Rosén, K., P. Gundersen, L. Tegnhammar, M. Johansson and T. Frogner, 1992. Nitrogen enrichment in Nordic forest ecosystems – The concept of critical loads. *Ambio* **21**:364–368.

Sverdrup, H. and P. Warfvinge, 1992. Calculating filed weathering rates using a mechanistic geochemical model – PROFILE. *Applied Geochemistry* **8**:273–283.

Sverdrup, H. and P. Warfvinge, 1993. *The effect of soil acidification on the growth of trees, grass and herbs as expressed by the (Ca+Mg+K)/Al ratio*. Reports in Ecology and Environmental Engineering 2:1993, Department of Chemical Engineering II, Lund University, Lund, Sweden, 177 pp.

UBA, 1996. *Manual on Methodologies and Criteria for Mapping Critical Levels/Loads and Geographical Areas Where They Are Exceeded,* Texte 71/96. Umweltbundesamt, Berlin, Germany, 144+lxxiv pp.

UN/ECE, 1994. *Protocol to the 1979 Convention on Long-range Transboundary Air Pollution on Further Reduction of Sulphur Emissions*. Document ECE/EB.AIR/40 (in English, French and Russian), New York and Geneva, 106 pp.

UN/ECE, 1995. Calculation of critical loads of nitrogen as a nutrient. *Summary report on the development of a library of default values*. Document EB.AIR/WG.1/R.108, Geneva.

Van der Salm, C., J. Kros, J.E. Groenenberg, W. de Vries and G.J. Reinds, 1995. Application of soil acidification models with different degrees of process description (SMART, RESAM and NUCSAM) on an intensively monitored spruce site. In: S.T. Trudgill (ed): *Solute Modelling in Catchment Systems*. John Wiley & Sons, pp.327–346.

Warfvinge, P., M. Holmberg, M. Posch and R.F. Wright, 1992. The use of dynamic models to set target loads. *Ambio* **21**:369–376.

Warfvinge, P. and H. Sverdrup, 1992. Calculating critical loads of acid deposition with PROFILE – A steady-state soil chemistry model. *Water, Air and Soil Pollution* **63**:119–143.

Warfvinge, P., U. Falkengren-Grerup, H. Sverdrup and B. Andersen, 1993. Modelling long-term cation supply in acidified forest stands. *Environmental Pollution* **80**:209–221.

Warfvinge, P., 1995. Basic principles of frequently used models. In: S.T. Trudgill (ed.): *Solute Modelling in Catchment Systems*. John Wiley & Sons, pp.57–72.

Warfvinge, P. and H. Sverdrup, 1995. *Critical loads of acidity to Swedish forest soils: Methods, data and results*. Reports in Ecology and Environmental Engineering 5:1995, Department of Chemical Engineering II, Lund University, Lund, Sweden, 104 pp.

White, C.C., M.S. Cresser and S.J. Langan, 1996. *The importance of marine-derived base cations and sulphur in estimating critical loads in Scotland*. The Science of the Total Environment 177:225–236.

Wright, R.F., M. Holmberg, M. Posch and P. Warfvinge, 1991. *Dynamic models for predicting soil and water acidification: Application to three catchments in Fenno-Scandinavia*. Acid Rain Research Report 25/1991, Norwegian Institute for Water Research, Oslo, Norway, 40 pp.

Appendix: List of variables

Symbol	Variable	Unit (e.g.)
X_{tot}	total amount of ion/element X in the soil*	eq/m^3
X_{dep}	total deposition flux of X	$eq/m^2/a$
X_{int}	flux of X generated in the soil	$eq/m^2/a$
X_{le}	leaching flux of X at bottom of root zone	$eq/m^2/a$
$[X]$	concentration of X in the soil solution	eq/m^3
f_X	fraction of ion X at the exchange complex	–
BC_w	weathering flux of base cations	$eq/m^2/a$
Bc_u	net growth uptake flux of base cations	$eq/m^2/a$
N_u	net growth uptake flux of nitrogen	$eq/m^2/a$
N_i	net immobilization flux of nitrogen	$eq/m^2/a$
N_{de}	denitrification flux	$eq/m^2/a$
f_{de}	denitrification fraction ($0 \leq f_{de} \leq 1$)	–
$N_{le,acc}$	acceptable leaching flux of nitrogen	$eq/m^2/a$
$ANC_{le,crit}$	critical leaching flux of ANC	$eq/m^2/a$
$(Bc/Al)_{crit}$	critical molar base cation to aluminium ratio	mol/mol
$SO_{4,ad}$	amount of adsorbed sulphate	meq/kg
K_{gibb}	gibbsite equilibrium constant	$(mol/l)^{-2}$
K_{AlBC}	selectivity constant for Al–BC exchange	mol/l
K_{HBC}	selectivity constant for H–BC exchange	$(mol/l)^{-1}$
K	$= K_{gibb}(K_{AlBC}/K_{HBC}^3)^{1/2}$	–
z	thickness of soil compartment (root zone)	m
ρ	bulk density of the soil	g/cm^3
Θ	volumetric water content of the soil	m/m
Q	percolation flux (precipitation–evapotranspiration)	m/a
CEC	cation exchange capacity of the soil	meq/kg
C/N	C:N ratio in soil organic matter	–
C/N_{min}	minimum C:N ratio in soil organic matter	–
C/N_{crit}	critical C:N ratio in soil organic matter	–
$CL_{nut}(N)$	critical load of nutrient nitrogen	$eq/m^2/a$
$CL(S)$	critical load of sulphur	$eq/m^2/a$
$CL(N)$	critical load of acidifying nitrogen	$eq/m^2/a$
$CL_{max}(S)$	maximum critical load of S (critical load of acidity)	$eq/m^2/a$
$CL_{min}(N)$	minimum critical load of acidifying nitrogen	$eq/m^2/a$
$CL_{max}(N)$	maximum critical load of acidifying nitrogen	$eq/m^2/a$

*X=H,Al,Ca,Mg,K,Na,NH_4,NO_3,SO_4,Cl,HCO_3,RCOO,N,S (Ca+Mg+K=Bc, Bc+Na=BC); charges are suppressed for convenience.

PROGRESS, FUTURE DIRECTIONS AND CONCLUDING REMARKS

S. J. LANGAN
Macaulay Land Use Research Institute, Craigiebuckler, Aberdeen, AB15 8QH, UK.

8.1 Introduction

The aim of this book and its constituent chapters is to provide an analysis and review of the work related to the sources and fate of atmospherically deposited nitrogen to natural and semi-natural ecosystems. In doing this, individual contributions have implicitly, or sometimes explicitly, highlighted the deficiencies in our understanding, data and existing monitoring strategies. This contribution provides a synthesis of the chapters and some of the recurrent themes together with consideration of more specific features or implications of this and other contemporary work. Thus the overview provides a basis from which future needs and research priorities can be identified and considered.

8.2 Emission Sources And Deposition

A precursor to understanding the impact of changing nitrogen deposition on ecosystems is an ability to quantify the variations in atmospheric inputs and the processes which drive this variability. The review of the available data on emissions and deposition in Chapter 2 shows that much of our current knowledge in this field is based on monitoring and modelling studies undertaken in Europe and to a lesser extent N. America.

In Europe the characterisation of emissions and the spatial disaggregation of nitrogen deposition is being pursued both through increased monitoring and concomitant model development. However, within this progression, Metcalfe *et al.*, suggest there are still significant problems and deficiencies left to be resolved. For the oxidised nitrogen species, emissions match areas of intensive industry and urban centres of population. With time, the total emissions from these sources have increased and they are currently higher than they have ever been. Against a background of stable emissions from natural sources and of declining anthropogenic SO_2 emissions in Europe and N. America, N contributes to a growing proportion of atmospheric input. However, the nature of the emissions is changing from a situation in which they have been dominated by relatively few large point sources at higher atmospheric levels to one in which inputs increasingly are from low level, mobile sources, namely transport. The inventory of sources and emissions for NO_x is relatively well characterised and Metcalfe *et al.* (Chapter 2) suggest the different methodologies for calculating these fluxes to the atmosphere are in good agreement (<10% difference).

The spatial characterisation of reduced nitrogen species in terms of their source and deposition fields represents a far greater challenge because of their inherent emission and reaction complexity over short distances. Emission patterns for NH_y contrast with those of NO_x because they are dominated by areas of agriculture, and, in particular, intensive livestock production. Source inventories for livestock are not as well spatially characterised and disaggregated as those for transport. Furthermore, the lifetime of ammonia in the atmosphere is a matter of hours which raises considerable problems for monitoring and modelling fluxes, particularly at the regional scale.

Perhaps of equal importance for the impacts based community and international policy makers are the limitations that these and other factors impose on the estimates of deposition. Emissions of NH_y from agricultural land are complicated by bi-directional fluxes with deposition and emissions occurring at the same place with time. In Europe, current models suggest that a high proportion (>80%) of NO_x emitted by any country is exported but, for reduced species, the very short atmospheric lifetime effectively reduces international export to around 40 percent within Europe.

8.2.1 FUTURE NEEDS – EMISSIONS AND DEPOSITION

In order to assist the development of abatement strategies based on a critical loads approach, there is a need to ensure that the dose–response relationships which underpin the concept are robust. Fundamental to this is the accurate characterisation of deposition loading at sites in which impacts have and have not been observed or predicted. In order to develop our understanding, there is a specific need to maintain and supplement existing monitoring networks to establish both the spatial and temporal variability in both wet and dry deposition. Such networks should also build up time-series which include short term, event inputs. To extend our knowledge base in the future, these data need to be integrated and used in the continued development of models. This will be particularly important with respect to the spatial variability in inputs which vary considerably with proximity to sources and terrain and where processes such as seeder–feeder enhancement may be important. Current transport and depositional models tend to be single layer constructs. A recognised problem of these models is the underestimation of near surface concentrations and hence dry deposition. For the impacts based research community this is likely to be highly significant in judging ecosystem responses to varying deposition inputs.

In terms of temporal variability and trends, it is vital that more attention is given to monitoring networks which encompass the change in N deposition over time. Without such information it will be difficult to assess and improve our understanding of the dynamic relationships between deposition input and ecosystem response.

These developments are particularly important with respect to ammonia deposition for which a high resolution modelling framework is required because of its high reactivity. The incorporation of these reactions provides a significant future challenge to progress our understanding and to derive deposition estimates at a regional scale. Related to this is the need to consider further the nature and quantification of the co-deposition of ammonia with sulphate. Ammonium sulphate has an atmospheric lifetime of 5–10 days and will therefore be highly significant in terms of long range transport and international abatement strategies.

Finally, in terms of the geography of emissions and deposition it is clear that much of the research and data presented are centred on Europe and to a lesser extent N. America. In Europe, abatement policies are being discussed but in N. America the situation is less clear, although nitrogen depositions are likely to be curtailed. The biggest changes in deposition and possible impact are likely to occur in S.E. Asia. In Europe and N. America, emissions of NO_x and ammonia are of similar magnitude. In S.E. Asia, ammonia is an order of magnitude greater than that for total oxidised nitrogen. In this context, Galloway (1995) makes the point that currently N. America and Europe with just 14% of the world's population have dominated the global release of SO_2 and NO_x (~70% of each species). For ammonia, the figures of Galloway are different with 40% from N. America and Europe and 90% of the remainder from Asia. By the year 2010 the situation will have changed significantly, as illustrated in Figure 8:1. The salient features of these changes are the decline in emissions in the developed countries compared to the increases (both in sulphur and nitrogen) for developing countries, most noticeably those in S.E. Asia. In the future more emphasis should be given to these regions of the world.

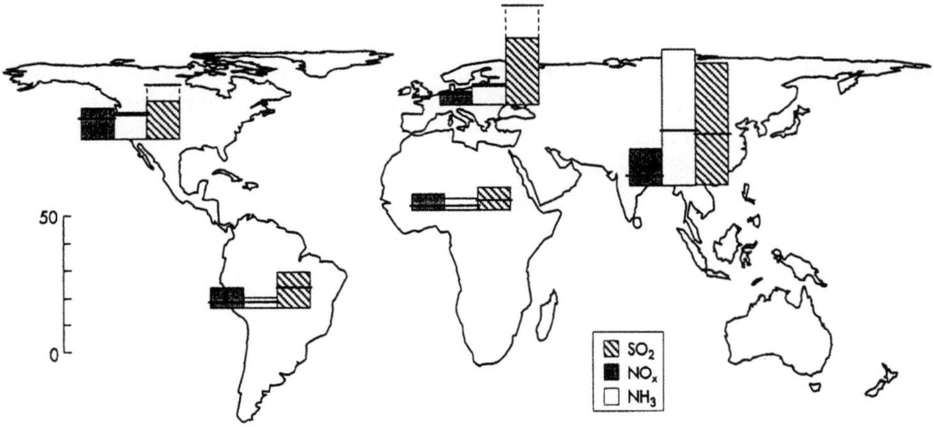

Figure 8.1: Continental emissions of NO_x, NH_3 and SO_2 projected for 2020, bar represents 1990 emissions (after Galloway, 1995).

8.3 Fate And Impact Assessment

8.3.1 PROCESSES AND RESPONSES

There are an increasing number of studies (and data) in which the fate of deposited nitrogen is being considered. These studies are aimed at improving both the identification and quantification of processes and also at our understanding of response indicators, in terms of thresholds for change. The data used are from a combination of manipulation experiments through inclusion, exclusion and relocation of ecological components. Further insights have come from observations at sites, compiled and compared, across pollutant gradients. The greatest attention (with the exception of the nutrient enrichment work in the Netherlands) has been on forested catchment ecosystems.

In the chapter by Williams and Anderson the extent of our understanding and quantification of the processes of nitrogen transformations in plants and soils in relation to changing atmospheric inputs is set out. Particular attention is paid to the (largely) European based manipulation experiments which indicate the importance of vegetation assimilation of nitrogen. In nutrient poor ecosystems in which there are multiple limitations to growth, it is not possible to isolate the impact of additional nitrogen. In this situation the C:N ratio of litter is reduced. The authors also discuss recent studies where increased inputs of nitrogen are associated with the release of dissolved organic matter. The chemical form and biological significance of this process and DON output have still to be determined. In terms of process identification and quantification, most work has been aimed at improving our understanding of the rates and controls on mineralisation, nitrification and denitrification processes.

In non-forest ecosystems, Aerts and Bobbink have provided an excellent review (in Chapter 4) of the range of processes and responses observed across a range of field experiments and manipulations involving heath, grass and wetland type communities and changes to their biomass allocation and mycorrhizal infection. In such communities the most obvious changes are in species composition which is the result of an increased rate of nitrogen cycling. In this situation, slow growing plants are typically replaced by fast growing plants with high rates of tissue turnover and high nutrient loss. Both observational and manipulation data suggest the biggest changes are associated with changes in ammonia sulphate inputs. At the individual plant level, N addition has been linked to changes in leaf and root weight ratios (see Indicators below). At the ecosystem level, further N additions give rise to such heathland ecosystems becoming phosphorus limited and nitrogen saturated. Under these conditions of increasing nitrogen supply the number of ectomycorrhizal infections decreases. Aerts and Bobbink also cite a limited number of studies which link these changes to secondary stress factors such as increased frost sensitivity and vulnerability to drought.

For forest ecosystems, Wilson and Emmett in Chapter 5 eloquently review the progress of research on the role of atmospheric N deposition and the controlling factors governing N saturation and nitrate leaching. In addition to reviewing indicators of a change in ecosystem functioning, the authors highlight the differing impact of nitrate as opposed to ammonia deposition and the role of site specific factors. Of particular importance in this respect are temperature and precipitation. Unfortunately, the changes in these environmental factors are frequently coincident with the pollution gradient and make them difficult to isolate. In the light of current climatic variability and future change such analysis is crucial to a better understanding of the interaction between pollutant input and ecosystem response. At the ecosystem level it is important that future research on modelling and prediction deals with spatial variability. For example, within forests, differences in uptake between deciduous and coniferous trees, the age of the canopy and presence or absence of a ground flora will be important in determining nitrate leaching.

In Chapter 6 Chapman and Edwards provide a comprehensive review of the available studies and data in which observed responses to increased nitrogen deposition to surface waters are judged against the conceptual framework provided by Stoddard (1994). The authors show that there are numerous studies from the U.S.A. and Europe to support the three stages of nitrogen saturation on the basis of nitrate leaching characteristics. Data from the Adirondacks, U.S.A. suggest that the spatial distribution of nitrate concentrations corresponds to the distribution of lake water acidity, whilst there is no such match with sulphate. In the same way data for Norway indicate the largest concentrations of nitrate are associated with lakes with the lowest pH. However, there are few data in which the long-term variability in relation to nitrogen deposition against variability in precipitation and temperature have been studied. Similarly for episodes, data are sparse. From the published data in a limited number of studies in streams and catchments the greatest changes have been recorded in nitrate concentrations over short time periods associated with snowmelt events. For rainfall driven events the response in surface waters tends to be smaller and, to a large extent, governed by the antecedent conditions in the catchment soils.

8.3.2 FUTURE NEEDS – RESPONSES

In order to improve the understanding of both the spatial and temporal dynamics of surface waters it will be necessary to continue monitoring at a range of sites at which deposition and ecosystem impact responses are recorded. Where networks are already in place it is important that they are assessed to ensure the range of sites cover the variability in the factors driving changes in water quality. Equally, data from these sites should be reviewed in order to assess temporal trends and used in the development of dynamic simulation models. Such models are essential in developing our ability to predict the implications of future change and its implications for ecosystem response. As observed in Chapters 4 and 6 the number of studies which have been maintained for more than ten years is very sparse. In the absence of such data it will be difficult to develop our understanding and prediction of ecosystem dynamics under changing environmental conditions. To date, little consideration has been given to characterising short term episodes and events both in terms of spatial input and ecosystem impact. Special attention should also be given to characterising deposition and potential impacts associated with differing inputs of ammonia. Currently our understanding of ammonia and its reaction products is poorly developed in comparison to oxidised atmospheric inputs. Given the characteristics, ammonia will be difficult to characterise (spatially) in terms of both emission and deposition. Furthermore its impact is complicated through the associated co-deposition with sulphur and its role in soil acidification and biological impact.

In terms of impacts based research there is a need to gain a better understanding of the dynamics of the ecosystems under threat and over what time scales change may occur. Work on the nature of the linkages between pollutant input and biological response, particularly with regards to secondary stresses such as drought and frost sensitivity are required. An important part of this work will be identifying the role of temperature and hydrological fluxes in order to predict responses under changing environmental conditions brought about by climate change. To answer some of these questions and to improve and provide a clear focus for monitoring programmes, further development of dynamic models will be necessary. As stated by Posch and De Vries, Chapter 7, the treatment of N processes in these models is rather simplistic. However, it is only through the use of such tools that scientists will be able to answer questions relating to the reversibility of change and indeed ecosystem recovery. Increasingly these issues will become the focus of attention in influencing decisions on the magnitude and timing of reductions in nitrogen deposition.

8.3.3 INDICATORS

Central to considering the environmental impact and fate of a pollutant is the ability to illustrate some change to the ecosystem functioning. Within the chapters reviewing the fate of atmospherically deposited nitrogen, many of the studies discuss the methods or corroboration of the methods by which the impact of increased nitrogen deposition are considered. A selection of the indicators used for assessing the impact of nitrogen (and total acidity) are summarised in Table 8:1.

Table 8.1: Selected indicators of ecosystem response to enhanced N deposition

Ecosystem component	Indicator	Reference
Vegetation effects	Change in species/community structure	Bobbink and Roelofs (1995)
	Foliar N content	DeVries (1993)
	Shoot–root ratio	Aerts and Bobbink (1998)
	NH_4/ K ratio	Relives *et al.* (1985)
Soil solution– vegetation effects	Ca:Al ratio	Ulrich and Matzner (1983)
		Sverdrup and Warfvinge (1993)
	Percent N and or C:N ratio	Gunderson (1998)
Surface Waters	Onset of NO_3 leaching	Dise and Wright (1995)
	Temporal trends in NO_3	Stoddard and Traeen (1995)
	NO_3:(NO_3+SO_4)	Henriksen

The table illustrates both the type and range of indicators which have been proposed and observed which link enhanced nitrogen deposition to ecosystem response. The indicators in the table are illustrative of more exhaustive lists available from, for example, the reviews of Nilsson and Grennfelt (1988), Sverdrup *et al.* (1990) and De Vries (1993). The available indicators tend to be relatively simple and often based on laboratory experiment data, and/or empirical field observation. For each indicator a threshold of N deposition is related to a change in ecosystem functioning, either through observation in biological change or, more commonly, observation of a changed chemical status in solution chemistry. These indicators are all based on the implicit assumption of steady state assessed by a sample (or samples) collected over a relatively short time period. Furthermore, field based observations are considered to reflect current deposition inputs and land management whilst there is a need to consider the role of past deposition (and/or past land management) histories in determining the observed response. In the absence of long term data sets this will remain a problem. Such a shortfall also has implications for critical loads and assessment of ecosystem recovery with emission abatement. The strength and assumptions underpinning indicators largely determine the efficacy on which critical loads are being developed and set.

8.3.4 FUTURE NEEDS – INDICATORS

Most of the indicators developed tend to use either a chemical change (e.g. soil solution ratios, surface waters NO_3 concentrations) or changes in biological composition. There are still relatively few indicators which link deposition, soil solution and biological response. Within soils the characterisation of the nitrogen and carbon pools is still rather simplistic in most effects based work. Thus the pools are frequently characterised in terms of C:N ratio and used to describe differences in soil organic materials and to predict the nitrification potential. This is a relatively crude measure but it is simple to determine and has proved robust. Further research may be able to produce more sophisticated indicators of soil organic matter from which it would produce better predictors of system response to pollutant N inputs. There are some areas in which there is apparent conflicting evidence, such as changes in forest ground flora. In such cases there is evidence to suggest that both present and historical differences in land-use may play an important role in the response. Equally, it is increasingly necessary to distinguish the past deposition history, both in terms of total loading but also the relative role of oxidised versus reduced species and any potential interaction with other compounding factors (e.g. co-deposition with sulphur, increased temperature or drought incidence).

For freshwaters the work of Dise and Wright (1995) suggests a threshold N loading above which enhanced nitrate leaching takes place. Is the threshold determined by the magnitude of annual inputs, a single years inputs or an aggregate input over time? Is the impact of 5 years N inputs at 20 kg ha yr the same as 10 years at 10 kg ha yr? The data presented by Yesmin *et al.* (1995) certainly suggests the history of N inputs at a given site is an important factor controlling response to further additions. Further work needs undertaking to identify the regional importance and severity of episodic acidification. It is possible that short term chronic effects, rather than long term accumulation of N are more important for the functioning of some ecosystem components. Under the same theme it is apparent that the importance of snowpack accumulation and elution of N have not been quantified. In tackling these issues not only will our knowledge and understanding of the underlying processes and factors be improved but the experiments established will also provide further data on the criteria against which nitrogen saturation can be assessed.

8.4 Critical Loads And The Policy Context

The synthesis of the environmental effects of atmospheric pollutant inputs to ecosystems in a dose–response type framework, such as that of critical loads, has already been introduced in Chapters 1, 2 and 7 of this book. The approaches range from simple empirical observational changes in vegetation communities through to complex process based formulation represented in steady state or time variant dynamic models. The former approach essentially deals with nitrogen as a nutrient, whereas the modelling approaches also consider the role of nitrogen inputs in soil acidification.

Empirical evidence suggests that at atmospheric deposition levels in excess of 15–20 kg ha^{-1} yr^{-1}, plant community changes and a loss of biodiversity will occur. Aerts and Bobbink (Chapter 4) suggest for more vulnerable species, often with higher conservation value, such change may occur at much lower deposition loadings of 7 kg ha^{-1} yr^{-1}. These values perhaps provide the currently best available upper limits for setting critical loads.

For acidification induced through enhanced nitrogen deposition, deriving critical loads is more complicated. This is reflected in the intricate presentation of the derivation of critical loads by Posch and De Vries in Chapter 7. Novel in this work is the consideration of the role of soil base cation status in providing a buffer against the onset of adverse effects. To date this mechanism has not been considered in setting critical loads.

Current models are centred on the representation on the central role of soils, both for vegetation and surface waters. The suite of models available use a chemical indicator, usually calcium to aluminium ratio in the soil solution to predict a 'harmful effect' to the relevant ecosystem. Within the range of models currently available there is a predilection to modelling forests. This is largely due to the large geographical expanse of forests in areas where natural and semi-natural ecosystems are under the threat from atmospheric pollutant inputs across Europe and N. America. The review of our ability to model system responses by Posch and De Vries amplifies the point made earlier on the lack of data and information on the heterogeneity of the governing processes.

In considering the potential role of pollutant deposition in the disruption of ecosystem function in terms of acidification it is impossible to isolate the role of sulphur from that of nitrogen. In terms of providing policy makers with a synthesis of the options available to reduce the impact from these sources, the principle of the critical load function has been developed in which the potential contribution and impact from eutrophication and acidity of these two pollutants can be conceptualised.

8.4.1 FUTURE NEEDS – CRITICAL LOADS

Many of the future requirements in the area of critical loads have been set out in the preceding sections of this chapter. Specifically for critical loads the issue is driven by the need to progress our understanding in order to reduce uncertainty in the formulation and quantification of the dose–response relationship. Equally there is a need to identify and predict the temporal changes in N deposition and ecosystem response.

At the process level, there is a need to capture information on the driving variables. Our current estimates of, for example, base cation depletion rates within soils, are subject to large uncertainties. Similar uncertainty lies in the linkage between the use of Ca:Al in soil solution as an adequate descriptor and linkage to variability in plant vitality. Much of the work to date has concentrated on forest ecosystems. In conservation terms there are other ecosystems of importance to which more effort should be considered. In terms of ecosystem and damage representation there are considerable sources of uncertainty in mapping heterogeneity. The key issue here would seem to be the way in which specific responses at a site represent the general overall situation. In order to develop meaningful damage scenarios, more emphasis should be placed on the 'level 2' approaches to assessing critical loads, through the use of dynamic modelling. Existing models need to incorporate the temporal dynamics of observed change, both in deposition, ecosystem response and the interaction with C cycling. Only in this manner will predictions of time sequences for damage and recovery under changing environmental scenarios be possible. Furthermore, work is needed on the nature of multi pollutant interactions and stresses and how these can be represented within the critical load function.

In the context of Europe and N. America these improvements need to be focused on the issues of reversibility in the light of stable or declining emissions. Whilst on a global scale the transboundary transport of pollutants may well change to consideration of the more localised (within country) but growing importance of ammonia.

Finally, perhaps the greatest uncertainty will be the characterisation and quantification of all the interactions against a changing climate. Wright and Schindler (1995) suggest that warming, the change in frequency of extremes and elevated CO_2 levels and the nature and intensity of their interactions with ecosystem response under long term climate change are largely unknown.

8.5 Concluding Remarks

Significant improvements have been made in our understanding of both processes and ecosystem response to the impact of atmospherically deposited nitrogen over the last ten years. The development and use of environmental thresholds as represented by critical loads has provided much of the stimulus for recent research. The development has provided a means of dialogue between scientist and policy maker. For the scientist this has meant an appraisal of the quantification of the driving processes and responses to nutrient enrichment and soil acidification. Of particular importance in this respect is the consideration of uncertainty. Within each chapter of this book the issue of uncertainty appears. Much of the discussion on uncertainty focuses on the representation of spatial and temporal variability at a scale which is appropriate to both the deposition input and ecosystem response. Central to this is the question:

Are the ecosystems and impacts identified stable against the driving forces for change from both deposition and climate?

Whilst these issues will remain, there is a range of ongoing research, illustrated throughout the book, in which the integration of monitoring and modelling strategies are rapidly developing and which will improve both our understanding of the processes and provide a framework to assess the variability at a range of scales.

In the light of these developments and current awareness there are moves to abate and lessen the impacts of nitrogen deposition in areas currently impacted, on the basis of a critical load approach. At the same time there is a growing need to assess those areas in which nitrogen deposition levels are set to continue increasing, noticeably SE Asia. It is less clear what the consequences of this change in deposition will be for some of these environments and against an uncertain future climate. However, the processes identified, methods and results presented throughout the book should provide a key from which to consider the potential impacts.

References

Bobbink, R and Roelofs, J. G. M. (1995) Nitrogen critical loads for natural and semi-natural ecosystems: The empirical approach. *Water, Air and Soil Pollution*, 85, 2413–2418.

De Vries, W. (1993) Average Critical Loads for Nitrogen and sulphur and its use in acidification abatement policy in the Netherlands. *Water, Air and Soil Pollution*, 68, 399–434.

Dise, N.B., and Wright, R.F., (1995) Nitrogen leaching from European forests in relation to Nitrogen deposition. *Forest Ecology and Management*, 71, 153–161.

Galloway, J.N. (1995) Acid deposition: Perspectives in time and space. *Water, Air and Soil Pollution*, 85, 15–24.

Henriksen, A. (1988) Critical Loads of Nitrogen to surface waters. In Nilsson, J., and Grennfelt, P., (eds.) 1988 *Critical loads for sulphur and nitrogen*. Report from a workshop held at Skokloster, Sweden March 1988, Published by Nordic Council of Ministers, Copenhagen.

Gunderson, P. (1998) Effects of enhanced nitrogen deposition in a spruce forest at Klosterhede, Denmark, examined by moderate NH_4NO_3 addition. *Forest Ecology and Management*, 101, 251–268.

Nilsson, J., and Grennfelt, P., (eds.) (1988) *Critical loads for sulphur and nitrogen*. Report from a workshop held at Skokloster, Sweden March 1988, Published by Nordic Council of Ministers, Copenhagen.

Roelofs, J.G.M., Kempers, A.J., Houdijk, A.L.F.M., and Jansen, J.(1985) *Plant and Soil*, 84, 45–00.

Stoddard, J.L., (1994) Long term changes in watershed retention of Nitrogen: its causes and aquatic consequences, in Environmental chemistry of lakes and reservoirs, *Advances in Chemistry* Series No. 237 (ed. L.A. Baker). American Chemical Society, pp.223–284.

Stoddard, J.L. and Traeen, Tor, S. (1995) The stages of Nitrogen saturation: Classification of catchments included in 'ICP' on waters, in *Mapping and modelling of critical loads for nitrogen: a workshop report* (eds. M. Hornung, M.A. Sutton and R.B. Wilson). Proceedings of Grange Over Sands Workshop 24–26 October 1994.

Sverdrup, H. And Warfvinge, P.U. (1993) *The effect of soil acidification on the growth of trees, grass and herbs as expressed by the (Ca+Mg+K)/Al ratio*. Reports in Ecology and Environmental Engineering 2. Department of Chemical Engineering II, Lund University, Lund, Sweden.

Ulrich, B. and Matzner, E. (1983) *Abiotische Folgewirkungen der weitraumigen Ausbreitung von Luftvenrunreinigung*. Umweltforschungengsplan der Mundermisniterium des Inneren. Forschungsbericht 10402615, Germany, 221p.

Wright, R.F. and Schindler, D.W. (1995) Interaction of acid rain and global changes: Effects on terrestrial and aquatic ecosystems. *Water, Air and Soil Pollution*, 85, 89–99.

Yesmin, L., Gammack, S.M., and Cresser, M.S. (1995) Impact of atmospheric N deposition on inorganic- and organic- N outputs in water draining from peat. *The Science of the Total Environment*, 166, 201–209.

INDEX

MIX
Papier aus verantwortungsvollen Quellen
Paper from responsible sources
FSC® C105338
FSC
www.fsc.org

If you have any concerns about our products,
you can contact us on
ProductSafety@springernature.com

In case Publisher is established outside the EU,
the EU authorized representative is:
Springer Nature Customer Service Center GmbH
Europaplatz 3, 69115 Heidelberg, Germany

Printed by Libri Plureos GmbH
in Hamburg, Germany